Graduate Texts in Physics

Graduate Texts in Physics

Graduate Texts in Physics publishes core learning/teaching material for graduate- and advanced-level undergraduate courses on topics of current and emerging fields within physics, both pure and applied. These textbooks serve students at the MS- or PhD-level and their instructors as comprehensive sources of principles, definitions, derivations, experiments and applications (as relevant) for their mastery and teaching, respectively. International in scope and relevance, the textbooks correspond to course syllabi sufficiently to serve as required reading. Their didactic style, comprehensiveness and coverage of fundamental material also make them suitable as introductions or references for scientists entering, or requiring timely knowledge of, a research field.

More information about this series at http://www.springer.com/series/8431

Gennady Stupakov · Gregory Penn

Classical Mechanics and Electromagnetism in Accelerator Physics

 Springer

Gennady Stupakov
SLAC National Accelerator Laboratory
Stanford University
Menlo Park, CA
USA

Gregory Penn
Lawrence Berkeley National Laboratory
Berkeley, CA
USA

ISSN 1868-4513   ISSN 1868-4521   (electronic)
Graduate Texts in Physics
ISBN 978-3-030-07956-7   ISBN 978-3-319-90188-6   (eBook)
https://doi.org/10.1007/978-3-319-90188-6

This Springer imprint is published by the registered company Springer International Publishing AG part of Springer Nature
The registered company address is: Gewerbestrasse 11, 6330 Cham, Switzerland

Preface

This book is a graduate-level textbook covering topics in classical mechanics and electromagnetic fields that are pertinent for the design and operation of particle accelerators. It was conceived as an intermediate step between general graduate university courses on these two subjects and specialized textbooks on accelerator physics. Our goal was to cover a selected number of subjects that we consider as an essential part of the knowledge that would help a student to better understand beam dynamics, collective effects and electromagnetic radiation of relativistic beams in accelerators.

The need for this type of course was first recognized by Richard Talman of Cornell University and Helmut Wiedemann of Stanford University, who taught this class in 2002 and 2004, respectively, at the US Particle Accelerator School (USPAS). The authors of this book offered this course as part of the USPAS program from 2007 to 2016 on a roughly biannual basis. It assumes an undergraduate-level background in mechanics and special relativity, and in electricity and magnetism. Matrix algebra, calculus, and complex variables are used throughout, again at an advanced undergraduate level.

In the presentation of the material, we made an effort to avoid lengthy derivations which can be found in existing textbooks. Instead, we tried to focus on major concepts and to connect those concepts to practical applications for accelerators. Fundamental notions of mechanics played a key role in the invention of the particle accelerator and continue to inspire new developments. Similarly, the electromagnetic fields produced by relativistic beams are both useful diagnostic signals and a key benefit of accelerator-based "light sources," while at the same time they impose significant constraints on the performance of accelerators.

The book consists of two parts. Part I is devoted to classical mechanics. In Chaps. 1–4, we first cover the basics of Lagrangian and Hamiltonian formalism, action-angle variables, and then linear and nonlinear oscillators. Starting from Chap. 5, we introduce specific features of accelerators. In Chaps. 5 and 6, we derive the Hamiltonian for a circular accelerator and formulate equations of motion, and in Chap. 7, we apply the action-angle formalism to the betatron oscillations in an accelerator. Having developed the basic machinery, we then apply it to the topics of field errors in Chap. 8 and nonlinear resonances in circular accelerators in Chap. 9. The first part of the book ends with Chap. 10, where we derive the kinetic equation governing the dynamics of beams and discuss some of its properties.

Part II of the book focuses on classical electromagnetism and begins with a discussion of the electromagnetic field of relativistic beams moving in free space in Chap. 11, and then, in Chap. 12, we introduce the effect of the material environment, obtaining the fields of a beam propagating inside a round pipe with resistive walls. In Chap. 13, we review plane electromagnetic waves and Gaussian beams as examples of the field propagating in free space. In Chap. 14, we expand the discussion to the modes in waveguides and radio-frequency cavities. A large fraction of the second part of the book deals, in Chaps. 15–20, with the radiation processes of relativistic beams in different conditions. We conclude this part with a discussion of laser-driven acceleration of charged particles in Chap. 21 and the radiation damping effect in Chap. 22. Selected references are included in some chapters for more detailed explanations of technical details and as a starting point for further reading.

We would like to thank our colleagues, who have been involved in the teaching of this class, and the organizers of the USPAS program. We are also grateful to the students who provided useful feedback in the development of this course. We are especially thankful to Chris Mayes who read and commented on several chapters of the book. We appreciate many valuable discussions with Alex Chao and Max Zolotorev on fundamental subjects of accelerator physics.

Cupertino, USA
Oakland, USA

Gennady Stupakov
Gregory Penn

Contents

Part I
Classical Mechanics

Chapter 1
The Basic Formulation of Mechanics: Lagrangian and Hamiltonian Equations of Motion

The Lagrangian and Hamiltonian formalisms are among the most powerful ways to analyze dynamic systems. In this chapter we will introduce Lagrange's equations of motion and discuss the transition from Lagrange's to Hamilton's equations. We write down the Lagrangian and Hamiltonian for a charged particle in an electromagnetic field, and introduce the Poisson bracket.

For more general information about classical mechanics, the authors recommend textbooks by Goldstein, Safko and Poole [1] and by Landau and Lifshiftz [2].

1.1 Lagrangian

There are many different types of dynamic systems. There can be imposed trajectories, or motion with constraints, or the basic $F = ma$ equations of motion. In accelerator systems particles move freely, and while control systems such as feedback are almost always used they have long time scales and are usually analyzed on a separate level. Thus, most accelerator systems involve applying forces to particles through prescribed fields, with the possibility of interactions among particles being important as well.

Systems with simple expressions for the acceleration can sometimes be solved for directly given the physical laws at work. For example, a simple harmonic oscillator satisfying the equation

$$\ddot{x} + \omega_0^2 x = 0 \tag{1.1}$$

has straightforward solutions of the form $a \cos(\omega_0 t + \phi)$, where a characterizes the amplitude of the motion and ϕ is a phase term that describes the timing.

© Springer International Publishing AG, part of Springer Nature 2018
G. Stupakov and G. Penn, *Classical Mechanics and Electromagnetism in Accelerator Physics*, Graduate Texts in Physics,
https://doi.org/10.1007/978-3-319-90188-6_1

Even slight generalizations, however, can be trickier to solve for and involve at the very least mathematical objects that are more esoteric than trigonometric functions. A common nonlinear example is the *pendulum equation*,

$$\ddot{\theta} + \omega_0^2 \sin\theta = 0, \tag{1.2}$$

where for a physical pendulum $\omega_0^2 = g/l$, with l being the length of the pendulum and g being acceleration due to gravity. We can still solve the pendulum equation exactly if we use energy conservation. Multiplying Eq. (1.2) by $\dot{\theta}$ gives

$$\frac{1}{2}\frac{d}{dt}\dot{\theta}^2 - \omega_0^2 \frac{d}{dt}\cos\theta = 0, \tag{1.3}$$

from which it follows that the quantity

$$E = \frac{1}{2\omega_0^2}\dot{\theta}^2 - \cos\theta = \text{const} \tag{1.4}$$

is conserved. We call E the energy of the system; each orbit is characterized by its own energy. For a given energy E we have

$$\dot{\theta} = \pm\omega_0\sqrt{2(E + \cos\theta)}, \tag{1.5}$$

and one more integration of this equation gives an explicit formula for the trajectory $\theta(t)$ (see details in Sect. 4.5).

How does one write equations of motion for more complicated mechanical systems, like the double pendulum shown in Fig. 1.1? The equations we would like to use are not easy to relate to the constraints representing, in this case, a joint. The Lagrangian formalism allows for easy formulation of such systems.

The first step in the Lagrangian formulation consists of choosing *generalized coordinates*, $q_1, q_2, \ldots, q_n$, which uniquely define a snapshot or configuration of the system at a particular time (these coordinates are said to define the *configuration space*). The number n is the number of degrees of freedom of our system. Each mechanical system possesses a Lagrangian function (or *Lagrangian* for short), which depends on the coordinates $q_1, q_2, \ldots, q_n$, velocities $\dot{q}_1, \dot{q}_2, \ldots, \dot{q}_n$ (with $\dot{q}_i = dq_i/dt$), and time t—for brevity, we will write the Lagrangian as $L(q_i, \dot{q}_i, t)$ instead of $L(q_1, q_2, \ldots, q_n, \dot{q}_1, \dot{q}_2, \ldots, \dot{q}_n, t)$.

The Lagrangian has the following property: the integral

$$S = \int_{t_1}^{t_2} L(q_i, \dot{q}_i, t)dt, \tag{1.6}$$

called the *action*, reaches an extremum along the true trajectory of the system when varied with fixed end points. The illustration in Fig. 1.2 gives an indication of what this condition means. This property can be used directly to find trajectories of a

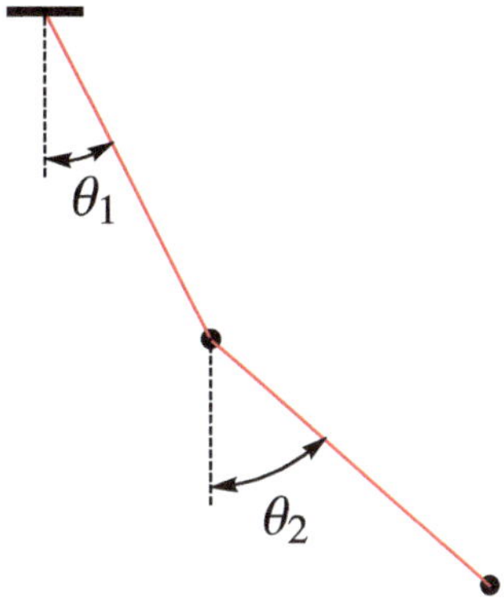

Fig. 1.1 A double pendulum

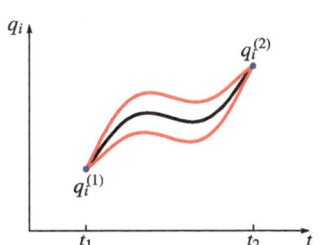

Fig. 1.2 Several trajectories have the same initial coordinates $q_i^{(1)}$ at t_1 and final coordinates $q_i^{(2)}$ at t_2. The integral of the Lagrangian reaches an extremum along the true trajectory

system by numerically minimizing the action S. It is however not very practical, in part because the varied trajectory is specified by its initial, $q_i(t_1)$, and final, $q_i(t_2)$, positions. In applications we would prefer to specify a trajectory by its initial position and velocity instead.

For mechanical systems, the Lagrangian is equal to the difference between the kinetic and the potential energies. The kinetic energy represents the energy from the particle motion alone, while the potential energy U defines a force acting on the particle through $F = -dU/dx$. For example, for a single pendulum with the equation of motion given by Eq. (1.2), with the angle θ chosen as a generalized coordinate q, the Lagrangian is

$$L(\theta, \dot{\theta}) = \frac{m}{2}l^2\dot{\theta}^2 + gml\cos\theta .\tag{1.7}$$

The most convenient approach to the problem of obtaining equations of motion for a given Lagrangian is based on the variational calculus. By direct minimization of the action integral, requiring

$$\delta \int_{t_1}^{t_2} L(q_i, \dot{q}_i, t)dt = 0 ,\tag{1.8}$$

one can get equations of motion in the following form:

$$\frac{\partial L}{\partial q_i} - \frac{d}{dt}\frac{\partial L}{\partial \dot{q}_i} = 0, \qquad i = 1, \ldots, n. \tag{1.9}$$

These are ordinary differential equations which are easier to solve than trying to directly minimize S.

Let us derive Eq. (1.9). Assume that $q_i(t)$ is a true orbit and the values $q_i(t_1)$ and $q_i(t_2)$ are fixed. Let $\delta q_i(t)$ be a deviation from this orbit; it has the property $\delta q_i(t_1) = \delta q_i(t_2) = 0$. Compute the variation of the action:

$$\delta \int_{t_1}^{t_2} L(q_i, \dot{q}_i, t)dt$$

$$= \int_{t_1}^{t_2} L(q_i + \delta q_i, \dot{q}_i + \dot{\delta}q_i, t)dt - \int_{t_1}^{t_2} L(q_i, \dot{q}_i, t)dt$$

$$= \int_{t_1}^{t_2} \sum_{i=1}^{n} \left(\frac{\partial L}{\partial q_i}\delta q_i + \frac{\partial L}{\partial \dot{q}_i}\delta\dot{q}_i\right) dt$$

$$= \int_{t_1}^{t_2} \sum_{i=1}^{n} \left(\frac{\partial L}{\partial q_i} - \frac{d}{dt}\frac{\partial L}{\partial \dot{q}_i}\right) \delta q_i dt, \tag{1.10}$$

where in the last step we integrated the term with $\delta\dot{q}_i$ by parts. Since $q_i(t)$ is a true orbit, the action reaches an extremum on it, and the variation of the action should be of second order, i.e., $\propto \delta q_i^2$. This means that the linear variation that we have found above vanishes for arbitrary δq_i, hence Eq. (1.9) must be satisfied.

The Lagrangian for a given system is not unique. There exist many Lagrangians for the same physical system that lead to identical equations of motion. This is easy to see from Eq. (1.8): adding to L any function $g(q_i, t)$ that is a total time derivative of an arbitrary function $f(q_i, t)$ of coordinate and time,

$$g(q_i, t) = \frac{df(q_i, t)}{dt} \equiv \frac{\partial f}{\partial t} + \sum_{i=1}^{n} \dot{q}_i \frac{\partial f}{\partial q_i}, \tag{1.11}$$

does not change the Eq. (1.9).

There are several advantages of using a Lagrangian as a starting point for the formulation of the equations of motion. First, we have complete freedom to choose generalized coordinates q_i. Second, the Lagrangian formalism is closely related to powerful variational principles, as indicated by Eq. (1.8). Finally, there is a connection between the symmetries of the Lagrangian and the conservation laws for the system. A simple example of such a connection is given by the case when L does not depend on q_i: as follows from Eq. (1.9), in this case the quantity $\partial L/\partial \dot{q}_i$ is conserved.

1.2 Lagrangian of a Relativistic Particle in an Electromagnetic Field

For particle accelerators, the primary interest is in the motion of relativistic charged particles in an electromagnetic field. The Lagrangian for such a particle is formulated in terms of the vector potential A and the scalar potential ϕ; the electric E and magnetic B fields are given by $E = -\nabla\phi - \partial A/\partial t$ and $B = \nabla \times A$. The Lagrangian has the following form:

$$L(r, v, t) = -mc^2\sqrt{1 - v^2/c^2} + ev \cdot A(r, t) - e\phi(r, t) \tag{1.12}$$

$$= -\frac{mc^2}{\gamma} + ev \cdot A(r, t) - e\phi(r, t),$$

where e is the electric charge of the particle, $\beta = v/c$, and $\gamma = (1 - \beta^2)^{-1/2}$ is the Lorentz factor. In a Cartesian coordinate system, $r = (x, y, z)$, and the Lagrangian is given as a function $L(x, y, z, \dot{x}, \dot{y}, \dot{z}, t)$, where, of course, $\dot{x} = v_x, \dot{y} = v_y, \dot{z} = v_z$.

As an example of applying the Lagrangian formalism, let us study in detail the motion of a particle in a uniform magnetic field using the above Lagrangian. We assume that the field is directed along the z-axis:

$$B = (0, 0, B_0), \tag{1.13}$$

with the corresponding vector potential

$$A = (-B_0 y, 0, 0). \tag{1.14}$$

This gives for the Lagrangian,

$$L = -mc^2\sqrt{1 - v^2/c^2} - eB_0 v_x y. \tag{1.15}$$

Let us first consider the z direction of motion along the magnetic field. Because L does not depend on z, from Eq. (1.9) we obtain $d(\partial L/\partial v_z)/dt = 0$ and it follows that

$$\frac{d\gamma v_z}{dt} = 0. \tag{1.16}$$

For the equation of motion in the x direction we have $\partial L/\partial x - d(\partial L/\partial v_x)/dt = 0$. Noting that L does not depend on x and that $\partial L/\partial v_x = mc^2\gamma v_x/c^2 - eB_0 y$, we obtain

$$-m\frac{d\gamma v_x}{dt} + eB_0 v_y = 0. \tag{1.17}$$

The Lagrange equation in the y direction similarly gives

$$-m\frac{d\gamma v_y}{dt} - eB_0 v_x = 0.$$
(1.18)

These three equations can also be written in the familiar vectorial form

$$\frac{d\boldsymbol{p}}{dt} = e\boldsymbol{v} \times \boldsymbol{B},$$
(1.19)

where the momentum $\boldsymbol{p} = m\gamma\boldsymbol{v}$. Because the magnetic force on the right-hand side of this equation is perpendicular to the velocity, it does not produce work on the particle, and the kinetic energy is conserved, $\gamma = \text{const}$. This can also be proven straightforwardly from Eqs. (1.16)–(1.18): multiplying each of the equations by v_z, v_x and v_y, respectively, and adding them gives

$$v_z\frac{d\gamma v_z}{dt} + v_y\frac{d\gamma v_y}{dt} + v_x\frac{d\gamma v_x}{dt} = 0,$$
(1.20)

which can also be written as $vd(\gamma v)/dt = 0$, with $v = \sqrt{v_x^2 + v_y^2 + v_z^2}$. Because γ is a function of v, from this equation it follows that $d\gamma/dt = 0$. Knowing that γ is constant, we can rewrite Eqs. (1.16)–(1.18) as

$$\dot{v}_z = 0, \quad \dot{v}_x = \omega_H v_y, \quad \dot{v}_y = -\omega_H v_x,$$
(1.21)

where we have introduced the *cyclotron frequency* ω_H,

$$\omega_H = \frac{eB_0}{\gamma m}.$$
(1.22)

Integrating Eq. (1.21) we find

$$v_x = v_0\cos(\omega_H t + \phi_0), \quad v_y = -v_0\sin(\omega_H t + \phi_0),$$
(1.23)

where v_0 and ϕ_0 are arbitrary transverse velocity and phase. With one more integration we arrive at

$$x = \frac{v_0}{\omega_H}\sin(\omega_H t + \phi_0) + x_0, \quad y = \frac{v_0}{\omega_H}\cos(\omega_H t + \phi_0) + y_0,$$
(1.24)

with arbitrary x_0 and y_0. Equation (1.24) represents a circular orbit with the center at x_0, y_0, and the radius

$$R = \frac{v_0}{\omega_H} = \frac{p}{eB_0}.$$
(1.25)

For a nonzero, constant v_z the full 3-dimensional orbit would be a helix.

1.3 From Lagrangian to Hamiltonian

The Hamiltonian approach has considerable advantages over the Lagrangian one. We will find that in the Hamiltonian approach it is simpler to change how a system is characterized in order to suit our needs, and the quantities which come out of this approach have clearer physical meanings as well. These advantages are especially useful in accelerator physics. A transition from the Lagrangian to the Hamiltonian description is made in three steps. First, we define the *generalized momenta* p_i:

$$p_i(q_k, \dot{q}_k, t) \equiv \frac{\partial L(q_k, \dot{q}_k, t)}{\partial \dot{q}_i}, \qquad i = 1, \ldots, n. \tag{1.26}$$

Second, from the n equations $p_i = p_i(q_k, \dot{q}_k, t)$, $i = 1, \ldots, n$ we express all the variables $\dot{q}_i$ in terms of $q_1, q_2, \ldots, q_n, p_1, p_2, \ldots, p_n$ and t,

$$\dot{q}_i = \dot{q}_i(q_k, p_k, t), \qquad i = 1, \ldots, n. \tag{1.27}$$

Finally, we construct the *Hamiltonian function H* as

$$H = \sum_{i=1}^{n} p_i \dot{q}_i - L(q_k, \dot{q}_k, t), \tag{1.28}$$

and express all $\dot{q}_i$ on the right-hand side through q_i, p_i and t using Eq. (1.27). This gives us the Hamiltonian as a function of variables q_i, p_i and t, namely $H(q_1, q_2, \ldots, q_n, p_1, p_2, \ldots, p_n, t)$. With the function H defined in this way, we claim that the equations of motion of our system become:

$$\dot{p}_i = -\frac{\partial H}{\partial q_i}, \qquad \dot{q}_i = \frac{\partial H}{\partial p_i}. \tag{1.29}$$

The partial derivatives $\partial/\partial q_i$ are here understood to be the derivatives taken while holding all p_i variables constant, in contrast to the derivatives $\partial/\partial q_i$ in Eq. (1.9) where the differentiation is carried out with constant $\dot{q}_i$. The variables p_i and q_i are called *canonically conjugate* pairs of variables.

Let us now prove Eq. (1.29) starting with the calculation of the partial derivative $-\partial H/\partial q_i$,

$$-\frac{\partial H}{\partial q_i} = -\frac{\partial}{\partial q_i}\left(\sum_{k=1}^{n} p_k \dot{q}_k - L\right) = \sum_{k=1}^{n}\left(-p_k \frac{\partial \dot{q}_k}{\partial q_i} + \frac{\partial L}{\partial \dot{q}_k}\frac{\partial \dot{q}_k}{\partial q_i}\right) + \frac{\partial L}{\partial q_i}. \tag{1.30}$$

Noting that, according to the definition Eq. (1.26), the momentum p_k is equal to $\partial L/\partial \dot{q}_k$, we conclude that the sum in the parentheses on the right-hand side cancels, and we are left with $\partial L/\partial q_i$. According to the Lagrangian equations of motion of Eq. (1.9), we then have

$$-\frac{\partial H}{\partial q_i} = \frac{\partial L}{\partial q_i} = \frac{d}{dt}\frac{\partial L}{\partial \dot{q}_i} = \frac{dp_i}{dt}. \tag{1.31}$$

This proves the first of Eq. (1.29). The second one is proven analogously.

As a simple exercise, let us derive the Hamiltonian for the pendulum. Starting from the Lagrangian of Eq. (1.7) we use Eq. (1.26) to find the generalized momentum p corresponding to the angular variable θ: $p = ml^2\dot{\theta}$. The Hamiltonian is then obtained directly from (1.28):

$$H(\theta, p) = \frac{p^2}{2ml^2} - \omega_0^2 ml^2 \cos\theta, \tag{1.32}$$

with $\omega_0 = \sqrt{g/l}$.

1.4 Hamiltonian of a Charged Particle in an Electromagnetic Field

Starting from the Lagrangian (1.12),

$$L(\boldsymbol{r}, \boldsymbol{v}, t) = -mc^2\sqrt{1 - v^2/c^2} + e\boldsymbol{v}\cdot\boldsymbol{A}(\boldsymbol{r}, t) - e\phi(\boldsymbol{r}, t),$$

we first need to find the canonical conjugate momentum for which we will use the vectorial notation $\boldsymbol{\pi} = (\pi_x, \pi_y, \pi_z)$, combining the three conjugate variables to the Cartesian coordinates $\boldsymbol{r} = (x, y, z)$:

$$\boldsymbol{\pi} = \frac{\partial L}{\partial \boldsymbol{v}} = m\frac{\boldsymbol{v}}{\sqrt{1 - v^2/c^2}} + e\boldsymbol{A} = m\gamma\boldsymbol{v} + e\boldsymbol{A}. \tag{1.33}$$

Here the derivative with respect to the vector $\boldsymbol{v}$ is understood as a collection of three derivatives, $(\partial/\partial v_x, \partial/\partial v_y, \partial/\partial v_z)$. Note that the conjugate momentum $\boldsymbol{\pi}$ differs from the kinetic momentum $m\gamma\boldsymbol{v}$ of the particle. Note also that as follows from the previous equation, $\gamma\boldsymbol{\beta} = (\boldsymbol{\pi} - e\boldsymbol{A})/mc$, and hence

$$\gamma^2\beta^2 = \frac{(\boldsymbol{\pi} - e\boldsymbol{A})^2}{m^2c^2}. \tag{1.34}$$

We will need this relation in what follows.

Now let us derive the Hamiltonian $H = \boldsymbol{v} \cdot \boldsymbol{\pi} - L$,

$$
\begin{aligned}
H &= \boldsymbol{v} \cdot \boldsymbol{\pi} + mc^2\sqrt{1 - v^2/c^2} - e\boldsymbol{v} \cdot \boldsymbol{A} + e\phi \\
&= m\gamma v^2 + \frac{mc^2}{\gamma} + e\phi \\
&= m\gamma c^2 + e\phi ,
\end{aligned} \tag{1.35}
$$

where we have used Eq. (1.33) and the relation $\gamma^2 = 1 + \gamma^2\beta^2$. Remarkably, the Hamiltonian is the sum of the particle energy γmc^2 and the potential energy associated with the electrostatic potential ϕ. It seems that the vector potential $\boldsymbol{A}$ does not enter this expression. This conclusion, however, is misleading — the vector potential is implicitly present in Eq. (1.35). To see this, remember that we need to express H in terms of the conjugate coordinates $\boldsymbol{r}$ and momenta $\boldsymbol{\pi}$. Again using $\gamma^2 = 1 + \gamma^2\beta^2$, this time in combination with Eq. (1.34), we express γ^2 through $\boldsymbol{\pi}$ and $\boldsymbol{A}$:

$$
\gamma^2 = 1 + \frac{(\boldsymbol{\pi} - e\boldsymbol{A})^2}{m^2c^2} . \tag{1.36}
$$

Taking the square root of this expression and substituting it for γ in the Hamiltonian of Eq. (1.35) gives

$$
H(\boldsymbol{r}, \boldsymbol{\pi}, t) = \sqrt{(mc^2)^2 + c^2(\boldsymbol{\pi} - e\boldsymbol{A}(\boldsymbol{r}, t))^2} + e\phi(\boldsymbol{r}, t) . \tag{1.37}
$$

In this form, the Hamiltonian is expressed in terms of the proper conjugate variables and explicitly contains both the scalar and vector potentials.

1.5 The Poisson Bracket

Let $f(q_i, p_i, t)$ be a function of the conjugate coordinates, momenta and time, and take any Hamiltonian trajectory $q_i(t)$ and $p_i(t)$. Then f becomes a function of time t only: $f(q_i(t), p_i(t), t)$. What is the derivative of this function with respect to time? Using the chain rule for computing the derivative of the composition of functions we find

$$
\frac{df}{dt} = \frac{\partial f}{\partial t} + \sum_i \left(\frac{\partial f}{\partial q_i}\dot{q}_i + \frac{\partial f}{\partial p_i}\dot{p}_i \right) . \tag{1.38}
$$

Often df/dt is referred to as the *convective*, or Lagrangian, derivative with respect to time. Substituting Eq. (1.29) into this equation gives

$$\frac{df}{dt}_{\text{conv}} = \frac{\partial f}{\partial t} + \sum_i \left(\frac{\partial f}{\partial q_i} \frac{\partial H}{\partial p_i} - \frac{\partial f}{\partial p_i} \frac{\partial H}{\partial q_i} \right)$$

$$= \frac{\partial f}{\partial t} + \{f, H\}, \tag{1.39}$$

where we have introduced the *Poisson bracket* of two functions f and g,

$$\{f, g\} \equiv \sum_i \left(\frac{\partial f}{\partial q_i} \frac{\partial g}{\partial p_i} - \frac{\partial f}{\partial p_i} \frac{\partial g}{\partial q_i} \right), \tag{1.40}$$

and specifically indicated with the subscript the convective time derivative of f.

Poisson brackets have many remarkable properties, which we will use in the next chapter. They are *anticommutative*: for two functions $f(q_i, p_i, t)$ and $g(q_i, p_i, t)$, changing the order of f and g in the Poisson bracket flips the sign,

$$\{g, f\} = -\{f, g\}. \tag{1.41}$$

From this property it immediately follows that

$$\{f, f\} = 0. \tag{1.42}$$

If $df/dt = 0$, then f is an *integral of motion*, meaning that its value remains constant along the orbit. Note that if f does not explicitly depend on time, $f = f(q_i, p_i)$, it is an integral of motion if and only if $\{f, H\} = 0$, as follows from Eq. (1.39). From this observation, we immediately conclude that a Hamiltonian that does not depend explicitly on time is an integral of motion, because the Poisson bracket of H with itself is always equal to zero.

Considering each coordinate q_k and momentum p_k as a function that only depends on itself, we can easily apply to them the partial derivative operators $\partial/\partial q_i$ and $\partial/\partial p_i$. The result is: $\partial q_k/\partial q_i = \partial p_k/\partial p_i = \delta_{ik}$ and $\partial q_k/\partial p_i = \partial p_k/\partial q_i = 0$. It is a simple exercise to substitute these relations into Eq. (1.40) and to obtain:

$$\{q_i, q_k\} = \{p_i, p_k\} = 0, \qquad \{q_i, p_k\} = \delta_{ik}. \tag{1.43}$$

Worked Examples

Problem 1.1 *Consider a pendulum of length l and mass m, supported by a pivot that oscillates in the vertical direction with frequency Ω, $Y(t) = a \sin(\Omega t)$, where a is the amplitude of oscillations. Obtain the Lagrangian and derive equations of motion for the pendulum.*

Solution: The pendulum position is given by $x = L \sin \theta$ and $y = Y - L \cos \theta$. The potential energy is then $V = mgy = mg[Y(t) - L \cos \theta]$, and the kinetic energy is

$$T = \frac{1}{2}m(\dot{x}^2 + \dot{y}^2)$$

$$= \frac{1}{2}m\left[(L\cos\theta)^2\dot{\theta}^2 + (\dot{Y} + \dot{\theta}L\sin\theta)^2\right]$$

$$= \frac{1}{2}m\left[L^2\dot{\theta}^2 + a^2\Omega^2\cos^2(\Omega t) + 2\dot{\theta}La\Omega\cos(\Omega t)\sin\theta\right].$$

We then find the Lagrangian

$$L = \frac{1}{2}m\left[L^2\dot{\theta}^2 + a^2\Omega^2\cos^2(\Omega t) + 2\dot{\theta}La\Omega\cos(\Omega t)\sin\theta\right] - mg\left[a\sin(\Omega t) - L\cos\theta\right],$$

and equations of motion from

$$\frac{\partial L}{\partial\theta} = -mgL\sin\theta + \dot{\theta}Lma\Omega\cos(\Omega t)\cos\theta,$$

$$\frac{d}{dt}\frac{\partial L}{\partial\dot{\theta}} = \frac{1}{2}m\left[2L^2\ddot{\theta} - 2a\Omega^2\sin(\Omega t)L\sin\theta + 2L\dot{\theta}a\Omega\cos(\Omega t)\cos\theta\right].$$

Combining, we get

$$L^2\ddot{\theta} + L[g - a\Omega^2\sin(\Omega t)]\sin\theta = 0,$$

so the gravitational term is modified by the acceleration, $g \to g - a\Omega^2\sin(\Omega t)$.

Problem 1.2 *Analyze a particle's motion in a rotating frame using the Lagrangian approach.*

Solution: For a free particle, $L = T$ so

$$L = \frac{1}{2}mv^2 = \frac{1}{2}m\left(\dot{x}^2 + \dot{y}^2 + \dot{z}^2\right).$$

For the rotating frame we introduce new coordinates x', y', and z' defined by

$$x = x'\cos\omega t - y'\sin\omega t,$$
$$y = x'\sin\omega t + y'\cos\omega t,$$
$$z = z',$$

giving

$$\dot{x} = \dot{x}'\cos\omega t - x'\omega\sin\omega t - \dot{y}'\sin\omega t - y'\omega\cos\omega t,$$
$$\dot{y} = \dot{x}'\sin\omega t + x'\omega\cos\omega t + \dot{y}'\cos\omega t - y'\omega\sin\omega t,$$
$$\dot{z} = \dot{z}'.$$

So we find

$$\dot{x}^2 + \dot{y}^2 = \left(\dot{x}' \cos \omega t - x'\omega \sin \omega t - \dot{y}' \sin \omega t - y'\omega \cos \omega t\right)^2$$
$$+ \left(\dot{x}' \sin \omega t + x'\omega \cos \omega t + \dot{y}' \cos \omega t - y'\omega \sin \omega t\right)^2$$
$$= \dot{x}'^2 + \dot{y}'^2 + 2\omega(\dot{y}x' - \dot{x}y') + \omega^2(x'^2 + y'^2),$$

and in the rotating frame we have the Lagrangian

$$L = \frac{1}{2}m(\dot{x}'^2 + \dot{y}'^2 + \dot{z}'^2) + m\omega(\dot{y}x' - \dot{x}y') + \frac{1}{2}m\omega^2(x'^2 + y'^2).$$

In addition to the kinetic energy term we have a Coriolis force term (second term) and a repulsive potential (final term, often identified as the centrifugal force).

Problem 1.3 *Derive equations of motion*

$$\frac{d\boldsymbol{p}}{dt} = e\boldsymbol{E} + e\boldsymbol{v} \times \boldsymbol{B},$$

from the Lagrangian (1.12). Give the appropriate definition of the relativistic momentum, $\boldsymbol{p}$.

Solution: The Langrangian for a charged particle in an electromagnetic field is

$$L(\boldsymbol{r}, \boldsymbol{v}, t) = -mc^2\sqrt{1 - v^2/c^2} + e\boldsymbol{v} \cdot \boldsymbol{A}(\boldsymbol{r}, t) - e\phi(\boldsymbol{r}, t).$$

To find the equations of motion we need the following quantities:

$$\frac{\partial L}{\partial \boldsymbol{r}} = e\nabla\left[\boldsymbol{v} \cdot \boldsymbol{A}(\boldsymbol{r}, t)\right] - e\nabla\left[\phi(\boldsymbol{r}, t)\right],$$

$$\frac{d}{dt}\frac{\partial L}{\partial \boldsymbol{v}} = \frac{d}{dt}\left(mc^2\frac{\boldsymbol{v}}{c^2\sqrt{1 - v^2/c^2}} + e\boldsymbol{A}(\boldsymbol{r}, t)\right)$$

$$= \frac{d}{dt}(\gamma m\boldsymbol{v}) + e\left(\frac{\partial \boldsymbol{A}}{\partial t} + (\boldsymbol{v} \cdot \nabla)\boldsymbol{A}\right),$$

where we used $d/dt = \partial/\partial t + \boldsymbol{v} \cdot \nabla$. Combining to find the equations of motion, we find

$$\frac{d}{dt}(\gamma m\boldsymbol{v}) = e\nabla(\boldsymbol{v} \cdot \boldsymbol{A}) - e\nabla\phi - e\left(\frac{\partial \boldsymbol{A}}{\partial t} + (\boldsymbol{v} \cdot \nabla)\boldsymbol{A}\right)$$

$$= \left(-e\nabla\phi - e\frac{\partial \boldsymbol{A}}{\partial t}\right) + e\left[(\boldsymbol{v} \cdot \nabla)\boldsymbol{A} + \nabla(\boldsymbol{v} \cdot \boldsymbol{A})\right]$$

$$= e\boldsymbol{E} + e(\boldsymbol{v} \times (\nabla \times \boldsymbol{A}))$$

$$= e\boldsymbol{E} + e\boldsymbol{v} \times \boldsymbol{B},$$

Fig. 1.3 The coordinate system ξ, s, z. The reference circle, shown in blue, has a radius equal to R. The coordinate ξ of a particle is defined as the difference between its polar radius r and the circle radius R

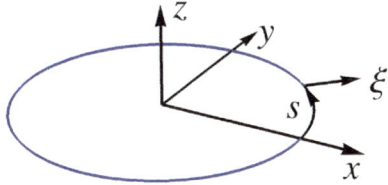

where we used the identity $\boldsymbol{a} \times (\nabla \times \boldsymbol{b}) = \nabla(\boldsymbol{a} \cdot \boldsymbol{b}) - (\boldsymbol{a} \cdot \nabla)\boldsymbol{b}$. The mechanical momentum is $\boldsymbol{p} = \gamma m \boldsymbol{v}$.

Problem 1.4 *Write the Lagrangian of a relativistic charge e moving in a uniform magnetic field B_0 with axis z directed along the magnetic field. Use the coordinate system (ξ, s, z) shown in Fig. 1.3. Derive the equations of motion.*

Solution: In the system of coordinates $\boldsymbol{r} = (\xi, s, z)$, the expression for the magnetic field in terms of the vector potential $\boldsymbol{A} = (A_\xi, A_s, A_z)$ is

$$\boldsymbol{B} = \nabla \times \boldsymbol{A}$$
$$= \left[\frac{1}{1+\xi/R} \frac{\partial A_z}{\partial s} - \frac{\partial A_s}{\partial z}, \; \frac{\partial A_\xi}{\partial z} - \frac{\partial A_z}{\partial \xi}, \; \frac{1}{1+\xi/R} \left(\frac{\partial(1+\xi/R)A_s}{\partial \xi} - \frac{\partial A_\xi}{\partial s} \right) \right].$$

We choose $A_\xi = -B_0 s \,(1 + \xi/R)$, with $A_z = A_s = 0$; a direct calculation shows that this vector potential corresponds to a uniform magnetic field $\boldsymbol{B} = (0, 0, B_0)$. We find that the Lagrangian is

$$L = -mc^2 \sqrt{1 - \frac{1}{c^2}\left[\dot{\xi}^2 + \left(1 + \frac{\xi}{R}\right)^2 \dot{s}^2 + \dot{z}^2 \right]} - e\dot{\xi} B_0 s \left(1 + \frac{\xi}{R}\right).$$

The derivatives of the Lagrangian are:

$$\frac{\partial L}{\partial \xi} = m\gamma \dot{s}^2 (1 + \xi/R)/R - e B_0 \dot{\xi} s/R, \qquad \frac{\partial L}{\partial s} = -e B_0 \dot{\xi}(1 + \xi/R), \qquad \frac{\partial L}{\partial z} = 0,$$

$$\frac{\partial L}{\partial \dot{\xi}} = m\gamma \dot{\xi} - e B_0 s(1 + \xi/R), \qquad \frac{\partial L}{\partial \dot{s}} = m\gamma \dot{s}\,(1 + \xi/R)^2, \qquad \frac{\partial L}{\partial \dot{z}} = m\gamma \dot{z},$$

where we simplified using $\gamma = 1/\sqrt{1 - v^2/c^2}$. We then use Eq. (1.9) to give the equations of motion:

$$m\gamma \ddot{\xi} = (1 + \xi/R)\left(e B_0 \dot{s} + \gamma m \dot{s}^2/R \right),$$
$$m\gamma(1 + \xi/R)^2 \ddot{s} = -2m\gamma(1 + \xi/R)\dot{\xi}\dot{s}/R - e B_0(1 + \xi/R)\dot{\xi},$$
$$\ddot{z} = 0.$$

Note one particular solution of these equations: $\xi = $ const, $\dot{s} = -eB_0R/\gamma m$, $z = $ const. This solution corresponds to circular motion about the origin with constant orbit radius $R + \xi$. Taking into account that the velocity is $|\dot{s}|(1 + \xi/R)$ it is easy to see that the radius of the orbit satisfies Eq. (1.25).

Problem 1.5 *The magnetic field as given in Eq. (1.13), $\boldsymbol{B} = (0, 0, B_0)$, can be represented not only by the vector potential (1.14) but also by $\boldsymbol{A} = \frac{1}{2}(-B_0 y, B_0 x, 0)$. Show that the equations of motion are the same as for the vector potential (1.14).*

Solution: With $\boldsymbol{A} = \frac{1}{2}(-B_0 y, B_0 x, 0)$, we have

$$L = -mc^2\sqrt{1 - (\dot{x}^2 + \dot{y}^2 + \dot{z}^2)/c^2} + eB_0\,(\dot{x}y/2 - \dot{y}x/2)\,.$$

We calculate:

$$\frac{\partial L}{\partial x} = -\frac{1}{2}B_0 e\dot{y}\,, \qquad \frac{\partial L}{\partial y} = \frac{1}{2}B_0 e\dot{x}\,, \qquad \frac{\partial L}{\partial z} = 0\,,$$

$$\frac{\partial L}{\partial \dot{x}} = \frac{1}{2}B_0 ey + m\gamma\dot{x}\,, \qquad \frac{\partial L}{\partial \dot{y}} = -\frac{1}{2}B_0 ex + m\gamma\dot{y}\,, \qquad \frac{\partial L}{\partial \dot{z}} = m\gamma\dot{z}\,.$$

We then use Eq. (1.9) to give the equations of motion,

$$m\gamma\ddot{x} = -B_0 e\dot{y}\,,$$
$$m\gamma\ddot{y} = B_0 e\dot{x}\,,$$
$$\ddot{z} = 0\,,$$

which are equivalent to Eqs. (1.16)–(1.18).

Problem 1.6 *Find conjugate momenta in cylindrical coordinates of a charged particle moving in an electromagnetic field.*

Solution: The Lagrangian is given by

$$L(\boldsymbol{r}, \boldsymbol{v}, t) = -mc^2\sqrt{1 - v^2/c^2} + e\boldsymbol{v} \cdot \boldsymbol{A}(\boldsymbol{r}, t) - e\phi(\boldsymbol{r}, t)\,.$$

For cylindrical coordinates ρ, ϕ, z, the velocity components are $v_\rho = \dot{\rho}$, $v_\phi = \rho\dot{\phi}$, $v_z = \dot{z}$, the total velocity is given by $v^2 = \dot{z}^2 + \dot{\rho}^2 + \rho^2\dot{\phi}^2$ and we find

$$L(\boldsymbol{r}, \boldsymbol{v}, t) = -mc^2\sqrt{1 - \frac{\dot{z}^2 + \dot{\rho}^2 + \rho^2\dot{\phi}^2}{c^2}} + e\left(\dot{\rho}A_\rho + \rho\dot{\phi}A_\phi + \dot{z}A_z\right) - e\phi(\boldsymbol{r}, t)\,.$$

The conjugate momenta for the cylindrical coordinates ρ, ϕ, and z are found by

$$\pi_\rho = \frac{\partial L}{\partial \dot{\rho}} = m\gamma\dot{\rho} + eA_\rho \,,$$

$$\pi_\phi = \frac{\partial L}{\partial \dot{\phi}} = m\gamma\rho^2\dot{\phi} + e\rho A_\phi \,,$$

$$\pi_z = \frac{\partial L}{\partial \dot{z}} = m\gamma\dot{z} + eA_z \,.$$

Problem 1.7 *The angular momentum M of a particle is defined as $M = r \times p$. Find the Poisson brackets $\{M_i, x_k\}$, $\{M_i, p_k\}$ and $\{M_i, M_k\}$, where the indices i and k take the values x, y and z.*

Solution: We write

$$M = r \times p = \left(p_z y - p_y z,\ p_x z - p_z x,\ p_y x - p_x y\right) \,,$$

or more conveniently

$$M_i = \sum_{a=1}^{3} \sum_{b=1}^{3} \epsilon_{iab} x_a p_b \,,$$

where ϵ_{iab} is the Levi-Civita symbol equal to $+1$ (-1) when (i, a, b) is an even (odd) permutation of $(1, 2, 3)$, and 0 if any index is repeated. We calculate the Poisson bracket (1.40) using

$$\frac{\partial x_a}{\partial x_i} = \frac{\partial p_a}{\partial p_i} = \delta_{ai} \,,$$

where δ_{ai} is the Kronecker delta.

To reduce the number of times needed to relabel indices, we will start by calculating the brackets of M_i with x_j and p_j:

$$\{M_i, x_j\} = -\frac{\partial M_i}{\partial p_j} = -\sum_{a=1}^{3} \epsilon_{iaj} x_a = \sum_{a=1}^{3} \epsilon_{ija} x_a \,,$$

$$\{M_i, p_j\} = \frac{\partial M_i}{\partial x_j} = \sum_{b=1}^{3} \epsilon_{ijb} p_b \,.$$

For $\{M_i, M_k\}$ we have

$$\{M_i, M_k\} = \sum_{j=1}^{3} \left(\frac{\partial M_i}{\partial x_j} \frac{\partial M_k}{\partial p_j} - \frac{\partial M_i}{\partial p_j} \frac{\partial M_k}{\partial x_j} \right)$$

$$= \sum_{j=1}^{3} \sum_{a=1}^{3} \sum_{b=1}^{3} \epsilon_{ija} \epsilon_{kjb} (x_a p_b - x_b p_a)$$

$$= x_i p_k - x_k p_i \,,$$

using the identity

$$\sum_{j=1}^{3} \epsilon_{ija} \epsilon_{kjb} = \delta_{ik} \delta_{ab} - \delta_{ib} \delta_{ak} \,.$$

The term including δ_{ab} drops out because $x_a p_b - x_b p_a$ must cancel whenever $a = b$. This result is equivalent to

$$\{M_i, M_k\} = \sum_{j=1}^{3} \epsilon_{ikj} M_j \,,$$

in other words it is the angular momentum component along the cross product of the directions corresponding to i and k.

Problem 1.8 *Simplify L and H in the nonrelativistic limit $v \ll c$.*

Solution: In the nonrelativistic limit, $v \ll c$, we can simplify the Lagrangian

$$L(\boldsymbol{r}, \boldsymbol{v}, t) = -mc^2 \sqrt{1 - v^2/c^2} + e\boldsymbol{v} \cdot \boldsymbol{A}(\boldsymbol{r}, t) - e\phi(\boldsymbol{r}, t)$$

$$\approx -mc^2 \left(1 - \frac{v^2}{2c^2} \right) + e\boldsymbol{v} \cdot \boldsymbol{A}(\boldsymbol{r}, t) - e\phi(\boldsymbol{r}, t) \,,$$

and the Hamiltonian

$$H = m\gamma c^2 + e\phi$$

$$\approx mc^2 \left(1 + \frac{(\boldsymbol{\pi} - e\boldsymbol{A})^2}{2m^2 c^2} \right) + e\phi \,,$$

where we used $\gamma = \sqrt{1 + (\boldsymbol{\pi} - e\boldsymbol{A})^2/m^2 c^2} \approx 1 + (\boldsymbol{\pi} - e\boldsymbol{A})^2/2m^2 c^2$, for γ close to 1.

References

1. H. Goldstein, J. Safko, C. Poole, *Classical Mechanics*, 3rd edn. (Wiley, New York, 1998)
2. L.D. Landau, E.M. Lifshitz, *Mechanics*, vol. 1, Course of Theoretical Physics (Elsevier Butterworth-Heinemann, Burlington, 1976). (translated from Russian)

Chapter 2
Canonical Transformations

One of the benefits of the Lagrangian approach to mechanical systems is that we can choose the generalized coordinates as we please. We have seen that once we select a set of coordinates q_i we can define the generalized momenta p_i according to Eq. (1.26) and form a Hamiltonian (1.28). We could also have chosen a different set of generalized coordinates $Q_i = Q_i(q_k, t)$, expressed the Lagrangian as a function of Q_i, used Eqs. (1.26) and (1.28), and obtained a different set of momenta P_i and a different Hamiltonian $H'(Q_i, P_i, t)$. Although mathematically different, these two representations are physically equivalent — they describe the same dynamics of our physical system. Understanding the freedom that we have in the choice of the conjugate variables for a Hamiltonian is important: a judicious choice of the variables could allow us to simplify the description of the system dynamics.

The above procedure shows us that we can rewrite our equations of motion for any coordinate system (even a moving one). But the Hamiltonian approach allows even more general changes in our choice of variables, as explained below. Let us assume that we have a set of canonical variables q_i, p_i and the corresponding Hamiltonian $H(q_i, p_i, t)$, and then make a transformation to new variables

$$Q_i = Q_i(q_k, p_k, t), \qquad P_i = P_i(q_k, p_k, t), \qquad i = 1 \ldots n. \qquad (2.1)$$

Can we find a new Hamiltonian $H'(Q_i, P_i, t)$ such that the dynamics as expressed in the new variables is also Hamiltonian? What are the requirements on the transformation (2.1) for such a Hamiltonian to exist?

These questions lead us to the notion of the *canonical transformation*.

© Springer International Publishing AG, part of Springer Nature 2018
G. Stupakov and G. Penn, *Classical Mechanics and Electromagnetism in Accelerator Physics*, Graduate Texts in Physics, https://doi.org/10.1007/978-3-319-90188-6_2

2.1 Canonical Transformations

We first consider a time-independent Hamiltonian $H(q_i, p_i)$, and later generalize the result for the case when H is a function of time. Instead of q_i, p_i we would like to use a new set of independent variables Q_i with the transformation from the old to new variables given by the following $2n$ equations,

$$Q_i = Q_i(q_k, p_k), \qquad P_i = P_i(q_k, p_k), \qquad i = 1 \ldots n, \tag{2.2}$$

see Fig. 2.1. An inverse transformation from Q_i, P_i to q_i, p_i is written as

$$q_i = q_i(Q_k, P_k), \qquad p_i = p_i(Q_k, P_k), \qquad i = 1 \ldots n. \tag{2.3}$$

It is obtained by considering Eq. (2.2) as $2n$ equations for the old variables and solving them for q_i, p_i.

Substituting (2.3) in $H(q_i, p_i)$, we can express our Hamiltonian in terms of the new variables:

$$H'(Q_k, P_k) = H(q_i(Q_k, P_k), p_i(Q_k, P_k)), \tag{2.4}$$

where we denote the new function by H'. Let us assume that we have solved the Hamiltonian equations of motion (1.29) and found a trajectory $q_i(t)$, $p_i(t)$. This trajectory is mapped, through the transformation (2.2), to an orbit in the phase space Q_i, P_i:

$$Q_i(t) = Q_i(q_k(t), p_k(t)), \qquad P_i(t) = P_i(q_k(t), p_k(t)). \tag{2.5}$$

We would like the trajectory defined by the functions $Q_i(t)$ and $P_i(t)$ to be a Hamiltonian orbit, which means that it has to satisfy the equations

$$\frac{dP_i}{dt} = -\frac{\partial H'(Q_k(t), P_k(t))}{\partial Q_i}, \qquad \frac{dQ_i}{dt} = \frac{\partial H'(Q_k(t), P_k(t))}{\partial P_i}. \tag{2.6}$$

The right-hand side of these equations is understood as follows: we first take a partial derivative of H' with respect to Q_i or P_i, holding all other Q and P variables constant,

Fig. 2.1 Transformation from the old variables q_i, p_i to the new variables Q_i, P_i: a point in the old phase space maps to a point in the new space, and an old orbit is transformed into a new one

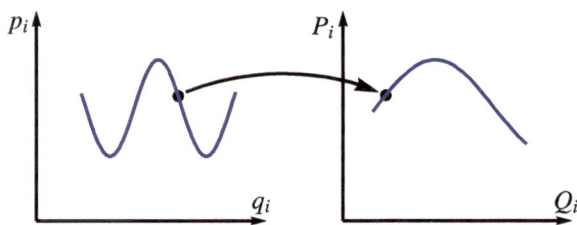

and then substitute for the arguments their values on the orbit $Q_k(t)$, $P_k(t)$. If the transformation (2.2) is such that Eq. (2.6) is satisfied for *every Hamiltonian H*, then it is called a *canonical transformation*.

Two simple examples of the canonical transformations are easily established by inspection:

$$Q_i = p_i, \qquad P_i = -q_i. \tag{2.7}$$

$$Q_i = -p_i, \qquad P_i = q_i. \tag{2.8}$$

In both cases, the transformation trivially renames the coordinates to momenta, and vice versa, with the proper change of signs. This example clearly shows that the conjugate variables in the Hamiltonian dynamics play an equal role, in contrast to the Lagrangian formalism where the coordinates q_i are fundamentally different from the velocities $\dot{q}_i$.

2.2 Poisson Brackets and Canonical Transformations

We will now show how to test that a given transformation (2.2) is canonical. The test is based on the invariance of the Poisson bracket with respect to the canonical transformations.

Let us assume that we have two functions of canonical variables, $f(q_i, p_i)$ and $g(q_i, p_i)$, and calculate their Poisson bracket:

$$\{f, g\}_{q,p} = \sum_i \left(\frac{\partial f}{\partial q_i} \frac{\partial g}{\partial p_i} - \frac{\partial f}{\partial p_i} \frac{\partial g}{\partial q_i} \right) \equiv J(q, p), \tag{2.9}$$

where, on the left-hand side, we now indicate the variables with respect to which the Poisson bracket is calculated. Using the inverse transformation (2.3) we can express our functions in terms of the new variables Q_i and P_i; the resulting new functions are denoted by $f'(Q_i, P_i)$ and $g'(Q_i, P_i)$. Let us also calculate the Poisson bracket of the new functions with respect to the new variables:

$$\{f', g'\}_{Q,P} = \sum_i \left(\frac{\partial f'}{\partial Q_i} \frac{\partial g'}{\partial P_i} - \frac{\partial f'}{\partial P_i} \frac{\partial g'}{\partial Q_i} \right) \equiv J'(Q, P). \tag{2.10}$$

We will now show that if $q_i, p_i \rightarrow Q_i, P_i$ is a canonical transformation, then rewriting $J'(Q, P)$ by expressing the new variables in terms of the old ones gives $J(q, p)$:

$$J'(Q_i(q_k, p_k), P_i(q_k, p_k)) = J(q_k, p_k). \tag{2.11}$$

To prove this statement we recall the relation between the Poisson bracket and the variation of a function along a trajectory, Eq. (1.39). Because the canonicity property is valid for all possible Hamiltonians, let us consider $g(q_i, p_i)$ as a Hamiltonian of a fictitious system. Denote by $q_i(t)$, $p_i(t)$ an orbit that satisfies this Hamiltonian. Then according to (1.39) the total time derivative df/dt along such an orbit is given by

$$\frac{df}{dt} = \{f, g\}_{q,p}. \tag{2.12}$$

When we change to the new variables Q_i and P_i in f, *the same time dependence* $f(t)$ is now given by $f(t) = f'(Q_i(t), P_i(t))$, where $Q_i(t)$, $P_i(t)$ is the orbit $q_i(t)$, $p_i(t)$ transformed into the new phase space. For the transformation to be canonical, the orbit $Q_i(t)$, $P_i(t)$ must satisfy Hamiltonian equations of motion with the new Hamiltonian $g'(Q_i, P_i) = g(q_i(Q_i, P_i), p_i(Q_i, P_i))$. Hence

$$\frac{df}{dt} = \{f', g'\}_{Q,P} \tag{2.13}$$

and with (2.12) we conclude that

$$\{f, g\}_{q,p} = \{f', g'\}_{Q,P}, \tag{2.14}$$

which proves Eq. (2.11) for arbitrary functions f and g.[1]

In practice, to establish that a transformation is canonical, we do not need to verify the equality (2.14) for all possible functions f and g; it is enough to make sure that it holds for a special set of $3n^2$ pairs of functions.[2] The first n^2 pairs are obtained by choosing two arbitrary indices i and k and setting $f = Q_i(q_l, p_l)$ and $g = Q_k(q_l, p_l)$. Using the fact that the Poisson brackets of the old coordinates are identically zero, Eq. (1.43), we conclude that the Poisson brackets of the new coordinates also have to vanish,

$$\{Q_i, Q_k\}_{q,p} = \{q_i, q_k\}_{q,p} = 0. \tag{2.15}$$

The second n^2 pairs are obtained by choosing $f = P_i(q_l, p_l)$ and $g = P_k(q_l, p_l)$ and requiring

$$\{P_i, P_k\}_{q,p} = \{p_i, p_k\}_{q,p} = 0. \tag{2.16}$$

Finally, choosing $f = Q_i(q_l, p_l)$ and $g = P_k(q_l, p_l)$ we obtain the third set of equations,

$$\{Q_i, P_k\}_{q,p} = \{q_i, p_k\}_{q,p} = \delta_{ik}. \tag{2.17}$$

---

[1] Strictly speaking we have only proved Eq. (2.14) for the points in the phase space along one particular trajectory $q_i(t)$, $p_i(t)$. To establish Eq. (2.14) in the whole domain of the transformation $q_i, p_i \to Q_i, P_i$ we need to consider a set of orbits in this domain with different initial conditions.

[2] Actually, due to the symmetry of the Poisson bracket with respect to transpositions, the number of independent, nontrivial equations is equal to $2n^2 - n$.

To summarize, we have proven that if Eq. (2.2) represents a canonical transformation, then for any pair of indices i and k the new variables satisfy Eqs. (2.15)–(2.17). These are the *necessary* conditions for a transformation to be canonical. It turns out (but we do not prove it here), that they also constitute a *sufficient* condition, that is if they are satisfied *for all* pairs i and k, the transformation is canonical.

2.3 Generating Functions

Poisson brackets are useful for establishing that a given transformation is canonical. They do not, however, provide a tool with which one can create canonical transformations. A technique that allows one to do that is based on the approach that uses so-called *generating functions*.

We will give a complete formulation of the method of generation functions in the next section. Here, we consider a particular case of a time-independent generating function of the so-called *first type*, F_1. Such a generating function depends on $2n$ variables: n old coordinates q_i and n new coordinates Q_i:

$$F_1(q_i, Q_i). \tag{2.18}$$

Having chosen an arbitrary function F_1, one can generate a transformation of variables (2.1) using the following set of equations:

$$p_k = \frac{\partial F_1(q_i, Q_i)}{\partial q_k}, \qquad P_k = -\frac{\partial F_1(q_i, Q_i)}{\partial Q_k}, \qquad k = 1 \ldots n. \tag{2.19}$$

The first n relations for p_k represent equations for n unknown Q_i. Solving these equations one finds n functions $Q_i(q_k, p_k)$, $i = 1 \ldots n$. Substituting these functions to the right-hand side of the second n relations for P_k gives n functions $P_k(q_i, p_i)$ in terms of the old variables. It turns out that the transformation of variables so obtained is canonical.

While this statement is valid for arbitrary n, we will only prove it here for a simple case of one degree of freedom, $n = 1$. In this case, we have two conjugate variables, q and p, and the canonical transformation (2.2) is defined by the following two equations,

$$Q = Q(q, p), \qquad P = P(q, p). \tag{2.20}$$

The generating function $F_1(q, Q)$ is a function of two variables, and Eq. (2.19) reduces to

$$p = \frac{\partial F_1(q, Q)}{\partial q}, \qquad P = -\frac{\partial F_1(q, Q)}{\partial Q}. \tag{2.21}$$

We need to show that from Eq. (2.21) follow Eqs. (2.15)–(2.17). Since in one dimension $i = k = 1$, Eqs. (2.15) and (2.16) are trivially satisfied, $\{Q, Q\} = \{P, P\} = 0$, and we only need to prove that

$$\{Q, P\}_{q,p} = \frac{\partial Q}{\partial q}\frac{\partial P}{\partial p} - \frac{\partial Q}{\partial p}\frac{\partial P}{\partial q} = 1. \tag{2.22}$$

From the second of Eq. (2.21) we find

$$\frac{\partial P}{\partial q} = -\frac{\partial^2 F_1}{\partial q \partial Q} - \frac{\partial^2 F_1}{\partial Q^2}\frac{\partial Q}{\partial q}, \qquad \frac{\partial P}{\partial p} = -\frac{\partial^2 F_1}{\partial Q^2}\frac{\partial Q}{\partial p}. \tag{2.23}$$

Substituting these equations into the Poisson bracket in (2.22) we obtain

$$\{Q, P\}_{q,p} = \frac{\partial Q}{\partial q}\frac{\partial P}{\partial p} - \frac{\partial Q}{\partial p}\frac{\partial P}{\partial q} = \frac{\partial Q}{\partial p}\frac{\partial^2 F_1}{\partial q \partial Q}. \tag{2.24}$$

The derivative $\partial Q/\partial p$ in this expression can be found when we differentiate the first of Eq. (2.21) with respect to p:

$$1 = \frac{\partial^2 F_1}{\partial Q \partial q}\frac{\partial Q}{\partial p}. \tag{2.25}$$

The right-hand sides of the previous two equations are identical, leading to the desired relation

$$\{Q, P\}_{q,p} = 1. \tag{2.26}$$

2.4 Transformations with a Time Dependence and Four Types of Generating Functions

Functions of the type (2.18) are not the only ones that generate canonical transformations. As it turns out, there are four different kinds of functions that can be used for this purpose. Moreover, all of these generating functions can be applied to time-dependent Hamiltonians, $H(q_i, p_i, t)$, and produce time-dependent canonical transformations as well. In this section, we will consider this most general case that includes a time dependence.

A time-dependent canonical transformation adds the time variable to the relations between the old and new variables:

$$Q_i = Q_i(q_k, p_k, t), \qquad P_i = P_i(q_k, p_k, t), \qquad i = 1 \ldots n. \tag{2.27}$$

They map the orbits from the old phase space to the new one as was discussed earlier, and the question is whether the orbits in the new phase space satisfy Hamiltonian equations. However, we will no longer insist on the particular relation (2.4) between the old and the new Hamiltonians, and allow for a broader class of functions $H'(Q_k, P_k, t)$.

The first type of the generating functions, generalized to include a time dependence, takes the form $F_1(q_i, Q_i, t)$. The relations between the old and the new variables are still given by Eq. (2.19), but the new Hamiltonian differs from the old one by the time derivative $\partial F_1/\partial t$:

$$p_i = \frac{\partial F_1}{\partial q_i}, \qquad P_i = -\frac{\partial F_1}{\partial Q_i}, \tag{2.28a}$$

$$H' = H + \frac{\partial F_1}{\partial t}. \tag{2.28b}$$

We remind the reader that the relations (2.28a) should be considered as a set of equations through which one finds the transformation (2.27), and the old variables on the right-hand side of (2.28b) are expressed through the new ones. As a consequence, while one can try any generating function to obtain a valid canonical transformation, it is not always straightforward to choose a generating function that yields a specific change in variables.

The second type of generating function depends on the old coordinates and new momenta, $F_2(q_i, P_i, t)$. The equations for the new variables and the new Hamiltonian are given by the following relations:

$$p_i = \frac{\partial F_2}{\partial q_i}, \qquad Q_i = \frac{\partial F_2}{\partial P_i}, \tag{2.29a}$$

$$H' = H + \frac{\partial F_2}{\partial t}. \tag{2.29b}$$

The third type is defined by a function $F_3(p_i, Q_i, t)$:

$$q_i = -\frac{\partial F_3}{\partial p_i}, \qquad P_i = -\frac{\partial F_3}{\partial Q_i}, \tag{2.30a}$$

$$H' = H + \frac{\partial F_3}{\partial t}, \tag{2.30b}$$

and the fourth type is generated by $F_4(p_i, P_i, t)$:

$$q_i = -\frac{\partial F_4}{\partial p_i}, \qquad Q_i = \frac{\partial F_4}{\partial P_i}, \tag{2.31a}$$

$$H' = H + \frac{\partial F_4}{\partial t}. \tag{2.31b}$$

Note that all of the generating functions depend on n old variables and n new ones.

Even for a time-dependent transformation the Poisson bracket can still be used for testing if a given transformation is canonical; the variable t in this case is considered as a parameter in calculations of partial derivatives.

Worked Examples

Problem 2.1 *In a later chapter, we will have need of a transformation that is not canonical. Show that the transformation $P_i = \lambda p_i$, $Q_i = q_i$, $H' = \lambda H$, where λ is a constant parameter, preserves the Hamiltonian structure of equations.*

Solution: The fact that the equations of motion in new variables are Hamiltonian is established by direct calculation:

$$\frac{dP_i}{dt} = \frac{d\lambda p_i}{dt} = -\lambda\frac{\partial H}{\partial q_i} = -\frac{\partial H'}{\partial Q_i},$$
$$\frac{dQ_i}{dt} = \frac{dq_i}{dt} = \frac{\partial H}{\partial p_i} = \frac{\partial H'}{\partial P_i}.$$

Note that this transformation is not canonical, because

$$\{Q_i, P_k\} = \lambda\{q_i, p_k\} = \lambda\delta_{ik}.$$

Problem 2.2 *Using the Poisson bracket, prove that the transformations Eqs. (2.7) and (2.8) are canonical.*

Solution: We start with the transformations

$$Q_i = p_i, \qquad P_i = -q_i.$$

We can then check $\{Q_i, Q_k\} = \{p_i, p_k\} = 0, \{P_i, P_k\} = \{-q_i, -q_k\} = \{q_i, q_k\} = 0$ and $\{Q_i, P_k\} = \{p_i, -q_k\} = \{q_k, p_i\} = \delta_{ik}$, so the transformation is canonical. We can easily see for the transformation

$$Q_i = -p_i, \qquad P_i = q_i,$$

that the same result follows.

Problem 2.3 *Find generating functions for the transformations (2.7) and (2.8).*

Solution: Again we start with the transformations

$$Q_i = p_i, \qquad P_i = -q_i.$$

For generating functions of the first type,

$$p_i = \frac{\partial F_1}{\partial q_i}, \qquad P_i = -\frac{\partial F_1}{\partial Q_i},$$

so plugging in for our transformation we find a generating function $F_1 = \sum_i Q_i q_i$. We could also have used

$$q_i = \frac{\partial F_4}{\partial p_i}, \qquad Q_i = -\frac{\partial F_4}{\partial P_i},$$

to find a generating function of the fourth type, $F_4 = -\sum_i p_i P_i$. For the transformation

$$Q_i = -p_i, \qquad P_i = q_i,$$

we find generating functions $F_1 = -\sum_i Q_i q_i$ or $F_4 = \sum_i P_i p_i$.

Problem 2.4 *Find the generating functions of the second and third type for the identity transformation*

$$Q_i = q_i, \qquad P_i = p_i.$$

This problem illustrates the fact that the choice of the type of the generating function is not unique.

Solution: The generating function of the second type for this transformation is

$$F_2 = \sum_i q_i P_i.$$

The generating function of the third type has to satisfy

$$q_i = -\frac{\partial F_3}{\partial p_i}, \qquad P_i = -\frac{\partial F_3}{\partial Q_i},$$

which is solved by $F_3 = -\sum_i p_i Q_i$.

Problem 2.5 *A change of coordinates which does not involve any dependence on the conjugate momenta is called a* point transformation*:*

$$Q_i = f_i(q_1, q_2, \ldots, q_n) \qquad i = 1 \ldots n, \tag{2.32}$$

(in general, there can be an explicit time dependence in the coordinate transformation, but for this problem we will not consider that). Find a generating function for a general time-independent point transformation (2.32).

Solution: The problem is solved by a generating function of the second type, $F_2 = \sum_i f_i(q_1, q_2, \ldots, q_n) P_i$:

$$Q_i = \frac{\partial F_2(q, P)}{\partial P_i} = f_i(q_1, q_2, \ldots, q_n)$$

$$p_i = \frac{\partial F_2(q, P)}{\partial q_i} = \sum_k \frac{\partial f_k}{\partial q_i} P_k = \sum_k \frac{\partial Q_k}{\partial q_i} P_k \,.$$

As an example, take the transformation from Cartesian to polar coordinates: $q_1 = x$, $q_2 = y \to Q_1 = r = \sqrt{x^2 + y^2}$, $Q_2 = \theta = \tan^{-1}(y/x)$. We then have the generating function $F_2 = P_1\sqrt{x^2 + y^2} + P_2 \tan^{-1}(y/x)$.

Problem 2.6 *For a simple harmonic oscillator with two degrees of freedom,*

$$H = p_x^2/2m + p_y^2/2m + m\omega_x^2 x^2/2 + m\omega_y^2 y^2/2 \,,$$

use a generating function of the second kind to change to variables $X = y$ and $Y = -x$. Find P_X, P_Y, and the new Hamiltonian H'.

Solution: The generating function of the second type is $F_2 = yP_x - xP_y$. Then the new coordinates are given by

$$X = \frac{\partial F_2}{\partial P_x} = y \,, \qquad Y = \frac{\partial F_2}{\partial P_y} = -x \,,$$

and we can trivially invert the expressions for the old momenta,

$$p_x = \frac{\partial F_2}{\partial x} = -P_y \,, \qquad p_y = \frac{\partial F_2}{\partial y} = P_x \,,$$

to obtain $P_x = p_y$ and $P_y = -p_x$.

Because F_2 has no explicit time dependence, the original Hamiltonian only has to be rewritten in terms of the new variables,

$$H' = P_x^2/2m + P_y^2/2m + m\omega_y^2 X^2/2 + m\omega_x^2 Y^2/2 \,.$$

Chapter 3
Action-Angle Variables and Liouville's Theorem

One of the most powerful uses of canonical transformations is to express the dynamics in terms of action-angle variables. These are phase space coordinates which provide a simple description of the Hamiltonian motion, and are widely used in particle dynamics. A geometrical view of the Hamiltonian flow in phase space leads us to the formulation of Liouville's theorem that is crucial for understanding the fundamental properties of large ensembles of beam particles in accelerators.

3.1 Canonical Transformation for a Linear Oscillator

We will now apply the general formalism of canonical transformations from the previous chapter to the particular example of the harmonic oscillator. The Hamiltonian for an oscillator with a unit mass is

$$H(x, p) = \frac{p^2}{2} + \frac{\omega_0^2 x^2}{2} \,, \tag{3.1}$$

where x is the coordinate, p is the conjugate momentum, and ω_0 is the oscillator frequency. Writing the equations of motion:

$$\dot{p} = -\frac{\partial H}{\partial x} = \omega_0^2 x \,, \qquad \dot{x} = \frac{\partial H}{\partial p} = p \,, \tag{3.2}$$

we easily find their solution,

$$x = a \cos(\omega_0 t + \phi_0) \,, \qquad p = -a\omega_0 \sin(\omega_0 t + \phi_0) \,, \tag{3.3}$$

where a is the amplitude and ϕ_0 is the initial phase of the oscillation. Both a and ϕ_0 are constant.

© Springer International Publishing AG, part of Springer Nature 2018
G. Stupakov and G. Penn, *Classical Mechanics and Electromagnetism
in Accelerator Physics*, Graduate Texts in Physics,
https://doi.org/10.1007/978-3-319-90188-6_3

We would like to find a canonical transformation from the old variables, x, p, to the new ones, ϕ, I, (where ϕ is the new coordinate and I is the new momentum), such that the transformation ϕ, $I \to x$, p takes the form

$$x = A(I) \cos \phi, \qquad p = -A(I) \omega_0 \sin \phi, \tag{3.4}$$

where $A(I)$ is an as-yet unknown function of I. The advantage of the new variables is clear: the new momentum I is a constant of motion (because the amplitude of the oscillations is constant), and the new coordinate evolves as a linear function of time, $\phi = \omega_0 t + \phi_0$.

To construct the canonical transformation (3.4) we will use the generating function $F_1(x, \phi)$ of the first type. First, we express p in terms of the old (x) and new (ϕ) coordinates by eliminating $A(I)$ from (3.4),

$$p = -\omega_0 x \tan \phi. \tag{3.5}$$

We then integrate the equation $(\partial F_1/\partial x)_\phi = p = -\omega_0 x \tan \phi$ to find

$$F_1(x, \phi) = \int p \, dx = -\frac{\omega_0 x^2}{2} \tan \phi. \tag{3.6}$$

The new momentum is obtained by differentiating F_1 with respect to the new coordinate ϕ,

$$I = -\frac{\partial F_1}{\partial \phi} = \frac{\omega_0 x^2}{2}(1 + \tan^2 \phi) = \frac{1}{2\omega_0}\left(\omega_0^2 x^2 + p^2\right), \tag{3.7}$$

where we have used Eq. (3.5) to express $\tan \phi$ in terms of x and p. The function $A(I)$ is then found when we substitute Eq. (3.4) into (3.7):

$$A(I) = \sqrt{\frac{2I}{\omega_0}}. \tag{3.8}$$

Equation (3.7) defines the new momentum in terms of the old variables. The new coordinate can be found from Eq. (3.5):

$$\phi = -\arctan \frac{p}{\omega_0 x}. \tag{3.9}$$

Because the canonical transformation does not depend on time, the new Hamiltonian is equal to the old one expressed in new variables. Comparing Eq. (3.1) with (3.7) we find that

$$H' = \omega_0 I. \tag{3.10}$$

We see that the new Hamiltonian does not depend on the new coordinate. The equations of motion in new variables are:

$$\dot{I} = -\frac{\partial H'}{\partial \phi} = 0, \qquad \dot{\phi} = \frac{\partial H'}{\partial I} = \omega_0. \tag{3.11}$$

They are easily integrated:

$$I = \text{const}, \qquad \phi = \omega_0 t + \phi_0. \tag{3.12}$$

Of course, this is the same dynamics as described by the original Eq. (3.3), but it is simpler because one of the coordinates, I, turns out to be an integral of motion and the other one, ϕ, is a simple linear function of time.

The (I, ϕ) pair is called the *action-angle* coordinates for this particular case. They are especially useful for building perturbation theory for more complicated systems that in the lowest approximation reduce to a linear oscillator. In the next section, we will derive the action-angle variables for a more general one-dimensional Hamiltonian system.

3.2 Action-Angle Variables in 1D

With a little more effort, we can generalize the action-angle variables introduced in the previous section for the harmonic oscillator to 1D periodic motion in an arbitrary but constant potential well $U(x)$. Assuming a unit mass, the Hamiltonian for this system is

$$H(x, p) = \frac{p^2}{2} + U(x). \tag{3.13}$$

The shape of the potential function is sketched in Fig. 3.1a with several trajectories in the phase space x, p shown in Fig. 3.1b. Each trajectory is defined by a constant value of the Hamiltonian, $H(x, p) = E$, where E is the energy. Both x and p for a given trajectory are periodic functions of time oscillating with the revolution frequency ω that depends on the energy, $\omega(E)$. The energy dependence of the frequency can be easily established using the relations $dx/dt = p = \sqrt{2(E - U(x))}$ and observing that half a period of the revolution is given by the following integral:

$$\frac{1}{2}T = \pi\omega^{-1} = \int_{x_1}^{x_2} \frac{dx'}{p(x')} = \int_{x_1}^{x_2} \frac{dx'}{\sqrt{2(E - U(x'))}}, \tag{3.14}$$

where $T = 2\pi/\omega$ is the period, and x_1 and x_2 are the turning points on the orbit (see Fig. 3.1b).

Fig. 3.1 Illustration of 1D
Hamiltonian: **a** the potential
energy $U(x)$ and **b** three
orbits with different
energies E

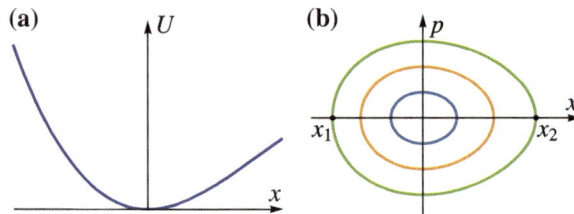

Let us try to make a canonical transformation to new variables choosing the new momentum to be equal to the energy E; the corresponding canonical conjugate coordinate is denoted by Q. We will use the generating function of the second type $F_2(x, E)$. Integrating the equation $\partial F_2/\partial x = p = \sqrt{2(E - U(x))}$, we obtain

$$F_2(x, E) = \int^x dx' \sqrt{2(E - U(x'))}. \qquad (3.15)$$

Of course, this integral cannot be taken analytically in general, but at least it can be computed numerically for any given function $U(x)$. Since this is a time-independent transformation, the new Hamiltonian H' is equal to the old one expressed in terms of the new variables:

$$H'(Q, E) = H = E, \qquad (3.16)$$

with the equations of motion for the new variables

$$\dot{Q} = \frac{\partial H'}{\partial E} = 1, \qquad \dot{E} = -\frac{\partial H'}{\partial Q} = 0. \qquad (3.17)$$

We see that the evolution of the variable Q is very simple,

$$Q = t + t_0. \qquad (3.18)$$

We can say that the conjugate variable to the energy is time.

The variable Q is not the most convenient one because in one revolution it increases by one period, $T = 2\pi/\omega(E)$, and this period changes with energy E. A better option would be to select a new coordinate, ϕ, in such a way that in one revolution it changes exactly by 2π — the same quantity for each trajectory. This coordinate is called the *angle*, and the corresponding generalized momentum, J, is the *action*.

As we will see, the action is a function of energy, $J(E)$, or, conversely, $E = E(J)$; this function will be found below. To find J and ϕ for the system (3.13) we need to go back and replace our canonical transformation (3.15) by a slightly modified one. The generating function $\tilde{F}_2(x, J)$ that accomplishes the transformation $(x, p) \to (\phi, J)$ is the same function F_2, in which we now replace E by $E(J)$,

$$\tilde{F}_2(x, J) = F_2(x, E(J)) \,. \tag{3.19}$$

With this arrangement, the new Hamiltonian $\tilde{H}$ is

$$\tilde{H}(\phi, J) = E(J) \,, \tag{3.20}$$

and the equation for ϕ reads

$$\dot{\phi} = \frac{\partial \tilde{H}}{\partial J} = \frac{dE}{dJ} \,. \tag{3.21}$$

We require that $\dot{\phi}$ be equal to $\omega(E)$, so that

$$\phi = \omega(E)t + \phi_0 \,. \tag{3.22}$$

With this time dependence, one orbital period corresponds to the change of variable ϕ by 2π, as desired. Combining (3.21) and (3.22) we obtain the differential equation for $E(J)$,

$$\frac{dE}{dJ} = \omega(E) \,. \tag{3.23}$$

Integrating this equation gives

$$J(E) = \int_{E_{\min}}^{E} \frac{dE'}{\omega(E')} \,, \tag{3.24}$$

where $E_{\min}$ is the energy corresponding to the bottom of the potential well U. The inverse function of $J(E)$ gives $E(J)$, which upon substitution into (3.19) in combination with (3.15) fully defines the generating function $\tilde{F}_2$ and finalizes the canonical transformation to ϕ, J.

The key features of action-angle coordinates are that the action is a constant of the motion, and the angle grows linearly in time, with periodic motion corresponding to a change in phase of 2π. The rate of change in the phase is generally different for different trajectories. The simple harmonic oscillator is a notable exception where all trajectories have the same period.

Strictly speaking, different Hamiltonians have different corresponding action-angle coordinates, but the resulting transformations will always be canonical. The transformation from a simple dynamical system can be useful as a first approximation to the action-angle coordinates in a more complicated system even when as a result the action is not exactly constant and the phase does not grow exactly linearly in time.

3.3   Hamiltonian Flow in Phase Space and Symplectic Maps

We will now take another look at the Hamiltonian motion, focusing on its geometrical aspect. Let us assume that for a given Hamiltonian $H(q_i, p_i, t)$, for every set of initial conditions q_i^0, p_i^0 from some domain, we can solve the equations of motion starting from initial time t_0 and find the values of q_i and p_i at time t. This gives us a *map* of the initial domain in the $2n$ dimensional phase space to a manifold in the same phase space at time t:

$$q_i = q_i(q_i^0, p_i^0, t_0, t), \qquad p_i = p_i(q_i^0, p_i^0, t_0, t). \tag{3.25}$$

Varying t in these equations moves each point (q_i, p_i) along its trajectory and the set of all trajectories starting from the initial domain constitutes a *Hamiltonian flow*, as illustrated by Fig. 3.2.

In an accelerator context one can associate, for example, each trajectory (3.25) with a different particle in a beam. Assume that there is a beam diagnostic at a location in the ring that measures coordinates of particles when the beam passes by at time t_0. On the next turn, at time $t = t_0 + T$, where T is the revolution period in the ring, it measures coordinates again. The relation between the new and old coordinates is given by Eq. (3.25) that connects the initial and final coordinates in a Hamiltonian flow.

A remarkable feature of the functional relations (3.25) is that, for a given t_0 and t, they constitute a canonical transformation from q_i^0, p_i^0 to q_i, p_i, which is also called a *symplectic* transfer map. While we are not going to prove the canonical properties of this map in the general case of n degrees of freedom, we will demonstrate it below for $n = 1$. In what follows, we drop the index i, using the notations q and p.

The proof is based on the calculation of the time derivatives of the Poisson brackets $\{q, p\}_{q^0, p^0}$, $\{p, p\}_{q^0, p^0}$ and $\{q, q\}_{q^0, p^0}$ and showing that they are equal to zero. Since at the initial time $t = t_0$ the transformation from q^0, p^0 to q, p is the identity transformation ($q = q^0$, $p = p^0$), it is clearly canonical. After we prove that the

Fig. 3.2 Hamiltonian flow in phase space. Orbits starting in the domain $\mathcal{M}_0$ at time t_0 end up in the domain $\mathcal{M}_1$ at time t

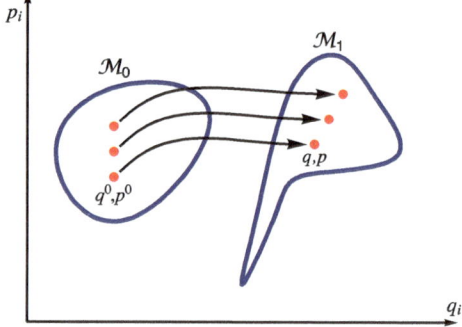

Poisson brackets do not change with time we will have verified that the map remains canonical for all values of t.

Let us calculate the time derivative of $\{q, p\}_{q_0, p_0}$:

$$\frac{d}{dt}\{q, p\}_{q_0, p_0} = \frac{d}{dt}\left(\frac{\partial q}{\partial q^0}\frac{\partial p}{\partial p^0} - \frac{\partial q}{\partial p^0}\frac{\partial p}{\partial q^0}\right) \tag{3.26}$$

$$= \frac{\partial p}{\partial p^0}\frac{\partial}{\partial q^0}\frac{dq}{dt} + \frac{\partial q}{\partial q^0}\frac{\partial}{\partial p^0}\frac{dp}{dt} - \frac{\partial p}{\partial q^0}\frac{\partial}{\partial p^0}\frac{dq}{dt} - \frac{\partial q}{\partial p^0}\frac{\partial}{\partial q^0}\frac{dp}{dt}$$

$$= \frac{\partial p}{\partial p^0}\frac{\partial}{\partial q^0}\frac{\partial H}{\partial p} - \frac{\partial q}{\partial q^0}\frac{\partial}{\partial p^0}\frac{\partial H}{\partial q} - \frac{\partial p}{\partial q^0}\frac{\partial}{\partial p^0}\frac{\partial H}{\partial p} + \frac{\partial q}{\partial p^0}\frac{\partial}{\partial q^0}\frac{\partial H}{\partial q}.$$

In the last transformation we have used the Hamiltonian equations of motion. Applying the chain rules,

$$\frac{\partial}{\partial p^0} = \frac{\partial p}{\partial p^0}\frac{\partial}{\partial p} + \frac{\partial q}{\partial p^0}\frac{\partial}{\partial q}, \qquad \frac{\partial}{\partial q^0} = \frac{\partial p}{\partial q^0}\frac{\partial}{\partial p} + \frac{\partial q}{\partial q^0}\frac{\partial}{\partial q}, \tag{3.27}$$

to the derivatives of the Hamiltonian, it is easy to show that all the terms on the right-hand side of (3.26) cancel and we obtain $d\{q, p\}_{q^0, p^0}/dt = 0$. For one degree of freedom, the other two Poisson brackets $\{q, q\}$ and $\{p, p\}$ are automatically always equal to 0 by the anticommutative property.

A somewhat different language of *symplectic maps* is often used in the literature in connection with canonical transformations (2.2) or (3.25). A symplectic map is defined with the help of the matrices J_{2n},

$$J_{2n} = \begin{pmatrix} J_2 & 0 & 0 & 0 \\ 0 & J_2 & 0 & 0 \\ 0 & 0 & \ddots & 0 \\ 0 & 0 & 0 & J_2 \end{pmatrix}, \tag{3.28}$$

where the right-hand side of this equation is a block $n \times n$ matrix with each element treated as 2×2 matrix. The diagonal elements are:

$$J_2 = \begin{pmatrix} 0 & -1 \\ 1 & 0 \end{pmatrix}, \tag{3.29}$$

and each zero in (3.28) is a 2×2 zero matrix. With the $2n \times 2n$ matrix J_{2n} defined in this way, one utilizes a uniform notation in which the conjugate coordinates and momenta q_i and p_i are replaced by $2n$ variables w_i: $w_{2i-1} = q_i$, $w_{2i} = p_i$, $i = 1, 2, \ldots, n$. Similarly, instead of the new coordinates and momenta Q_i and P_i one uses W_i: $W_{2i-1} = Q_i$, $W_{2i} = P_i$, $i = 1, 2, \ldots, n$. For example, for $n = 2$ we have $w_1 = q_1$, $w_2 = p_1$, $w_3 = q_2$, $w_4 = p_2$, and $W_1 = Q_1$, $W_2 = P_1$, $W_3 = Q_2$, $W_4 = P_2$. The transformation from the old to new variables (2.2) is then replaced by $2n$ functions

$$W_i = W_i(w_k), \qquad i, k = 1, 2, \ldots, 2n . \tag{3.30}$$

It is a matter of a straightforward calculation to show that the requirement that all possible Poisson brackets satisfy Eqs. (2.15)–(2.17) (which, as we know, is equivalent to the requirement for the transformation to be canonical) can be concisely written as

$$M J_{2n} M^T = J_{2n} , \tag{3.31}$$

where M is the Jacobian matrix of the transformation with the matrix elements

$$M_{ij} = \frac{\partial W_i}{\partial w_j}, \tag{3.32}$$

and the superscript T denotes the transposition of a matrix.

3.4   Liouville's Theorem

Let us consider a general canonical transformation (2.2) that maps a $2n$-dimensional domain $\mathcal{M}_0$ in the phase space (q_i, p_i) to a manifold $\mathcal{M}_1$ in the phase space (Q_i, P_i). The volume V_1 of $\mathcal{M}_0$ is given by the integral

$$V_1 = \int_{\mathcal{M}_0} dq_1 dq_2 \ldots dq_n dp_1 dp_2 \ldots dp_n , \tag{3.33}$$

where the integration goes over the manifold $\mathcal{M}_0$. The manifold $\mathcal{M}_1$ has a volume V_2,

$$V_2 = \int_{\mathcal{M}_1} dQ_1 dQ_2 \ldots dQ_n dP_1 dP_2 \ldots dP_n . \tag{3.34}$$

It turns out that the new volume is equal to the old one, $V_2 = V_1$. The proof follows from the following statement from calculus: the ratio of infinitesimal volumes in a transformation of variables is equal to the absolute value of the determinant of the Jacobian matrix of the transformation M,

$$\frac{dQ_1 dQ_2 \ldots dQ_n dP_1 dP_2 \ldots dP_n}{dq_1 dq_2 \ldots dq_n dp_1 dp_2 \ldots dp_n} = |\det M| . \tag{3.35}$$

At this point one can see the advantage of using the uniform notations of old, w_i, and new, W_i, variables introduced at the end of the previous section: the Jacobian matrix is simply given by Eq. (3.32). For a canonical transformation, this matrix satisfies the Eq. (3.31). Taking the determinant of the right-hand and left-hand sides of (3.31) and noting that the determinant of a product of matrices is equal to the product of the determinants, with the observation that $\det J_{2n} = 1$, we conclude that $|\det M| = 1$. From the conservation of the infinitesimal volumes then follows the conservation of integral volumes for arbitrary manifolds that are mapped by a canonical transformation.

Applied to the case of the Hamiltonian flow discussed in the previous section, Liouville's theorem guarantees that the phase space volume occupied initially by a beam remains the same through its Hamiltonian evolution with time. In particular, it forbids increasing the phase space density of a beam by applying to its component particles electromagnetic fields that do not destroy the Hamiltonian nature of their motion. This fact has a fundamental importance for beam properties in accelerators.

3.5 Non-conservative Forces in Hamiltonian Dynamics

The dynamics of some physical systems do not fit into the framework of the Hamiltonian and Lagrangian formalisms that we have been focusing on. In particular, damping and arbitrary externally-applied forces lead to equations of motion that do not quite match those which we have looked at previously. We can consider such terms to be corrections to the equations of motion. This will be most useful when the non-conservative forces are small perturbations on time scales of a single oscillation.

The simplest place to start is with Eq. (1.9) from Chap. 1, and generalize it to:

$$\frac{d}{dt}\frac{\partial L}{\partial \dot{q}_i} - \frac{\partial L}{\partial q_i} = F_i , \qquad i = 1, \ldots, n , \tag{3.36}$$

where $F_i = F_i(q_k, \dot{q}_k)$ is a generalized force (for example, friction). Sometimes this force can be defined in terms of a potential-like term $R(q_k, \dot{q}_k)$, called the *Rayleigh dissipation function*, as $F_i = -\partial R/\partial \dot{q}_i$. Although R does not represent a true potential or relate to any conserved quantity, it is convenient because, in contrast to F_i, it does not change under coordinate transformations. Therefore, when changing coordinates from q_i to Q_i, it is sufficient to replace $R(q_k, \dot{q}_k)$ with

$$\tilde{R}(Q_i, \dot{Q}_i) = R\left(q_k(Q_i), \sum_{j=1}^{n} \frac{\partial q_k}{\partial Q_j}\dot{Q}_j \right) ,$$

where the functions $q_k(Q_i)$ express the old coordinates in terms of the new ones. In other words, R preserves the covariance of the Lagrangian equations, even though it breaks some of the conservation rules [1].

The switch to the Hamiltonian formalism uses the exact same definitions as for the case where F (and R) vanish, with conjugate momentum $p_i = \partial L/\partial \dot{q}_i$ and

$$H = \sum_{i=1}^{n} p_i \dot{q}_i - L(q_k, \dot{q}_k, t) . \tag{3.37}$$

Repeating the derivation from Sect. 1.3 yields

$$\frac{dp_i}{dt} = -\frac{\partial H}{\partial q_i} + F_i \,,$$
$$\frac{dq_i}{dt} = \frac{\partial H}{\partial p_i} \,. \tag{3.38}$$

In these equations F_i is now understood as a function of the Hamiltonian variables q_k and p_k which is obtained by expressing $\dot{q}_k$ in the arguments of F_i through these variables, see Eq. (1.27). The form of the F_i may change dramatically after a canonical transformation. Finally, when we calculate the total time derivative of a general function $f(q_i, p_i, t)$ in Hamiltonian coordinates, which in this context is called the *convective* derivative, we find that

$$\frac{df}{dt}_{\text{conv}} = \frac{\partial f}{\partial t} + \{f, H\} + \sum_i F_i \frac{\partial f}{\partial p_i} \,. \tag{3.39}$$

This equation generalizes Eq. (1.39) for the case of non-conservative forces F_i. In particular, the Hamiltonian evolves as

$$\frac{dH}{dt} = \frac{\partial H}{\partial t} + \sum_i F_i \frac{\partial H}{\partial p_i} = \frac{\partial H}{\partial t} + \sum_i F_i \dot{q}_i = \frac{\partial H}{\partial t} - \sum_i \dot{q}_i \frac{\partial R}{\partial \dot{q}_i} \,, \tag{3.40}$$

where the last expression is for a frictional force corresponding to a potential R, but the derivatives of this potential and the $\dot{q}_i$ terms should be viewed as functions of the q_i and p_i coordinates. Here, it is more obvious that the numerical value of R does not have to change for a simple change in coordinates.

We can also use this formalism for one degree of freedom to calculate the impact of frictional forces on the dynamic flow maps defined by $(q_0, p_0) \rightarrow (q(t), p(t))$. If we simplify Eq. (3.26) as far as possible without using the Hamiltonian equations of motion, instead only using the definition of the Poisson bracket, we find that for one degree of freedom

$$\frac{d}{dt}\{q, p\}_{q_0, p_0} = \{q, p\}_{q_0, p_0} \left(\frac{\partial \dot{q}}{\partial q} + \frac{\partial \dot{p}}{\partial p} \right) \,. \tag{3.41}$$

Using the expressions from Eq. (3.38), we see that the partial derivatives of H cancel in the last bracket, leaving only the correction $\partial F / \partial p$. We can integrate this equation to find

$$\{q, p\}_{q_0, p_0} = \exp \left[\int_{t_0}^{t} \left(\frac{\partial F}{\partial p} \right) dt' \right] \,, \tag{3.42}$$

where the integration goes along the trajectory that starts at time t_0 with the initial conditions q_0, p_0, and we used the fact that at the initial coordinates $\{q_0, p_0\}_{q_0, p_0} = 1$. This directly gives us the extent to which phase space volumes are changing, instead of remaining constant as in Liouville's theorem.

For multiple degrees of freedom, the generalization of the above is more complex and no simple expression can be derived for the evolution of the Poisson brackets $\{q_i, p_k\}$ in the system governed by Eq. (3.36). However, if we focus on Liouville's theorem, we only need to consider a single scalar, det M. We will show that a simple expression can be obtained for the time derivative of det M. In our derivation we will use Jacobi's formula [2] for the time derivative of the determinant of an invertible matrix $M(t)$:

$$\frac{d}{dt} \det M = \det(M) \operatorname{tr}\left(M^{-1}\frac{dM}{dt}\right), \qquad (3.43)$$

where the trace operator, tr, denotes the sum of the diagonal elements.

To simplify the calculation we first look at the time derivative at $t = t_0$. In this case, the starting matrix $M(t = t_0)$ is the Jacobian of the identity map and hence is a unit $2n \times 2n$ matrix, $M(t = t_0) = I_{2n}$. Substituting this matrix to the right-hand side of Eq. (3.43) we find

$$\frac{d}{dt} \det M\bigg|_{t=t_0} = \operatorname{tr}\left(\frac{dM}{dt}\right)\bigg|_{t=t_0} = \sum_i \left(\frac{\partial \dot{q}_i}{\partial q_i} + \frac{\partial \dot{p}_i}{\partial p_i}\right) = \sum_i \left(\frac{\partial F_i}{\partial p_i}\right)_{q_0, p_0},$$
$$(3.44)$$

where we used the definition (3.32) to calculate the diagonal elements of M, and Eq. (3.38) for non-conservative forces. To get the derivative at any time t we separate M into two matrices, one for a fixed map from $t = t_0$ to $t_1 < t$, $M(t_0 \to t_1)$, and another mapping from the t_1 coordinates to t, which we denote by $M(t_1 \to t)$. In calculation of the time derivative of det M we will use the fact that the determinant of a product of two matrices is equal to the product of the determinants,

$$\frac{d}{dt} \det M = \det M(t_0 \to t_1)\frac{d}{dt} \det M(t_1 \to t). \qquad (3.45)$$

In the limit when t_1 approaches t, for the time derivative on the right-hand side of this expression we can use the result Eq. (3.44) in which t_0 is replaced by t_1. The result is:

$$\frac{d}{dt} \det M = \det M(t) \sum_i \left(\frac{\partial F_i}{\partial p_i}\right)_{q(t), p(t)}. \qquad (3.46)$$

Although it is not necessary to know all the details of the map corresponding to the dynamic flow, the right-hand side in this equation does have to be evaluated along particle trajectories.

Because we are only extracting a scalar, it is straightforward to integrate this equation:

$$\det M(t) = \exp\left[\int_{t_0}^t \sum_i \left(\frac{\partial F_i}{\partial p_i}\right)_{q(t'),p(t')} dt'\right]$$

$$= \prod_i \exp\left[\int_{t_0}^t \left(\frac{\partial F_i}{\partial p_i}\right)_{q(t'),p(t')} dt'\right]. \tag{3.47}$$

In some very simple examples for one degree of freedom the damping force reduces to $F = -\gamma p$ for constant γ, and we find that $\det M = e^{-\gamma(t-t_0)}$.

Worked Examples

Problem 3.1 *Find the action-angle variables for the system with the following potential*

$$U(x) = \begin{cases} \infty, & x < 0 \\ Fx, & x > 0 \end{cases}.$$

Solution: The Hamiltonian $H = (1/2)p^2 + U(x)$ defines the energy E, and in this case excludes particles from occupying $x < 0$ so particles "bounce" at $x = 0$. The period is given by

$$\pi\omega^{-1} = \int_{x_1}^{x_2} \frac{dx'}{\sqrt{2(E - U(x'))}}$$

$$= \int_0^{E/F} \frac{dx'}{\sqrt{2(E - Fx')}} = \frac{1}{F}\sqrt{2(E - Fx')}\Big|_{x'=0}^{E/F}$$

$$= \frac{1}{F}\sqrt{2E},$$

with limits 0 and E/F chosen at the turning points. So we find

$$\omega = \frac{F\pi}{\sqrt{2E}},$$

and then

$$I(E) = \int_{E_{\min}}^E \frac{dE'}{\omega(E')} = \frac{1}{\pi F}\int_0^E \sqrt{2E'}dE'$$

$$= \frac{2\sqrt{2}}{3\pi F}E'^{3/2}\Big|_{E'=0}^E = \frac{2\sqrt{2}}{3\pi F}E^{3/2}$$

$$= \frac{2\sqrt{2}}{3\pi F}\left(\frac{p^2}{2} + Fx\right)^{3/2}.$$

We can then find the generating function $\tilde{F}_2(x, E)$ corresponding to the new momentum equal to the energy E,

$$\tilde{F}_2(x, E) = \int_0^x dx' \sqrt{2(E - Fx')} = -\frac{1}{3F} \left([2(E - Fx)]^{3/2} - (2E)^{3/2} \right) .$$

The generating function for the action I is $F_2(x, I) = \tilde{F}_2(x, E(I))$. Inverting our equation for $I(E)$ we find

$$E = \frac{1}{2} (3\pi F I)^{2/3} ,$$

so

$$F_2(x, I) = -\frac{1}{3} \left(\frac{1}{F} \left[(3\pi F I)^{2/3} - 2Fx \right]^{3/2} - 3\pi I \right) .$$

We then find the angle variable from

$$\begin{aligned}
\phi &= \frac{\partial F_2(x, E)}{\partial I} = -\frac{1}{3} \left(\frac{1}{F} \left[(3\pi F I)^{2/3} - 2Fx \right]^{1/2} \frac{3}{2} (3\pi F)^{2/3} \frac{2}{3} I^{-1/3} - 3\pi \right) \\
&= -\frac{1}{3F} \left(\left[(3\pi F I)^{2/3} - 2Fx \right]^{1/2} (3\pi F) (3\pi F I)^{-1/3} \right) + \pi \\
&= \pi - \frac{\pi p}{\sqrt{p^2 + 2Fx}} ,
\end{aligned}$$

using the identity $(3\pi F I)^{2/3} = 2E = p^2 + 2Fx$. The π phase offset comes from the starting point of the integral used to calculate $F_2(x, E)$. Starting at $x = x_{max}$ instead of $x = 0$ would have removed this phase term.

Problem 3.2 *Derive Eq. (3.31) for $n = 2$.*

Solution: For $n = 2$, we have the matrix J_4 defined in terms of 2×2 submatrix components as

$$J_{2n} = \begin{pmatrix} J_2 & 0 \\ 0 & J_2 \end{pmatrix} ,$$

where J_2 is given by Eq. (3.29). It is convenient to break down the matrix M in the same way, as

$$M = \begin{pmatrix} A_{11} & A_{12} \\ A_{21} & A_{22} \end{pmatrix} ,$$

where

$$A_{11} = \begin{pmatrix} \partial Q_1/\partial q_1 & \partial Q_1/\partial p_1 \\ \partial P_1/\partial q_1 & \partial P_1/\partial p_1 \end{pmatrix}, \qquad A_{12} = \begin{pmatrix} \partial Q_1/\partial q_2 & \partial Q_1/\partial p_2 \\ \partial P_1/\partial q_2 & \partial P_1/\partial p_2 \end{pmatrix},$$

$$A_{21} = \begin{pmatrix} \partial Q_2/\partial q_1 & \partial Q_2/\partial p_1 \\ \partial P_2/\partial q_1 & \partial P_2/\partial p_1 \end{pmatrix}, \qquad A_{22} = \begin{pmatrix} \partial Q_2/\partial q_2 & \partial Q_2/\partial p_2 \\ \partial P_2/\partial q_2 & \partial P_2/\partial p_2 \end{pmatrix}.$$

This readily generalizes to more degrees of freedom. While we do not know much about any single term M_{ij}, we have constraints related to combinations that correspond to Poisson brackets, because we are looking at canonical transformations.

We evaluate

$$M J_4 M^T = \begin{pmatrix} A_{11} J_2 A_{11}^T + A_{12} J_2 A_{12}^T & A_{11} J_2 A_{21}^T + A_{12} J_2 A_{22}^T \\ A_{21} J_2 A_{11}^T + A_{22} J_2 A_{12}^T & A_{21} J_2 A_{21}^T + A_{22} J_2 A_{22}^T \end{pmatrix}.$$

Now we can consider these 2×2 terms individually. We will only calculate the left column, the quantities in the right column can be found in the same way just by switching $1 \longleftrightarrow 2$ in the partial derivative terms.

We begin by looking at

$$\begin{aligned} A_{11} J_2 A_{11}^T &= \begin{pmatrix} M_{11} & M_{12} \\ M_{21} & M_{22} \end{pmatrix} \begin{pmatrix} 0 & -1 \\ 1 & 0 \end{pmatrix} \begin{pmatrix} M_{11} & M_{21} \\ M_{12} & M_{22} \end{pmatrix} \\ &= \begin{pmatrix} -M_{11}M_{12} + M_{12}M_{11} & -M_{11}M_{22} + M_{12}M_{21} \\ -M_{21}M_{12} + M_{11}M_{22} & -M_{21}M_{22} + M_{22}M_{21} \end{pmatrix} \\ &= (M_{11}M_{22} - M_{12}M_{21}) \begin{pmatrix} 0 & -1 \\ 1 & 0 \end{pmatrix} = (\det A_{11}) J_2, \end{aligned}$$

where $\det A_{11}$ is the determinant of that 2×2 matrix. Repeating this calculation for $A_{12} J_2 A_{12}^T$ gives $A_{12} J_2 A_{12}^T = (\det A_{12}) J_2$ and the top left corner of the matrix $M J_4 M^T$ becomes

$$A_{11} J_2 A_{11}^T + A_{12} J_2 A_{12}^T = (\det A_{11} + \det A_{12}) J_2.$$

Writing the elements of A_{11} and A_{12} as partial derivatives yields

$$\begin{aligned} \det A_{11} + \det A_{12} &= \frac{\partial Q_1}{\partial q_1} \frac{\partial P_1}{\partial p_1} - \frac{\partial Q_1}{\partial p_1} \frac{\partial P_1}{\partial q_1} + \frac{\partial Q_1}{\partial q_2} \frac{\partial P_1}{\partial p_2} - \frac{\partial Q_1}{\partial p_2} \frac{\partial P_1}{\partial q_2} \\ &= \sum_i \left(\frac{\partial Q_1}{\partial q_i} \frac{\partial P_1}{\partial p_i} - \frac{\partial Q_1}{\partial p_i} \frac{\partial P_1}{\partial q_i} \right) \equiv \{Q_1, P_1\}_{q,p} = 1. \end{aligned}$$

This verifies that the top left corner is J_2.

Now it only remains to show that the bottom left corner of $M J_4 M^T$ is equal to 0. The individual terms in each off-diagonal submatrix are:

$$A_{21} J_2 A_{11}^T = \begin{pmatrix} -M_{31} M_{12} + M_{32} M_{11} & -M_{31} M_{22} + M_{32} M_{21} \\ -M_{41} M_{12} + M_{42} M_{11} & -M_{41} M_{22} + M_{42} M_{21} \end{pmatrix}$$

$$A_{22} J_2 A_{12}^T = \begin{pmatrix} -M_{33} M_{14} + M_{34} M_{13} & -M_{33} M_{24} + M_{34} M_{23} \\ -M_{43} M_{14} + M_{44} M_{13} & -M_{43} M_{24} + M_{44} M_{23} \end{pmatrix} .$$

Expressing these quantities as partial derivatives, we see that all of the terms in $A_{21} J_2 A_{11}^T$ give derivatives with respect to q_1 and p_1, while all the terms in $A_{22} J_2 A_{12}^T$ give derivatives with respect to q_2 and p_2. The 2×2 submatrix can be written as

$$A_{21} J_2 A_{11}^T + A_{22} J_2 A_{12}^T = \begin{pmatrix} \{Q_1, Q_2\}_{q,p} & \{P_1, Q_2\}_{q,p} \\ \{Q_1, P_2\}_{q,p} & \{P_1, P_2\}_{q,p} \end{pmatrix} .$$

All of these Poisson brackets must equal 0 for a canonical transformation, and the identity

$$A_{21} J_2 A_{11}^T + A_{22} J_2 A_{12}^T = 0$$

is established.

References

1. E. Minguzzi, Rayleigh's dissipation function at work. Eur. J. Phys. **36**, 035014 (2016)
2. J.H. Hubbard, B.B. Hubbard, *Vector Calculus, Linear Algebra, and Differential Forms: A Unified Approach*, 2nd edn. (Prentice Hall, Upper Saddle River, 2001)

Chapter 4
Linear and Nonlinear Oscillators

The linear oscillator is a simple model that lies at the foundation of many physical phenomena and plays a crucial role in accelerator dynamics. Many systems can be viewed as an approximation to a set of independent linear oscillators. In this chapter, we will review the main properties of the linear oscillator including its response to resonant excitations, slowly varying forces, random kicks, and parametric variation of the frequency. We will discuss the impact of damping terms as well as how small, nonlinear terms in the oscillator equation modify the oscillator frequency and lead to nonlinear resonance.

4.1 Harmonic Oscillator Without and with Damping

We have already encountered Eq. (1.1) for an ideal harmonic oscillator without damping. It has the general solution,

$$x(t) = a \cos(\omega_0 t + \phi_0), \tag{4.1}$$

where a is the amplitude and ϕ_0 is the initial phase of oscillations.

Damping due to a friction force that is proportional to the velocity $\dot{x}$ adds a term with the first derivative into the differential equation,

$$\ddot{x} + \gamma \dot{x} + \omega_0^2 x = 0, \tag{4.2}$$

where γ is the damping constant and has the dimension of frequency. When the damping is not too strong, $\gamma < 2\omega_0$, the general solution to this equation is

$$x(t) = a e^{-\gamma t/2} \cos(\omega_1 t + \phi_0), \tag{4.3}$$

© Springer International Publishing AG, part of Springer Nature 2018
G. Stupakov and G. Penn, *Classical Mechanics and Electromagnetism
in Accelerator Physics*, Graduate Texts in Physics,
https://doi.org/10.1007/978-3-319-90188-6_4

with

$$\omega_1 = \omega_0\sqrt{1 - \frac{\gamma^2}{4\omega_0^2}} \, . \tag{4.4}$$

If $\gamma \ll \omega_0$, the frequency ω_1 is close to ω_0, $\omega_1 \approx \omega_0$. The damping effect is often quantified by the so-called *qualtity factor* Q defined as $Q = \omega_0/2\gamma$; the regime of weak damping is characterized by $Q \gg 1$.

If the oscillator is driven by an external force we have the following equation for $x(t)$:

$$\ddot{x} + \gamma\dot{x} + \omega_0^2 x = f(t) \, , \tag{4.5}$$

where $f(t)$ is the force divided by the oscillator mass. The general solution to this equation can be found by the method of variation of parameters. Below we will need the special case of this solution for $\gamma = 0$,

$$x(t) = x_0 \cos\omega_0 t + \frac{\dot{x}_0}{\omega_0}\sin\omega_0 t + \frac{1}{\omega_0}\int_0^t \sin[\omega_0(t - t')]f(t')dt' \, , \tag{4.6}$$

where x_0 and $\dot{x}_0$ are the initial values of the coordinate and velocity of the oscillator at $t = 0$ (see a derivation in Problem 4.1).

We can apply the formalism of the Rayleigh dissipation function from Sect. 3.5 to the harmonic oscillator with friction. The Hamiltonian of an ideal oscillator (3.1) does not have an explicit dependence on time. For a linear oscillator the conjugate momentum $p = \dot{x}$, and when the friction force $F = -\gamma\dot{x}$ we find that $R = \gamma p^2/2$ and $dH/dt = -\gamma p^2$. For a quadratic potential, we know that the average of p^2 is H in the absence of damping. Thus at least for the weak damping case, $\gamma \ll \omega_0$, we can expect to be able to take an average over a single oscillation (denoted below by the angular brackets) to find the approximate relation

$$\frac{d}{dt}\langle H \rangle \simeq -\gamma\langle H \rangle \, , \tag{4.7}$$

so the Hamiltonian decays as $H \simeq H_0 e^{-\gamma t}$. Because the Hamiltonian scales as the square of the amplitude of motion, this is consistent with Eq. (4.3).

More general forms of the frictional force can also be considered. While the above expression for H is not a replacement for the exact solution when the damping term is strong, it is useful in the case of weak damping, as will be seen in later chapters.

4.2 Resonance in a Damped Oscillator

Consider now an oscillator driven by a sinusoidal force with the frequency ω, $f(t) = f_0 \cos \omega t$. For this particular case, it is convenient to represent $x(t)$ as the real part of a complex function $\xi(t)$, $x(t) = \text{Re}\, \xi(t)$. The differential equation for $\xi(t)$,

$$\ddot{\xi} + \gamma \dot{\xi} + \omega_0^2 \xi = f_0 e^{-i\omega t} , \qquad (4.8)$$

is written in such a way that taking its real part gives Eq. (4.5) with $f(t) = f_0 \cos \omega t$. Due to the linearity of the problem, the real part of a solution to (4.8) is also a solution to (4.5).

Let us seek a particular solution to (4.8) in the form $\xi(t) = \xi_0 e^{-i\omega t}$ where ξ_0 is a complex number, $\xi_0 = |\xi_0| e^{i\phi_0}$. For such $\xi(t)$ we have $x(t) = \text{Re}\, (|\xi_0| e^{-i\omega t + i\phi_0}) = |\xi_0| \cos(\omega t - \phi_0)$, hence $|\xi_0|$ is the amplitude of the driven oscillation and ϕ_0 is its phase. Substituting $\xi(t)$ into (4.8) and solving it for ξ_0 we find

$$\xi_0 = \frac{f_0}{\omega_0^2 - \omega^2 - i\omega\gamma} . \qquad (4.9)$$

Taking the absolute value squared of ξ_0 we obtain

$$|\xi_0|^2 = \frac{f_0^2}{(\omega_0^2 - \omega^2)^2 + \omega^2 \gamma^2} . \qquad (4.10)$$

When the damping factor γ is small, $\gamma \ll \omega_0$, the dependence of $|\xi_0|^2$ versus the frequency ω exhibits *resonant* behavior: the amplitude of the oscillations increases when the driving frequency approaches the resonant frequency ω_0, reaching its maximum at $\omega \approx \omega_0$ with the resonant amplitude $|\xi_0| = f_0/\omega_0\gamma$. The *resonant width* $\Delta\omega_{\text{res}}$ is defined as a characteristic width of the resonant curve; a crude estimate for $\Delta\omega_{\text{res}}$ is $\Delta\omega_{\text{res}} \sim \gamma$.

A plot of the amplitude of the oscillations versus the frequency ω for several values of the parameter γ is shown in Fig. 4.1.

4.3 Random Kicks

What happens to an oscillator if the external force is a sequence of random kicks? Let us assume that the external force is given by the following expression,

$$f(t) = \sum_i f_i \delta(t - t_i) , \qquad (4.11)$$

Fig. 4.1 Resonant curves for various values of the damping parameter γ. The ratio γ/ω_0 takes the values of 0, 0.1, 0.2 and 0.5 with larger values corresponding to flatter curves. The curve for $\gamma = 0$ tends to infinity at $\omega = \omega_0$

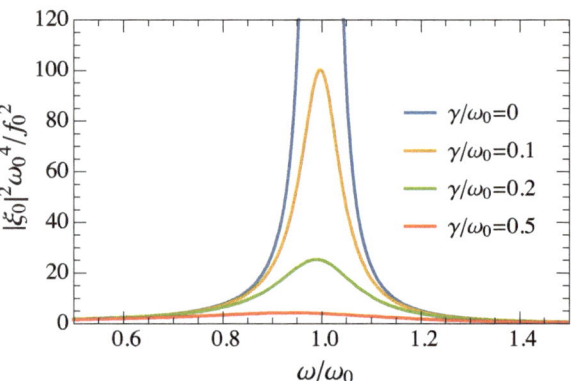

where t_i are random moments of time, and the kick amplitudes f_i take random values with zero average value, $\langle f_i \rangle = 0$. The formal solution to this problem (for $\gamma = 0$) is given by Eq. (4.6):

$$x(t) = \frac{1}{\omega_0} \int_0^t \sin[\omega_0(t - t')] f(t') dt' = \sum_i \frac{f_i}{\omega_0} \sin[\omega_0(t - t_i)], \qquad (4.12)$$

where we have assumed that at time $t = 0$ the oscillator was at rest, $x_0 = \dot{x}_0 = 0$. The result is a random process whose particular values are determined by the specific sequence of f_i and t_i. For any given realization of the random numbers f_i and t_i, the solution (4.12) would exhibit wild fluctuations in amplitude and phase. It makes sense then to consider the amplitude squared *averaged* over many random realizations of the random force with the same statistical properties. Note that for free oscillations, as it follows from (4.1), the square of the amplitude is given by $x^2 + \omega_0^{-2}\dot{x}^2$, so we associate with the averaged amplitude squared the following quantity:

$$\langle x(t)^2 + \omega_0^{-2}\dot{x}^2(t) \rangle =$$
$$= \omega_0^{-2} \sum_{i,j} \langle f_i f_j \{ \sin[\omega_0(t - t_i)] \sin[\omega_0(t - t_j)] + \cos[\omega_0(t - t_i)] \cos[\omega_0(t - t_j)] \} \rangle$$
$$= \omega_0^{-2} \sum_{i,j} \langle f_i f_j \cos[\omega_0(t_i - t_j)] \rangle, \qquad (4.13)$$

where the angular brackets denote the averaging. Let us assume that t_i and t_j are statistically independent random numbers, and they are not correlated with the kick amplitudes f_i. Then the averaging of $f_i f_j$ on the last line of Eq. (4.13) can be split from the averaging of the cosine functions, and using $\langle f_i f_j \rangle = f^2 \delta_{ij}$ we arrive at the estimate

$$\langle x(t)^2 + \omega_0^{-2}\dot{x}^2(t) \rangle = \frac{f^2}{\omega_0^2} N(t), \qquad (4.14)$$

where $N(t)$ is the average number of kicks in the interval $[0, t]$. The latter can be estimated as $N(t) \approx t/\Delta t$, where Δt is the averaged time between the kicks. We see that the square of the oscillation amplitude grows linearly with time which is a characteristic feature of the *diffusion* process. Hence the random uncorrelated kicks lead to diffusion-like behavior of the oscillation amplitude with time.

Note that in the limit $\omega_0 \to 0$, which corresponds to a free particle, we can neglect the term $x(t)^2$ on the left-hand side to obtain

$$\langle \dot{x}^2(t) \rangle = f^2 N(t) \, . \tag{4.15}$$

This is a well known result for the velocity diffusion of a free particle caused by uncorrelated random kicks.

4.4 Parametric Resonance and Slow Variation of the Oscillator Parameters

Let us now consider what happens if the frequency $\omega_0(t)$ of the linear oscillator varies with time,

$$\ddot{x} + \omega_0^2(t)x = 0 \, , \tag{4.16}$$

and moreover, $\omega_0(t)$ is a periodic function of time. Unfortunately, there is no general solution to this equation for an arbitrary function $\omega_0(t)$, so we will narrow our scope and limit our analysis to the case where $\omega_0^2(t)$ is given by the following formula,

$$\omega_0^2(t) = \Omega^2[1 - h\cos(\nu t)]. \tag{4.17}$$

Equation (4.16) with this time-dependent frequency is called the Mathieu equation; its solutions are given by the Mathieu functions.

Naively, one might think that if h is small, solutions will be close to those of the harmonic oscillator with the frequency approximately equal to Ω. This is true, however, only on a limited time interval while for some combinations of the parameters h, Ω and ν, Eq. (4.16) may exhibit unstable solutions that grow exponentially with time. For small values of h, oscillations become unstable if the ratio of the frequencies Ω/ν is close to $n/2$ where n is an integer or, in other words, for $\nu \approx 2\Omega$, Ω, $\frac{2}{3}\Omega$, $\frac{1}{2}\Omega$, ... The exact pattern of stable and unstable regions in the plane of Ω, h parameters is rather complicated; it is shown in Fig. 4.2. The unstable gaps between the stable regions become exponentially narrow when $h \lesssim 1$ and Ω/ν increases. This means that for a slow modulation, $\nu \ll \Omega$, the region $h \lesssim 1$ can be considered as a practically stable area. This is the region of *adiabatically* slow variation of the oscillator parameters.

Fig. 4.2 Stability regions for the Mathieu equation (4.16), (4.17) as functions of amplitude modulation h. The stable regions are shadowed

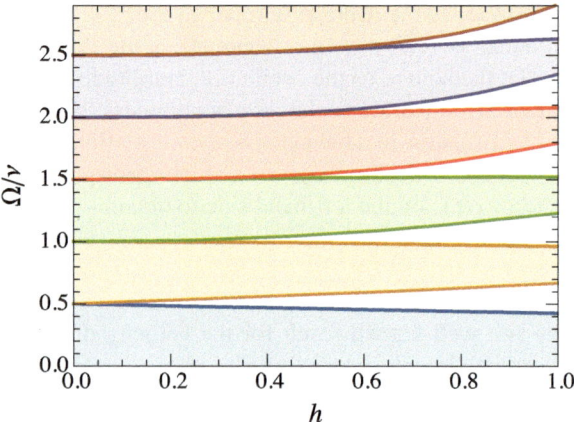

Remarkably, the adiabatic regime allows for a simple analytical treatment even for an arbitrary dependence $\omega_0(t)$. An adiabatically slow variation means that

$$\omega_0^{-2}\left|\frac{d\omega_0}{dt}\right| \ll 1 \,, \tag{4.18}$$

which also means that the relative change of the frequency ω_0 over time ω_0^{-1} is small. We seek a solution to Eq. (4.16) as a real part of the complex function $\xi(t)$, with $\xi(t)$ given by the following expression:

$$\xi(t) = A(t)\exp\left(-i\int_0^t \omega_0(t')dt' + i\phi_0\right), \tag{4.19}$$

where $A(t)$ is the slowly varying amplitude of the oscillations and ϕ_0 is the initial phase. The integral over time in this expression properly accounts for the accumulation of the phase when ω_0 is a function of time. Substituting this into Eq. (4.16) yields

$$\ddot{A} - 2i\omega_0\dot{A} - i\dot{\omega}_0 A = 0 \,. \tag{4.20}$$

Since we expect that the amplitude A is a slow function of time, we neglect the second derivative $\ddot{A}$ in this equation in comparison with the second term, which gives

$$2\omega_0\dot{A} + \dot{\omega}_0 A = 0 \,. \tag{4.21}$$

This equation can also be written as

$$\frac{d}{dt}\ln(A^2\omega_0) = 0 \,, \tag{4.22}$$

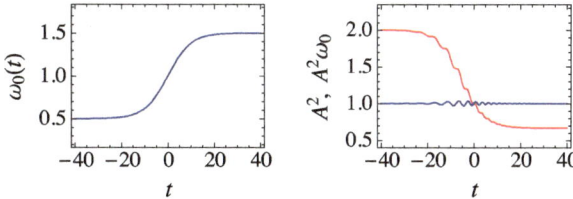

Fig. 4.3 Illustration of the adiabatic invariance of $A^2\omega_0$: the left panel shows the function $\omega_0(t)$; on the right panel the red curve shows the quantity $x(t)^2 + \dot{x}^2(t)/\omega_0^2(t)$ (which is close to the amplitude squared, A^2) while the blue curve shows the product of this quantity with $\omega_0(t)$

and then easily integrated: $A(t)^2\omega_0(t) = $ const. We found an *adiabatic invariant* for the harmonic oscillator, from which it follows that, indeed, $A(t)$ varies on the same time scale as $\omega_0(t)$, and hence is a slow-varying function as was assumed in the derivation. The value of the constant is defined by the initial values of A and ω_0; at a later time the amplitude of the oscillation varies as $A \propto 1/\sqrt{\omega_0}$.

In Fig. 4.3 we show an example of numerical integration of Eq. (4.16) in an adiabatic regime that demonstrates an approximate conservation of the product $A^2\omega_0$. It is a common feature of adiabatic invariants that once $\omega_0(t)$ stops changing, the invariant settles back to a value very close to its original value.

4.5 Nonlinear Oscillator and Nonlinear Resonance

The linear oscillator is only the lowest order approximation for a system in which the potential energy depends on the coordinate x in a fashion more general than the quadratic dependence $\omega_0^2 x^2/2$. Accounting for higher order terms in the potential energy, Eq. (1.1) for $x(t)$ is replaced by

$$\ddot{x} = -\omega_0^2 x + \alpha x^2 + \beta x^3 + \cdots, \tag{4.23}$$

where the terms on the right side of the equation are obtained through the Taylor expansion of the potential energy close to the equilibrium position. The oscillator is *weakly* nonlinear if the nonlinear terms with α and β are small in comparison with the linear ones. One of the most important properties of the nonlinear oscillator — in contrast to the linear one — is that its frequency depends on the amplitude.

Instead of trying to solve for the nonlinear oscillator (4.23) we will first analyze a particularly important example for particle accelerators, which is the *pendulum equation* (1.2),

$$\ddot{\theta} + \omega_0^2 \sin\theta = 0. \tag{4.24}$$

Fig. 4.4 Phase space orbits
for the pendulum with the
red curve showing the
separatrix

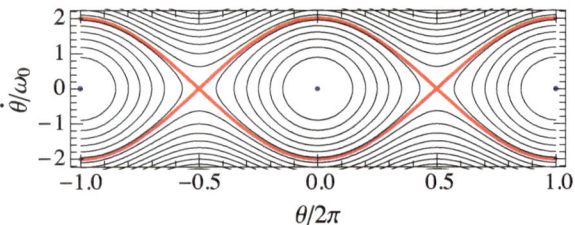

Note that for small amplitudes, $|\theta| \ll 1$, we have

$$\sin \theta \approx \theta - \frac{1}{6}\theta^3 \, , \tag{4.25}$$

and we recover Eq. (4.23) with $\alpha = 0$ and $\beta = \omega_0^2/6$. The linear approximation for
the pendulum equation is obtained if we neglect the cubic term in this expansion.

As was outlined in Sect. 1.1, using the integral of motion that characterizes the
energy, $E = \frac{1}{2}\omega_0^{-2}\dot{\theta}^2 - \cos \theta$, we obtain the following equation for the time derivative $\dot{\theta}$:

$$\dot{\theta} = \pm\omega_0\sqrt{2(E + \cos \theta)} \, . \tag{4.26}$$

This equation allows us to graph the *phase portrait* of the system where we plot
trajectories in the plane $(\theta, \dot{\theta}/\omega_0)$ using contours of constant E, see Fig. 4.4. This
plot shows the *stable points* at $\theta = 2\pi n$, $\dot{\theta} = 0$ and *unstable points* at $\theta = 2\pi n + \pi$,
$\dot{\theta} = 0$, where n is an integer number. The trajectories that pass through the unstable
points are called *separatrices*. Pendulum oscillations correspond to the values of
E such that $-1 < E < 1$; these trajectories exhibit *bounded* motion and occupy
a limited extension in θ. Orbits with $E > 1$ describe unbounded motion with the
pendulum rotating about the pivot point — the angle θ on these trajectories varies
without limits. The separatrices are the orbits with energy $E = 1$.

We now calculate the period of the pendulum T (and hence the frequency $\omega = 2\pi/T$) as a function of the oscillation amplitude. Inside the separatrix, for a given
energy E, the pendulum swings between $-\theta_0$ and θ_0, where θ_0 is defined by the
equation $\cos \theta_0 = -E$. To find a half period of the oscillations we need to integrate
$dt = d\theta/\dot{\theta}$ from $-\theta_0$ to θ_0:

$$\frac{1}{2}T\omega_0 = \omega_0 \int_{-\theta_0}^{\theta_0} \frac{d\theta}{\dot{\theta}} = \frac{1}{\sqrt{2}} \int_{-\theta_0}^{\theta_0} \frac{d\theta}{\sqrt{\cos \theta - \cos \theta_0}} \, . \tag{4.27}$$

This integral can be expressed in terms of the complete elliptic integral of the first
kind K,[1]

---

[1] Here we use the definition of the complete elliptic integral following the convention of the software
package Mathematica [1], $K(m) = \int_0^{\pi/2} \left(1 - m \sin^2 \theta\right)^{-1/2} d\theta$.

Fig. 4.5 Period T for the pendulum as a function of the amplitude angle θ_0 in the range $0 < \theta_0 < \pi$

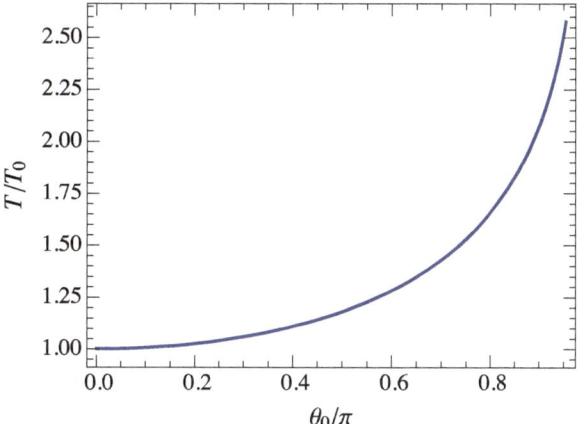

$$\frac{T}{T_0} = \frac{2}{\pi} K \left[\sin^2 \left(\frac{\theta_0}{2} \right) \right] = \frac{2}{\pi} K \left(\frac{1+E}{2} \right) , \tag{4.28}$$

where $T_0 = 2\pi/\omega_0$ is the period in the linear approximation. A plot of this function is shown in Fig. 4.5.

For small values of the argument, the Taylor expansion of the elliptic function is $(2/\pi)K(x) \approx 1 + x/4$. This means that for small amplitudes the oscillation frequency is given by

$$\omega \approx \omega_0 \left(1 - \frac{\theta_0^2}{16} \right) . \tag{4.29}$$

This frequency decreases with the amplitude θ_0.

Returning to the general case of the weakly nonlinear oscillator (4.23) and using our knowledge from the analysis of the pendulum in the limit of small amplitude (4.29), we can expect a correction to the frequency ω_0,

$$\omega(a) \approx \omega_0 + \nu a^2 , \tag{4.30}$$

where a is the amplitude and ν is a constant. Indeed, as detailed calculations show [2], for given nonlinearity parameters α and β, the constant ν is given by the following expression:

$$\nu = -\frac{3\beta}{8\omega_0} - \frac{5\alpha^2}{12\omega_0^3} . \tag{4.31}$$

The approximation (4.30) is valid if the nonlinear correction to the frequency is much smaller than ω_0, $\omega_0 \gg |\nu| a^2$. Another interesting property of a weakly nonlinear oscillation is *anharmonicity* — its Fourier spectrum contains not only the fundamental frequency $\omega(a)$, but also small contributions from higher harmonics $n\omega(a)$, where n is an integer.

Nonlinearity also changes the resonance effect. We saw in Sect. 4.2 that for a linear oscillator an external force at the resonant frequency can drive the oscillator amplitude to very large values, if the damping is small. The situation is different for a nonlinear oscillator for a reason that is easy to understand: when the amplitude grows, the frequency of the oscillator drifts from its initial value, detuning the oscillator from the resonance. The unlimited growth of the amplitude ceases when the amplitude reaches some value a_* which depends on the strength of the external force and the nonlinearity.

It is easy to get a rough estimate of the maximum value of a_* using Eq. (4.10) for the square of the amplitude in which we set $\gamma = 0$ assuming no damping. We replace ω_0 by $\omega_0 + \nu a_*^2$ and then set $\omega = \omega_0$ (choosing the frequency of the driving force to be equal to that of the linear oscillator). Using the smallness of νa_*^2 we find

$$a_*^2 \approx \frac{f_0^2}{(2|\nu|\omega_0 a_*^2)^2} \,. \tag{4.32}$$

This relation should be considered as an equation for a_*, from which we find

$$a_* \approx \left(\frac{f_0}{2|\nu|\omega_0} \right)^{1/3} \,. \tag{4.33}$$

We see that due to the nonlinearity, even at exact resonance the amplitude of the oscillations is finite; it is proportional to $|\nu|^{-1/3}$. For drive frequencies different from ω_0, this dependence will change and the peak amplitude may even increase for a forcing term having the same magnitude.

Worked Examples

Problem 4.1 *Prove that Eq. (4.6) gives a solution to Eq. (4.5) with $\gamma = 0$,*

$$\ddot{x} + \omega_0^2 x = f(t) \,. \tag{4.34}$$

Verify that the initial conditions are satisfied. Generalize the solution Eq. (4.6) for the case when $\gamma \neq 0$.

Solution: Calculating the first and the second time derivatives of Eq. (4.6) we find

$$\frac{dx}{dt} = -x_0\omega_0 \sin \omega_0 t + \dot{x}_0 \cos \omega_0 t + \int_0^t \cos [\omega_0(t - t')] f(t') dt' \,,$$

and

$$\frac{d^2 x}{dt^2} = f(t) - x_0 \omega_0^2 \cos \omega_0 t - \dot{x}_0 \omega_0 \sin \omega_0 t - \omega_0 \int_0^t \sin [\omega_0 (t - t')] f(t') dt' .$$

Substitution into Eq. (4.34) verifies the equation. To verify initial conditions, the evaluation of Eq. (4.6) at $t = 0$ yields

$$x(0) = x_0 , \qquad \dot{x}(0) = \dot{x}_0 . \tag{4.35}$$

It is instructive to understand the origin of different terms in the solution (4.6). The first two terms, $x_0 \cos \omega_0 t + \dot{x}_0 \omega_0^{-1} \sin \omega_0 t$, is a solution of the homogeneous equation (with $f = 0$) with the initial conditions (4.35). The third, integral term is a convolution of the force with the function $\omega_0^{-1} \sin(\omega_0 t)$. The latter is also a solution of Eq. (4.34) with $f(t) = \delta(t)$ and the initial conditions $x(t < 0) = 0$ and $\dot{x}(t < 0) = 0$. This solution is called the *Green function* for the harmonic oscillator. Alternatively, $\omega_0^{-1} \sin(\omega_0 t)$ can be understood as a solution of the homogeneous equation (4.34) with the initial conditions $x(0) = 0$ and $\dot{x}(0) = 1$.

The Green function approach can also be used for Eq. (4.5). Using the general solutions (4.3) of the homogeneous equation we first find a combination that satisfies the initial conditions (4.35):

$$e^{-\gamma t/2} \left[x_0 \cos \omega_1 t + \left(\frac{x_0 \gamma}{2\omega_1} + \frac{\dot{x}_0}{\omega_1} \right) \sin \omega_1 t \right] . \tag{4.36}$$

We then find the Green function as a solution of the homogeneous Eq. (4.5) with the initial conditions $x(0) = 0$ and $\dot{x}(0) = 1$:

$$\frac{1}{\omega_1} e^{-\gamma t/2} \sin (\omega_1 t) . \tag{4.37}$$

Summing Eq. (4.36) with the convolution of (4.37) with $f(t)$ gives

$$x(t) = e^{-\gamma t/2} \left[x_0 \cos \omega_1 t + \left(\frac{x_0 \gamma}{2\omega_1} + \frac{\dot{x}_0}{\omega_1} \right) \sin \omega_1 t \right]$$
$$+ \frac{1}{\omega_1} \int_0^t e^{-\gamma(t-t')/2} \sin [\omega_1 (t - t')] f(t') dt' .$$

Similar to what we did above for the solution (4.6) it is now straightforward to verify that this solution satisfies Eq. (4.5).

Problem 4.2 *The function $f(t)$ is shown in Fig. 4.6: it is equal to zero for $t < -\Delta t$, and is constant for $t > \Delta t$ with a smooth transition in between. Describe the behavior of the linear oscillator driven by this force in the limits $\Delta t \ll \omega_0^{-1}$ and $\Delta t \gg \omega_0^{-1}$.*

Fig. 4.6 A sketch of the
function $f(t)$

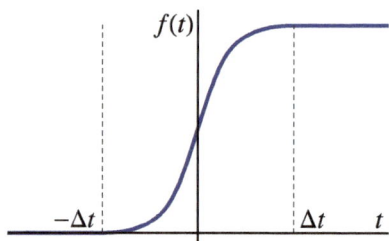

Solution: Assume $\gamma = 0$, and use the solution (4.5) with initial conditions $x_0 = \dot{x}_0 = 0$. Over time Δt, the function $f(t)$ goes from 0 to f_0. We have motion given by

$$x = \frac{1}{\omega_0} \int_{-\Delta t}^{t} \sin \omega_0(t - t') f(t') dt'$$

$$= \frac{1}{\omega_0} \left[\frac{\cos \omega_0(t - t')}{\omega_0} f(t') \Big|_{t'=-\Delta t}^{t} - \int_{-\Delta t}^{t} \frac{\cos \omega_0(t - t')}{\omega_0} \dot{f}(t') dt' \right]$$

$$= \frac{1}{\omega_0^2} \left[f(t) - \int_{-\Delta t}^{t} \cos \omega_0(t - t') \dot{f}(t') dt' \right].$$

For the case of $\Delta t \ll \omega_0^{-1}$, $f(t)$ has a steep jump at $t = 0$, and can be approximated by the Heaviside function, $f(t) \approx f_0 h(t)$, where $h = 0$ for $t < 0$ and $h = 1$ for $t > 1$. Then $\dot{f} \approx f_0 \delta(t)$ and for $t > \Delta t$ we find

$$x \approx \frac{f_0}{\omega_0^2} (1 - \cos \omega_0 t).$$

So a sudden kick from f leads to oscillations centered around f/ω_0^2 (with f the force per unit mass to give correct units). In the opposite limit, $\Delta t \gg \omega_0^{-1}$, we ignore $\dot{f}$ as small, giving

$$x \approx \frac{f(t)}{\omega_0^2}.$$

In this adiabatic limit, the position changes smoothly from $x = 0$ to $x = f_0/\omega_0^2$.

Problem 4.3 Assume $\gamma = 0$ in Eq. (4.8). Show that if $\omega \gg \omega_0$ then one can neglect the term $\omega_0^2 \xi$ in the equation. In other words, the oscillator responds to the driving force as a free particle. This fact explains why the dielectric response of many media to x-rays can be computed neglecting the binding of electrons to nuclei.

Solution: For $\gamma = 0$ and $\omega \gg \omega_0$ from Eq. (4.9) we find

$$\xi_0 = \frac{f_0}{\omega_0^2 - \omega^2} \approx -\frac{f_0}{\omega^2} \,.$$

But this approximate solution, $\xi(t) = -(f_0/\omega^2)e^{-i\omega t}$, solves the original equation (4.8) without the ω_0 term (and $\gamma = 0$):

$$\ddot{\xi} = f_0 e^{-i\omega t} \,.$$

This is an equation of motion for a free particle under the influence of the sinusoidal force with the frequency ω.

Problem 4.4 *Draw the phase portrait of a linear oscillator with and without damping.*

Solution: The phase portraits are shown in Fig. 4.7. The derivative of the position is scaled to ω_0. The left image is for the special case of no damping, $\gamma = 0$, the center image is a typical example of weak damping, $0 < \gamma < 2\omega_0$, and the final image is an example of strong damping, $\gamma > 2\omega_0$.

When there is no damping, the trajectories form periodic orbits, each corresponding to a fixed energy. With small damping ($\gamma < 2\omega_0$), the trajectories both oscillate and lose energy, and so spiral in to the center. The two trajectories shown both slowly converge to the center. When $\gamma > 2\omega_0$, ω_1 is imaginary and we do not have oscillatory behavior, only rapid decay towards the center. Multiple trajectories are shown, all converging to this point.

Problem 4.5 *Derive Eq. (4.29) directly from Eq. (4.27).*

Solution: The period is given by

$$\frac{1}{2}T\omega_0 = \frac{1}{\sqrt{2}} \int_{-\theta_0}^{\theta} \frac{d\theta}{\sqrt{\cos\theta - \cos\theta_0}} \,.$$

For small angles we can use the Taylor expansion $\cos\theta \approx 1 - \theta^2/2! + \theta^4/4!$ and approximate this as

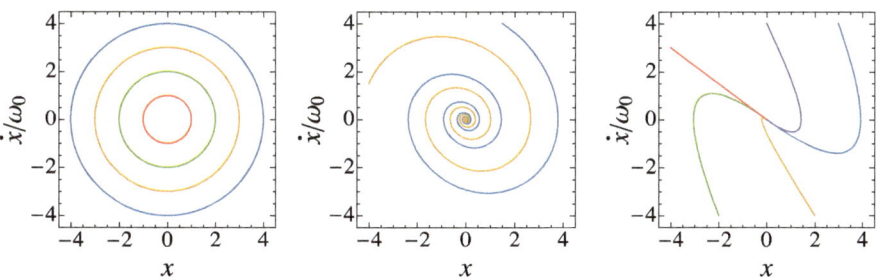

Fig. 4.7 Linear oscillator phase portraits for $\gamma = 0$ (left image), $\gamma = 0.3\,\omega_0$ (center image), and $\gamma = 2.1\,\omega_0$ (right image)

Fig. 4.8 Dependence of $\dot\theta$ versus time for a pendulum trajectory

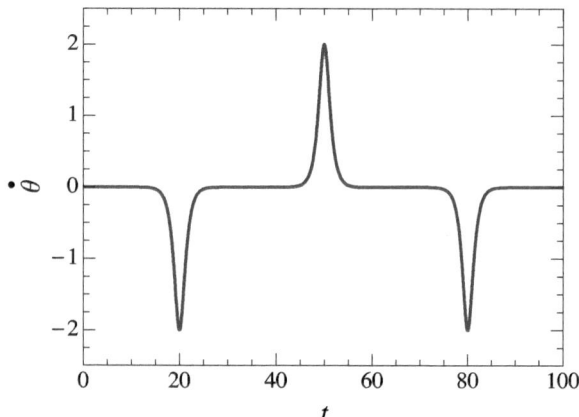

$$\frac{1}{2}T\omega_0 \approx \frac{1}{\sqrt{2}} \int_{-\theta_0}^{\theta_0} d\theta \left[\left(\frac{\theta_0^2}{2!} - \frac{\theta_0^4}{4!} \right) - \left(\frac{\theta^2}{2!} - \frac{\theta^4}{4!} \right) \right]^{-1/2}$$

$$= \frac{1}{\sqrt{2}} \int_{-\theta_0}^{\theta_0} d\theta \left\{ \left(\frac{\theta_0^2}{2} - \frac{\theta^2}{2} \right) \left[1 - \frac{1}{6} \left(\frac{\theta_0^2}{2} + \frac{\theta^2}{2} \right) \right] \right\}^{-1/2}$$

$$\approx \int_{-\theta_0}^{\theta_0} \frac{d\theta \left[1 + (\theta_0^2 + \theta^2)/24 \right]}{\sqrt{\theta_0^2 - \theta^2}}$$

$$= \int_{-1}^{1} \frac{dx}{\sqrt{1 - x^2}} \left[1 + \frac{\theta_0^2}{24} \left(1 + x^2 \right) \right] ,$$

with $x \equiv \theta/\theta_0$. This integral can be found as

$$\frac{1}{2}T\omega_0 = \sin^{-1} x \Big|_{-1}^{1} + \frac{\theta_0^2}{24} \left(\frac{3}{2} \sin^{-1} x - \frac{1}{2} x \sqrt{1 - x^2} \right) \Big|_{-1}^{1}$$

$$= \pi \left(1 + \frac{\theta_0^2}{16} \right) .$$

So with $T \equiv 2\pi/\omega$, we find

$$\omega \approx \omega_0 \left(1 - \frac{\theta_0^2}{16} \right) .$$

Problem 4.6 *Figure. 4.8 shows a numerically computed function $\dot\theta(t)$ for a pendulum with $\omega_0 = 1$. Try to figure out what is the energy E for this trajectory and explain qualitatively the shape of the curve.*

Solution: Near the separatrix, that corresponds to $\theta_0 = \pi$ and $E = 1$, the period of oscillations becomes very large. The argument of the elliptic integral K from

Eq. (4.28) will also be close to unity. Denoting $E = 1 - 2\epsilon$ we can use an approximation for the function K:

$$K(1 - \epsilon) \approx \frac{1}{2} \ln\left(\frac{16}{\epsilon}\right),$$

valid in the limit $\epsilon \ll 1$. This gives us an estimate for the period T near the separatrix:

$$\frac{T}{T_0} \approx \frac{1}{\pi} \ln \frac{32}{1 - E}.$$

As we see, the period diverges logarithmically as E approaches its value at the separatrix. The particle motion comes almost to a stop as it moves close to the 'x' points of the separatrix, which leads to the long periods of nearly zero $\dot{\theta}$ corresponding to almost reaching the highest possible position. Eventually, the particle falls back down the way it came, undergoing a nearly full cycle until it approaches the highest position from the other side.

Inverting the above expression yields

$$E \doteq 1 - 32 \exp\left(-\frac{\pi T}{T_0}\right).$$

In the figure, the period of motion is about 60, while the low-amplitude period is $T_0 = 2\pi/\omega_0 = 2\pi$ because $\omega_0 = 1$. Therefore, the energy must be $E \simeq 1 - 32e^{-30}$; it takes 12 significant digits to see a difference between this value and 1.

Problem 4.7 *Verify that Eq. (4.31) gives the result (4.29) for the pendulum.*

Solution: For the pendulum, $\alpha = 0$ and $\beta = \omega_0^2/6$, so

$$a = -\frac{3\beta}{8\omega_0} - \frac{5\alpha^2}{12\omega_0^3} = -\frac{\omega_0}{16},$$

and

$$\omega(A) \approx \omega_0 + aA^2 = \omega_0\left(1 - \frac{A^2}{16}\right),$$

where the amplitude $A = \theta_0$ is the maximum amplitude of the motion.

References

1. Wolfram Research, Inc. Mathematica, Version 11.2, 2017
2. L.D. Landau, E.M. Lifshitz, *Course of Theoretical Physics*, 3rd edn., Mechanics (Elsevier Butterworth-Heinemann, Burlington MA, 1976). (translated from Russian)

Chapter 5
Coordinate System and Hamiltonian for a Circular Accelerator

In this chapter, we derive the Hamiltonian for a particle moving in a circular accelerator. Our derivation uses several simplifying assumptions. First, we assume that there is no electrostatic field, $\phi = 0$, and the magnetic field does not vary with time. Second, the magnetic field is arranged in such a way that there is a *closed reference* (or *nominal*) *orbit* for a particle with a nominal momentum p_0—this is achieved by a proper design of the magnetic lattice of the ring. We will also assume that this reference orbit is a *plane curve* lying in the horizontal plane. Our goal is to describe the motion in the vicinity of this reference orbit of particles having energies (or, equivalently, momenta) that can slightly deviate from the nominal one.

There are very practical reasons for wanting a circular accelerator to at least approximate these properties. Accelerators can be very large and generally need to lie flat on the ground. Horizontal bending is obviously necessary to form the ring, while significant vertical bending will usually lead to coupling of the horizontal and vertical degrees of freedom unless extraordinary measures are taken in designing the beamline.

More general consideration of the issues related to the derivation of the Hamiltonian of a charged particle in an accelerator can be found in Refs. [1, 2]. The *Handbook of Accelerator Physics and Engineering* [3] is an excellent reference text for accelerators in general, with extensive references.

5.1 Coordinate System

A segment of the reference orbit is shown in Fig. 5.1. It is specified by the vector function $\mathbf{r}_0(s)$, where s is the arclength measured along the orbit *in the direction of motion*. In connection with this orbit, we will define three unit vectors. The first vector $\hat{\mathbf{s}}$ is tangential to the orbit, $\hat{\mathbf{s}} = d\mathbf{r}_0/ds$. The second vector, $\hat{\mathbf{x}}$, is perpendicular to $\hat{\mathbf{s}}$ and lies in the plane of the orbit. The third vector, $\hat{\mathbf{y}}$, is perpendicular to the plane of the orbit, $\hat{\mathbf{y}} = \hat{\mathbf{s}} \times \hat{\mathbf{x}}$. The three vectors $\hat{\mathbf{x}}$, $\hat{\mathbf{y}}$, and $\hat{\mathbf{s}}$ constitute a right-hand oriented

© Springer International Publishing AG, part of Springer Nature 2018
G. Stupakov and G. Penn, *Classical Mechanics and Electromagnetism in Accelerator Physics*, Graduate Texts in Physics, https://doi.org/10.1007/978-3-319-90188-6_5

Fig. 5.1 A segment of a
plane reference orbit (thick
blue curve), with the global
coordinate system X, Y, Z,
and a local Cartesian
coordinate system x, y and s.
For this orbit, the bending
radius ρ is positive. A
particle is located at point P

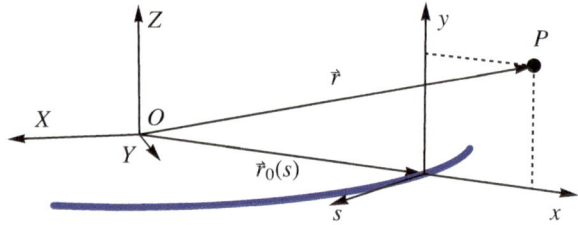

Fig. 5.1 A segment of a plane reference orbit (thick blue curve), with the global coordinate system X, Y, Z, and a local Cartesian coordinate system x, y and s. For this orbit, the bending radius ρ is positive. A particle is located at point P

base for the local coordinate system. The coordinate x is measured along $\hat{x}$, and the coordinate y is along $\hat{y}$.

Note that the direction of the vector $\hat{x}$ in our definition is not uniquely specified: it can either be directed along, or against, the curvature vector of the orbit. It is customary to direct $\hat{x}$ away from the interior of the closed orbit, as shown in Fig. 5.1. Vector $\hat{y}$, however, is unique after both $\hat{s}$ and $\hat{x}$ are established.

The three vectors $\hat{s}$, $\hat{x}$, and $\hat{y}$, being defined locally, vary with s. Frenet-Serret formulas from differential geometry [4] establish the following relations between their derivatives:

$$\frac{d\hat{s}}{ds} = -\frac{\hat{x}}{\rho(s)}, \tag{5.1a}$$

$$\frac{d\hat{x}}{ds} = \frac{\hat{s}}{\rho(s)}, \tag{5.1b}$$

$$\frac{d\hat{y}}{ds} = 0, \tag{5.1c}$$

where $\rho(s)$ is the radius of curvature of the reference orbit. Since we have made the simplifying assumption that there is no vertical bending, $d\hat{x}/ds$ is always parallel to the longitudinal coordinate $\hat{s}$ and the reference orbit lies in a plane. Therefore, the curvature sufficiently defines the local properties of the orbit. Furthermore, the magnetic field can only have y (vertical field) and/or s (solenoidal field) components. The absolute value of the radius ρ is given by the equation (see Eq. (1.25))

$$|\rho(s)| = \frac{p_0}{|eB_y(s)|}, \tag{5.2}$$

where p_0 is the kinetic momentum of the reference particle and B_y is the vertical component of the magnetic field. In the more general case of orbits that move out of a single plane, the expressions for the derivatives above would have additional terms related to *torsion*, and the magnetic field could also have an x component, but those will not be considered here. Under the above constraints, the torsion is always zero.

Beam particles deviate from the reference orbit, but move close to it. A particle position is defined in the local coordinate system as shown in Fig. 5.1, where a radius vector r is represented by coordinates s, x, and y such that

$$r = r_0(s) + x\hat{x}(s) + y\hat{y}.$$ (5.3)

Below we will need to carry out various differential operations in curvilinear coordinates. Here are some useful formulae for the gradient of a scalar function $\phi(x, y, s)$, and for the curl and divergence of a vector function $A = (A_x(x, y, s), A_y(x, y, s), A_s(x, y, s))$:

$$\nabla\phi = \hat{x}\frac{\partial\phi}{\partial x} + \hat{y}\frac{\partial\phi}{\partial y} + \hat{s}\frac{1}{1 + x/\rho}\frac{\partial\phi}{\partial s},$$ (5.4a)

$$(\nabla \times A)_x = -\frac{1}{1 + x/\rho}\frac{\partial A_y}{\partial s} + \frac{\partial A_s}{\partial y},$$ (5.4b)

$$(\nabla \times A)_y = -\frac{1}{1 + x/\rho}\frac{\partial A_s(1 + x/\rho)}{\partial x} + \frac{1}{1 + x/\rho}\frac{\partial A_x}{\partial s},$$ (5.4c)

$$(\nabla \times A)_s = -\frac{\partial A_x}{\partial y} + \frac{\partial A_y}{\partial x},$$ (5.4d)

$$\nabla \cdot A = \frac{1}{1 + x/\rho}\frac{\partial A_x(1 + x/\rho)}{\partial x} + \frac{\partial A_y}{\partial y} + \frac{1}{1 + x/\rho}\frac{\partial A_s}{\partial s}.$$ (5.4e)

5.2 Hamiltonian in Curvilinear Coordinate System

The general Hamiltonian for a charged particle is given by Eq. (1.37), which for $\phi = 0$ takes the form

$$H = \sqrt{(mc^2)^2 + c^2(\pi - eA)^2}.$$ (5.5)

This Hamiltonian was derived for a Cartesian coordinate system. We now want to use it in the coordinate system related to the reference orbit described in the previous section. This requires a canonical transformation from the old to new coordinates that we will carry out with the help of generating functions introduced in Sect. 2.4.

As a first step, we choose the local coordinates s, x, and y as coordinate variables for our new Hamiltonian. For the transformation from the original Cartesian coordinates X, Y, and Z (see Fig. 5.1) to the new ones, we use the generating function of the third type:

$$F_3(\pi, x, y, s) = -\pi \cdot (r_0(s) + x\hat{x}(s) + y\hat{y}).$$ (5.6)

In this equation π is the *old* momentum and x, y and s are the new coordinates. The new canonical momentum is denoted by Π; it is given by Eq. (2.30),

$$\Pi_x = -\frac{\partial F_3}{\partial x} = \boldsymbol{\pi} \cdot \hat{\boldsymbol{x}} = \pi_x,$$

$$\Pi_y = -\frac{\partial F_3}{\partial y} = \boldsymbol{\pi} \cdot \hat{\boldsymbol{y}} = \pi_y,$$

$$\Pi_s = -\frac{\partial F_3}{\partial s} = \boldsymbol{\pi} \cdot \left(\frac{d\boldsymbol{r_0}}{ds} + x\frac{d\hat{\boldsymbol{x}}}{ds}\right) = \boldsymbol{\pi} \cdot \left(\hat{\boldsymbol{s}} + \frac{x}{\rho}\hat{\boldsymbol{s}}\right) = \pi_s\left(1 + \frac{x}{\rho}\right), \quad (5.7)$$

where we have used Eq. (5.1b). Note that

$$(\boldsymbol{\pi} - e\boldsymbol{A})^2 = (\pi_x - eA_x)^2 + (\pi_y - eA_y)^2 + (\pi_s - eA_s)^2$$

$$= (\Pi_x - eA_x)^2 + (\Pi_y - eA_y)^2 + \left(\frac{\Pi_s}{1 + x/\rho} - eA_s\right)^2, \quad (5.8)$$

and our Hamiltonian can be written as

$$H = c\left[m^2c^2 + (\Pi_x - eA_x)^2 + (\Pi_y - eA_y)^2 + \left(\frac{\Pi_s}{1 + x/\rho} - eA_s\right)^2\right]^{1/2}. \quad (5.9)$$

Here we have introduced the notation $A_x = \boldsymbol{A} \cdot \hat{\boldsymbol{x}}$, $A_y = \boldsymbol{A} \cdot \hat{\boldsymbol{y}}$, and $A_s = \boldsymbol{A} \cdot \hat{\boldsymbol{s}}$[1].

Equation (5.9) is our new Hamiltonian as a function of the new coordinates x, y, s and the new conjugate momenta Π_x, Π_y and Π_s.

5.3 Using s as Time Variable

It was assumed at the beginning of this chapter that our Hamiltonian does not depend on time t and hence is a constant of motion. Because a particle has three degrees of freedom, the Hamiltonian depends on three pairs of conjugate variables. It turns out that, using conservation of H, one can transform the equations of motion in such a way that they are described by a Hamiltonian with two pairs of conjugate variables, effectively lowering the number of degrees of freedom from three to two. This is achieved by changing the independent variable from time t to the longitudinal coordinate s.

Let us assume that we have solved the equations of motion and found all the Hamiltonian variables as functions of time, $x(t)$, $y(t)$, $s(t)$, etc. Then the form of, say, $x(s)$ is obtained in the following way. From the equation $s = s(t)$ we find the inverse function $t(s)$ and substitute it into the argument of x: $x(t) \rightarrow x(t(s))$. The latter becomes a function of s: $x(s) = x(t(s))$. Repeating the same procedure for the coordinate y and the components of the momentum vector $\boldsymbol{\Pi}$ we obtain $y(s)$ and $\boldsymbol{\Pi}(s)$.

---

[1] In some textbooks the reader can find a different definition of A_s, $A_s = (1 + x/\rho)\boldsymbol{A} \cdot \hat{\boldsymbol{s}}$.

It turns out that the dependence of x, y, Π_x, and Π_y versus s can be found directly by solving Hamiltonian equations with two degrees of freedom, without invoking the intermediate step of reversing the function $s(t)$. We will first show how to construct the new Hamiltonian, and then will prove that the new Hamiltonian equations are equivalent to the original ones.

We start from equating the Hamiltonian (5.9) to a constant value of h,

$$h = H(x, \Pi_x, y, \Pi_y, s, \Pi_s), \tag{5.10}$$

and then treat this relation as an equation for Π_s. Solving it, we find Π_s as a function of the two pairs of conjugate variables plus h and s,

$$\Pi_s = \Pi_s(x, \Pi_x, y, \Pi_y, h, s), \tag{5.11}$$

where h is understood as a constant; recalling Eq. (1.35) and remembering that $\phi = 0$, we see that the value of this constant is equal to γmc^2. The crucial next step is to introduce a new Hamiltonian, which we denote by K, equal to the negative function Π_s,

$$K(x, \Pi_x, y, \Pi_y, h, s) = -\Pi_s(x, \Pi_x, y, \Pi_y, h, s). \tag{5.12}$$

Here x, Π_x, y, Π_y are considered as canonical conjugate variables, s is an independent "time-like" variable, and h is a (constant) parameter. As we see, the Hamiltonian K has two pairs of conjugate variables, and hence describes motion of a system which has two degrees of freedom. However, this Hamiltonian depends on s and hence, in contrast to the original H, is not a conserved quantity. The loss of the Hamiltonian constancy is the price paid for lowering the number of degrees of freedom from three to two.

Let us now prove that the s-evolution of the functions $x(s)$, $\Pi_x(s)$, $y(s)$, and $\Pi_y(s)$ are governed by the Hamiltonian (5.12). Because $x(s)$ is obtained from $x(t)$ and $s(t)$ by eliminating the variable t, we have for dx/ds,

$$\frac{dx}{ds} = \frac{dx/dt}{ds/dt} = \frac{\partial H/\partial \Pi_x}{\partial H/\partial \Pi_s}, \tag{5.13}$$

where we have used the Hamiltonian equations of motion for dx/dt and ds/dt. On the other hand, the derivative $\partial K/\partial \Pi_x$ can be calculated as a derivative of an implicit function by differentiating Eq. (5.10) with respect to Π_x, yielding:

$$\frac{\partial K}{\partial \Pi_x} = -\left(\frac{\partial \Pi_s}{\partial \Pi_x}\right)_h = \frac{\partial H/\partial \Pi_x}{\partial H/\partial \Pi_s}. \tag{5.14}$$

We see that

$$\frac{dx}{ds} = \frac{\partial K}{\partial \Pi_x},$$

(5.15)

which is a Hamiltonian equation for dx/ds with the Hamiltonian K. The same approach works for Π_x,

$$\frac{d\Pi_x}{ds} = \frac{d\Pi_x/dt}{ds/dt} = \frac{-\partial H/\partial x}{\partial H/\partial \Pi_s} = \left(\frac{\partial \Pi_s}{\partial x}\right)_h = -\frac{\partial K}{\partial x}.$$

(5.16)

Similarly one can calculate dy/ds and $d\Pi_y/ds$ and find out that they are equal to $\partial K/\partial \Pi_y$ and $-\partial K/\partial y$, respectively.

Although time is now eliminated from the equations, the time dependence of s, and hence all of the variables, can be easily recovered. Indeed, in the original Hamiltonian equations we had $ds/dt = \partial H/\partial \Pi_s$. Taking the reciprocal gives the equation for $t(s)$:

$$\frac{dt}{ds} = \frac{1}{\partial H/\partial \Pi_s} = \frac{\partial \Pi_s}{\partial h} = -\frac{\partial K}{\partial h}.$$

(5.17)

The second equality is found by substituting Eq. (5.11) into Eq. (5.10) and then differentiating with respect to h. Integrating this equation over s we find $t = t(s)$, with the inverse function defining $s(t)$.

5.4 Small Amplitude Approximation

We will now accept s as an independent variable and switch from the Hamiltonian H (5.9) to the Hamiltonian K given by Eq. (5.12). Solving Eq. (5.9) for Π_s, we obtain the following expression for K:

$$K = -\left(1 + \frac{x}{\rho}\right)\left[\frac{1}{c^2}h^2 - (\Pi_x - eA_x)^2 - (\Pi_y - eA_y)^2 - m^2c^2\right]^{1/2}$$
$$- eA_s\left(1 + \frac{x}{\rho}\right).$$

(5.18)

As we will see in the next chapter, in most cases of interest a single component A_s is sufficient to describe the magnetic field in an accelerator, so we set $A_x = A_y = 0$ in Eq. (5.18). With this choice of the vector potential, Π_x and Π_y are equal to the kinetic momenta, $\Pi_x = p_x = m\gamma v_x$ and $\Pi_y = p_y = m\gamma v_y$ (see Eqs. (5.7) and (1.33)), and we can use p_x and p_y instead of Π_x and Π_y:

$$K = -\left(1 + \frac{x}{\rho}\right)\left(\frac{1}{c^2}h^2 - p_x^2 - p_y^2 - m^2c^2\right)^{1/2} - eA_s\left(1 + \frac{x}{\rho}\right).$$

(5.19)

Particles in an accelerator beam typically move at small angles relative to the reference orbit. This means that p_x and p_y are small in comparison with the total momentum p, and the square root in (5.19) can be expanded as a Taylor series. This is the main advantage of the local coordinate system, in which x and y are measured relative to the reference orbit. Expanding K in p_x and p_y and keeping only the lowest order terms, we obtain:

$$K \approx -p \left(1 + \frac{x}{\rho}\right)\left(1 - \frac{p_x^2}{2p^2} - \frac{p_y^2}{2p^2}\right) - eA_s\left(1 + \frac{x}{\rho}\right), \qquad (5.20)$$

where $p(h) = \sqrt{h^2/c^2 - m^2c^2}$ is the total kinetic momentum of the particle (which together with the energy is a conserved quantity in a constant magnetic field).

Instead of using the variables p_x and p_y it is convenient to introduce dimensionless quantities $P_x = p_x/p_0$ and $P_y = p_y/p_0$, where p_0 is the nominal momentum in the ring. The transformation from x, p_x, y, p_y to x, P_x, y, P_y is not canonical, but it does not change the Hamiltonian structure of the equations of motion. A simple analysis shows that the new Hamiltonian, which we denote by $\mathcal{H}$, is obtained by dividing K by p_0 (see Problem 2.1 on p. 33):

$$\mathcal{H}(x, P_x, y, P_y) = \frac{K}{p_0}$$
$$= -\frac{p}{p_0}\left(1 + \frac{x}{\rho}\right)\left[1 - \frac{1}{2}P_x^2\left(\frac{p_0}{p}\right)^2 - \frac{1}{2}P_y^2\left(\frac{p_0}{p}\right)^2\right] - \frac{e}{p_0}A_s\left(1 + \frac{x}{\rho}\right).$$
$$(5.21)$$

As mentioned at the beginning of this chapter, we are interested here in the case when the energy and the total momentum of the particles deviate only slightly from the nominal one, that is

$$\frac{p}{p_0} = 1 + \eta, \qquad (5.22)$$

with $\eta \ll 1$. This somewhat simplifies $\mathcal{H}$:

$$\mathcal{H}(x, P_x, y, P_y)$$
$$\approx -(1 + \eta)\left(1 + \frac{x}{\rho}\right)\left(1 - \frac{1}{2}P_x^2 - \frac{1}{2}P_y^2\right) - \frac{e}{p_0}A_s\left(1 + \frac{x}{\rho}\right), \qquad (5.23)$$

where we have replaced $(p_0/p)^2$ by unity in small quadratic terms proportional to P_x^2 and P_y^2.

Finally, we note that our dimensionless momenta P_x, P_y are approximately equal to the slopes $x' \equiv dx/ds$ and $y' \equiv dy/ds$, respectively. Indeed, we have

$$x' \equiv \frac{dx}{ds} = \frac{v_x}{v_s} = \frac{p_x}{p_s} \approx P_x , \tag{5.24}$$

with a similar expression for y'. Some authors use the notations x' and y' instead of P_x, P_y for the canonical momenta conjugate to x and y — when using that notation, one has to be careful to avoid confusion between a canonical variable, e.g., P_x, with the rate of change of its conjugate, dx/ds.

5.5 Time-Dependent Hamiltonian

While we emphasized above that for a time-independent Hamiltonian the transition from t to s as an independent variable eliminates one degree of freedom, the requirement of being independent of time is actually not needed. Without changes, our derivation in Sect. 5.3 can be easily generalized to the case when H is also a function of t. Writing down Eq. (5.10) we now have $H(x, \Pi_x, y, \Pi_y, s, \Pi_s, t)$ on the right-hand side and, correspondingly, the new Hamiltonian K becomes a function of time,

$$K(x, \Pi_x, y, \Pi_y, t, h, s) = -\Pi_s(x, \Pi_x, y, \Pi_y, t, h, s). \tag{5.25}$$

Time t in this Hamiltonian should now be understood as a third coordinate (in addition to x and y). Indeed, Eq. (5.17) is a Hamiltonian equation for $t(s)$, if $-h$ is associated with the momentum conjugate to t. For such an association to be valid, we also need the complementary Hamiltonian equation,

$$\frac{dh}{ds} = \frac{\partial K}{\partial t}. \tag{5.26}$$

This equation is easily proven if one takes into account that $dh/ds = (1/\dot{s})\partial H/\partial t$, expresses the time derivative of H in terms of the partial derivatives of K, $\partial H/\partial t = -(\partial K/\partial t)/(\partial K/\partial h)$, and uses Eq. (5.17). So we conclude that for a time-dependent Hamiltonian H, in the transformation to the Hamiltonian K, one has to treat time and the negative energy as a canonically conjugate pair of variables.

One important physical effect that requires a time-dependent Hamiltonian for its description is acceleration of charged particles by radio-frequency (RF) electromagnetic fields. In a simple model which assumes that the field is localized in a short RF cavity which is powered to voltage V, an additional term that needs to be added to the Hamiltonian (5.18) is [1]

$$\frac{eV}{\omega_{\mathrm{RF}}} \delta(s - s_0) \sin(\omega_{\mathrm{RF}} t + \phi) , \tag{5.27}$$

where ω_{RF} is the RF frequency, ϕ is the RF phase, and s_0 is the coordinate of the cavity location in the ring. The cavity length in this equation is assumed to be infinitesimally small, so that its action is localized at one point in the ring. Using Eq. (5.26), it is easy to calculate that a passage through the point s_0 at time t changes the kinetic energy h of the particle by $eV \cos(\omega_{RF} t + \phi)$, which reaches the maximum value of $\Delta h = eV$ when the argument of the cos function is equal to a multiple of 2π. Because the Hamiltonian (5.23) is obtained from K through division by p_0, when adding the RF term to $\mathcal{H}$ we also need to divide it by the same factor p_0.

Worked Examples

Problem 5.1 *Check that Eq. (5.1) holds for a circular orbit.*

Solution: A circular reference orbit of radius ρ_0 is parametrized in Cartesian coordinates by

$$\boldsymbol{r}_0 = (\rho_0 \cos(s/\rho_0), \rho_0 \sin(s/\rho_0), 0) \, ,$$

where we assumed that the orbit lies in the plane $Z = 0$. Using $\hat{\boldsymbol{x}} = \boldsymbol{r}_0/\rho_0$ we find that

$$\frac{d\boldsymbol{r}_0}{ds} = (-\sin(s/\rho_0), \cos(s/\rho_0), 0) = \hat{\boldsymbol{s}} \, ,$$

$$\frac{d\hat{\boldsymbol{s}}}{ds} = \frac{d}{ds} (-\sin(s/\rho_0), \cos(s/\rho_0), 0)$$

$$= \left(-\frac{1}{\rho_0} \cos(s/\rho_0), -\frac{1}{\rho_0} \sin(s/\rho_0), 0\right) = -\frac{\boldsymbol{r}_0}{\rho_0^2} = -\frac{\hat{\boldsymbol{x}}}{\rho_0} \, ,$$

$$\frac{d\hat{\boldsymbol{x}}}{ds} = \frac{1}{\rho_0} \frac{d\boldsymbol{r}_0}{ds} = \frac{\hat{\boldsymbol{s}}}{\rho_0} \, ,$$

$$\frac{d\hat{\boldsymbol{y}}}{ds} = \frac{d}{ds}(0, 0, 1) = (0, 0, 0) \, .$$

Problem 5.2 *Verify that if we extend Eq. (5.2) to define a specific choice for the sign of ρ,*

$$\rho(s) = \frac{p_0}{q B_y(s)} \, , \tag{5.28}$$

this choice of sign is correct within Eq. (5.1) for arbitrary sign of the charge q and of the direction of the motion in the reference orbit.

Solution: Because $\hat{\boldsymbol{s}}$ is directed along $\boldsymbol{v}$ we can write $\boldsymbol{v} = v\hat{\boldsymbol{s}}$. Substituting this velocity into the magnetic force on the charged particle $\boldsymbol{F} = q\boldsymbol{v} \times \boldsymbol{B}$ we obtain

Fig. 5.2 Electron trajectory
in a chicane with the color
rectangles showing the
positions of the four dipole
magnets. Assume that the y
axis is directed out of the
page

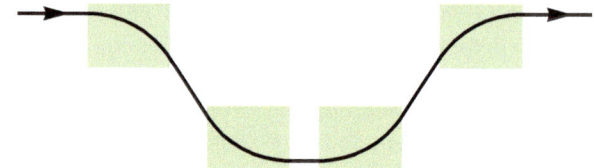

$F = qvB_y\hat{s} \times \hat{y} = -qvB_y\hat{x}$, where we have used the relation $\hat{s} \times \hat{y} = -\hat{x}$ for a right-handed coordinate system x, y, s. Substituting Eq. (5.28) into Eq. (5.1a) we find that $d\hat{s}/ds = -qB_y\hat{x}/p_0$, which means that $d\hat{s}/ds$ is co-linear with the force F, and the trajectory turns in the direction of the force. Thus, this choice of sign is consistent with the sign in Eqs. (5.1a, 5.1b, 5.1c) for either sign of q or B_y. We emphasize that for this consistency to hold true it is necessary to align $\hat{s}$, the direction of increasing s, with the velocity of the particle.

Problem 5.3 *Figure 5.2 shows the electron trajectory in a four-dipole chicane (typically used for bunch compressions). Indicate the direction of axis x assuming that the y axis is directed out of the page toward you. Determine the sign of the orbit radius ρ and the magnetic field direction of each of four dipoles along the orbits. What happens with this sign if the particle is moving in the direction opposite to the one shown in the figure?*

Solution: For the chicane, taking the y axis out of the plane and s to the right, the x axis is pointing up. Tracking the rotation of $\hat{x}$ through the chicane, it is easy to see that in the first and the last magnets $\hat{x}$ rotates in the $+\hat{s}$ direction. In the second and third magnets it rotates in the $-\hat{s}$ direction. From Eq. (5.1b) we then conclude that ρ is positive in the first and the last magnets and negative in the second and third ones. The direction of the magnetic field is then determined from Eq. (5.28) taking into account that $q = -e$ negative for electrons: B_y is negative in the first and the last magnets and positive in the second and the third.

If the electron follows the same path in the opposite direction, then all of the B fields change direction to preserve the bending direction. If we keep $\hat{y}$ in the same direction as before, the signs of ρ are flipped. Because the direction of the electron trajectory is reversed, the geometry requires $\hat{x}$ to be reversed as well, which is another way of seeing that ρ changes sign even though the vector form of the radius of curvature is a function of the curve alone.

Problem 5.4 *Verify that Eqs. (5.4a)–(5.4e) hold for a circular orbit.*

Solution: For a circular orbit, the curvilinear coordinate system (x, y, s) reduces to the standard cylindrical coordinates (r, θ, z) if we establish the following correspondence between the unit vectors:

$$\hat{r} = \hat{x}, \qquad \hat{\theta} = \hat{s}, \qquad \hat{z} = -\hat{y},$$

while for the coordinates

$$r = \rho + x, \qquad \theta = \frac{s}{\rho}, \qquad z = -y.$$

The minus sign in the $y \rightarrow z$ correspondence is needed to have the (r, θ, z) coordinate system be right-handed. Changing the notation $A_x \rightarrow A_r$, $A_y \rightarrow -A_z$ and $A_s \rightarrow A_\theta$ it is then a straightforward calculation to check that Eqs. (5.4a)–(5.4e) are the same as the corresponding differential operators in the cylindrical coordinates.

Problem 5.5 *Find the Hamiltonian K for the following model Hamiltonian H:*

$$H(x, \Pi_x, s, \Pi_s) = \frac{\Pi_x^2}{2} + \omega^2(s)\frac{x^2}{2} + v\Pi_s,$$

where v is a constant. Prove that both Hamiltonians describe the same dynamics.

Solution: Solving equation $H(x, \Pi_x, s, \Pi_s) = h$ for Π_s we find

$$K(x, \Pi_x, s, h) = -\Pi_s = \frac{1}{v}\left(-h + \frac{\Pi_x^2}{2} + \omega^2(s)\frac{x^2}{2}\right).$$

The equations of motion that follow from the old Hamiltonian are:

$$\dot{x} = \frac{\partial H}{\partial \Pi_x} = \Pi_x$$

$$\dot{\Pi}_x = -\frac{\partial H}{\partial x} = -x\omega^2$$

$$\dot{s} = \frac{\partial H}{\partial \Pi_s} = v$$

$$\dot{\Pi}_s = -\frac{\partial H}{\partial s} = -x^2\omega\frac{d\omega}{ds}.$$

The new Hamiltonian K has two conjugate variables, x and Π_x, and the two equations of motion are

$$\frac{dx}{ds} = \frac{\partial K}{\partial \Pi_x} = \frac{\Pi_x}{v}$$

$$\frac{d\Pi_x}{ds} = -\frac{\partial K}{\partial x} = -\frac{x\omega^2}{v}.$$

These equations are complemented by Eq. (5.17),

$$\frac{dt}{ds} = -\frac{\partial K}{\partial h} = \frac{1}{v}.$$

The last equation that we need relates $d\Pi_s/ds$ to the rate of change of the Hamiltonian K,

$$\frac{d\Pi_s}{ds} = -\frac{dK}{ds} = -\frac{\partial K}{\partial s} = -\frac{x^2\omega}{v}\frac{d\omega}{ds}.$$

We can then compare the two sets of equations, and see that $ds/dt = v$ and $dt/ds = 1/v$ are consistent with each other. The remaining equations differ only by a factor of $1/v$, as we expect because we are taking derivatives with respect to s instead of t. Multiplying by $ds/dt = v$ makes these equations identical.

References

1. J.M. Jowett, Introductory statistical mechanics for electron storage rings. AIP Conf Proc **153**, 864–970 (1987)
2. K.R. Symon, Derivation of Hamiltonians for accelerators. Report ANL/APS/TB-28, Argonne National Laboratory, 1997
3. A.W. Chao, M. Tigner, *Handbook of Accelerator Physics and Engineering*, 3rd edn. (World Scientific Publishing, Singapore, 2006)
4. E. Kreyszig, *Differential Geometry* (Dover Publications, 1991)

Chapter 6
Equations of Motion in Accelerators

A typical accelerator uses a sequence of various types of magnets separated by sections of free space (so-called *drifts*) to control the motion of the particle beam. To specify the Hamiltonian (5.23) we need to know the vector potential, A_s, for these magnets. We evaluate the fields and Hamiltonian for the major magnet types assuming that the field profiles are uniform over their length. Often in analysis and simulations, one has to take into account that at the end points of the magnets different field geometries appear, called *fringe fields*. The impact of these fields are usually treated as highly localized corrections which are calculated separately from the bulk of the magnet, and involve higher order terms that we will simply neglect in this chapter. When fringe fields are weak they can be treated as field errors, which are covered in Chap. 8.

6.1 Vector Potential for Different Types of Magnets

There are several types of magnets that are used in accelerators and each of them is characterized by a specific dependence of the longitudinal component of the vector potential, A_s, versus x and y. In this section, we will list several magnet types and write down approximate expressions for $A_s(x, y)$. In our approximations, we will use the fact that we are only interested in fields near the reference orbit, $|x|, |y| \ll |\rho|$, so we can neglect higher powers of the ratios x/ρ and y/ρ.

We first consider the *dipole* magnets that are used to bend the orbit and, in circular accelerators, to eventually make it close on itself. Assuming that the field is directed along y and, in the lowest approximation, neglecting its variation in the transverse plane (that is, neglecting its dependence on x and y), we have

$$\boldsymbol{B} = \hat{\boldsymbol{y}} B(s) . \tag{6.1}$$

© Springer International Publishing AG, part of Springer Nature 2018
G. Stupakov and G. Penn, *Classical Mechanics and Electromagnetism in Accelerator Physics*, Graduate Texts in Physics,
https://doi.org/10.1007/978-3-319-90188-6_6

The function $B(s)$ characterizes the longitudinal variation of the field, and vanishes outside of the magnets. Within the dipole the field can be represented by the following vector potential:

$$A_s = -B(s)x \left(1 - \frac{x}{2\rho} \right) . \tag{6.2}$$

Indeed, using Eq. (5.4c) we obtain

$$\begin{aligned}
B_y &= -\frac{1}{1 + x/\rho} \frac{\partial A_s (1 + x/\rho)}{\partial x} \\
&\approx B(s) \left(1 - \frac{x}{\rho} \right) \frac{\partial}{\partial x} \left[x \left(1 - \frac{x}{2\rho} \right) \left(1 + \frac{x}{\rho} \right) \right] \\
&\approx B(s) + O \left(\frac{x^2}{\rho^2} \right) .
\end{aligned} \tag{6.3}$$

This is an approximation in which we only keep terms to the first order in $|x/\rho|$.

The second type is a *quadrupole* magnet. It is used to focus off-orbit particles so that they remain close to the reference orbit. It has two components of the field, B_x and B_y, that linearly increase with the distance from the axis:

$$B = G(s)(\hat{y}x + \hat{x}y) , \tag{6.4}$$

where the function $G(s)$ again isolates the longitudinal variation of the field. A picture of quadrupole field lines is shown in Fig. 6.1a.

Note that the field on the axis is zero, which means that the reference orbit is a straight line, $\rho = \infty$. It is straightforward to verify that the corresponding vector potential is

$$A_s = \frac{1}{2}G(s) \left(y^2 - x^2 \right) . \tag{6.5}$$

A *skew quadrupole* is a normal quadrupole rotated by 45°:

$$B = G_{sq}(s)(-\hat{y}y + \hat{x}x) , \tag{6.6}$$

with

$$A_s = G_{sq}(s)xy . \tag{6.7}$$

Finally, we also consider a *sextupole magnet*. Sextupoles are used to correct some properties of the transverse oscillations of the beam particles around the reference orbit. This magnet has a nonlinear dependence of the magnetic field with the transverse coordinates:

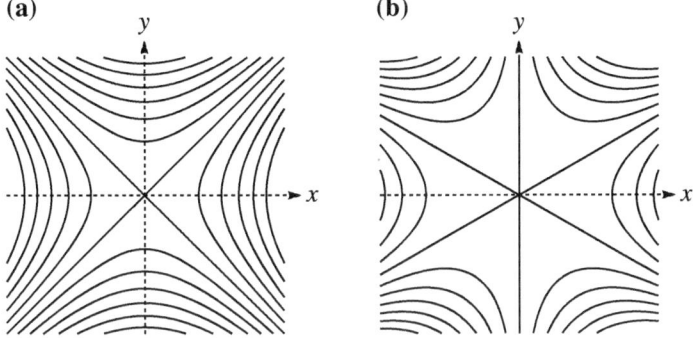

Fig. 6.1 Quadrupole (**a**) and sextupole (**b**) field lines

$$\boldsymbol{B} = S(s)\left[\frac{1}{2}\hat{\boldsymbol{y}}(x^2 - y^2) + \hat{\boldsymbol{x}}xy\right].$$ (6.8)

The vector potential corresponding to this field is

$$A_s = S(s)\left(\frac{1}{2}xy^2 - \frac{1}{6}x^3\right).$$ (6.9)

The field lines of a sextupole are shown in Fig. 6.1b. There is also a skew version of the sextupole, and for higher order magnets as well.

6.2 Taylor Expansion of the Hamiltonian

We are now in position to specify the Hamiltonian (5.23) for a circular accelerator that has dipole and quadrupole magnets in the ring. Replacing A_s with the sum of the vector potentials (6.2) and (6.5) we obtain

$$
\begin{aligned}
\mathcal{H} &= -(1+\eta)\left(1+\frac{x}{\rho}\right)\left(1 - \frac{1}{2}P_x^2 - \frac{1}{2}P_y^2\right) \\
&\quad - \frac{e}{p_0}\left[-B(s)x\left(1 - \frac{x}{2\rho}\right) + \frac{1}{2}G(s)\left(y^2 - x^2\right)\right]\left(1 + \frac{x}{\rho}\right) \\
&\approx -1 - \eta - \eta\frac{x}{\rho} + \frac{1}{2}P_x^2 + \frac{1}{2}P_y^2 + \frac{x^2}{2\rho^2} - \frac{e}{p_0}\frac{1}{2}G(s)\left(y^2 - x^2\right),
\end{aligned}
$$ (6.10)

where we have made use of $\rho = p_0/eB(s)$ and neglected terms of the third and higher orders (assuming that η, as well as all four canonical variables x, P_x, y and P_y are of the first order). In what follows, we drop the constant first term -1 on the last line of Eq. (6.10).

Our main interest in the following chapters will be the case of on-momentum particles, $\eta = 0$. In this case, the Hamiltonian is the sum of two terms corresponding to the vertical and horizontal degrees of freedom,

$$\mathcal{H} = \mathcal{H}_x + \mathcal{H}_y , \tag{6.11}$$

with

$$\mathcal{H}_x = \frac{1}{2}P_x^2 + \frac{x^2}{2\rho^2} + \frac{1}{2}\frac{e}{p_0}G(s)x^2 \tag{6.12}$$

and

$$\mathcal{H}_y = \frac{1}{2}P_y^2 - \frac{1}{2}\frac{e}{p_0}G(s)y^2 . \tag{6.13}$$

The fact that the Hamiltonian is split into two pieces each of which involves only variables corresponding to one degree of freedom means that the horizontal and vertical motion are decoupled. The skew quadrupole and sextupole magnets which have been left out of this example can in practice be used to correct unintended coupling as needed.

The quadrupole magnetic field acts in opposite ways in x and y: positive G means that the effective potential energy in $\mathcal{H}_x$ has a minimum on axis $x = 0$ and leads to stable oscillations around this equilibrium point. At the same time, the effective potential energy in $\mathcal{H}_y$ has a maximum at $y = 0$, which is unstable. Negative G changes the sign of the effect in x and y. In the following sections, we will show that notwithstanding the defocusing effect in one of the transverse directions, a sequence of quadrupoles with alternating polarities can make the transverse motion stable in both directions and confine it near the reference orbit. As a result, a particle near the equilibrium orbit executes stable *betatron* oscillations.

Note that in the absence of quadrupoles, there is a focusing force in the horizontal direction inside dipole magnets due to the second term in Eq. (6.12) for $\mathcal{H}_x$. Being inversely proportional to ρ^2, this term is typically small and does not play a big role in the beam dynamics (it is referred to as the *weak* focusing effect).

To study general properties of the motion in both transverse planes, in the next section we will use a generic Hamiltonian

$$\mathcal{H}_0(x, P_x, s) = \frac{1}{2}P_x^2 + \frac{1}{2}K(s)x^2 , \tag{6.14}$$

where $K = \rho^{-2} + eG/p_0$ for the horizontal plane, and $K = -eG/p_0$ for the vertical plane.

6.3 Hill's Equation, Betatron Function and Betatron Phase

From the Hamiltonian (6.14) we find the following equation of motion in the transverse plane:

$$x''(s) + K(s)x(s) = 0,\qquad(6.15)$$

where the prime denotes the derivative with respect to s. In a circular accelerator, $K(s)$ is a *periodic* function of s with a period that we denote by L (which may be equal to the ring circumference or a fraction of it). Equation (6.15) with a periodic K is called Hill's equation; it describes the so-called *betatron* oscillations around the reference orbit. Note that we have encountered the same equation, Eq. (4.16), in the discussion of the parametric resonance, with the only difference that we now have s as an independent variable instead of t. We know that this equation can have both stable and unstable solutions. Of course, to successfully contain particles inside the vacuum chamber of an accelerator, one has to design a magnetic system that avoids unstable solutions.

While we cannot explicitly solve Eq. (6.15) in the general case, some of its fundamental properties can be studied without specifying the function $K(s)$. Let us seek its solution in the following form,

$$x(s) = Aw(s)\cos\psi(s),\qquad(6.16)$$

where A is an arbitrary constant, and the functions $w(s)$ and $\psi(s)$ will be determined by the requirement that (6.16) satisfies (6.15). Note that the function $w(s)$ is not uniquely defined: we can always multiply it by an arbitrary factor w_0 and redefine the amplitude $A \to A/w_0$, so that $x(s)$ is not changed. It turns out that if the particle motion is stable, we can require that $w(s)$ and $d\psi/ds$ be periodic functions of s with the period L. The requirement on the function ψ, which is called the *betatron phase*, can equivalently be formulated as $\psi(s + L) = \psi(s) + \psi_0$, where ψ_0 is a constant. These two properties of the functions w and ψ are guaranteed by the Floquet theory of differential equations with periodic coefficients [1]. Introducing the two unknown functions $w(s)$ and $\psi(s)$ instead of one $x(s)$ gives us the freedom to impose a constraint of our choice on the functions w and ψ to obtain an optimal parametrization of the solution.

Substituting (6.16) into (6.15) we obtain

$$\left[w'' - w\psi'^2 + K(s)w\right]\cos\psi - \left(2w'\psi' + w\psi''\right)\sin\psi = 0.\qquad(6.17)$$

We now use the freedom mentioned above and set to zero each of the terms that multiply $\cos\psi$ and $\sin\psi$. The motivation here is to have a solution which is insensitive to the constant part of the phase ψ, and instead only depends on its derivatives. This gives us two equations:

$$w'' - w\psi'^2 + K(s)w = 0,$$
$$-2w'\psi' - w\psi'' = 0. \tag{6.18}$$

The last equation can be written as

$$\frac{1}{w}(\psi'w^2)' = 0. \tag{6.19}$$

At this point we integrate (6.19) and introduce the *beta function*, $\beta(s) = w^2(s)$,

$$\psi' = \frac{a}{\beta(s)}, \tag{6.20}$$

where a is an arbitrary constant of integration. Without loss of generality, we can assume that $a > 0$; if this is not the case, we can always change its sign by redefining the angle $\psi \rightarrow -\psi$, which does not change $x(s)$ in (6.16). As was pointed out above, the function w can be multiplied by an arbitrary constant factor. Choosing this factor equal to $\sqrt{a}$ and replacing $\beta \rightarrow a\beta$ eliminates a from (6.20) and converts it into

$$\psi' = \frac{1}{\beta(s)}. \tag{6.21}$$

The first equation in (6.18) now becomes

$$w'' - \frac{1}{w^3} + K(s)w = 0. \tag{6.22}$$

Finally, substituting $w(s) = \sqrt{\beta(s)}$ into (6.22) yields

$$\frac{1}{2}\beta\beta'' - \frac{1}{4}\beta'^2 + K(s)\beta^2 = 1. \tag{6.23}$$

We have derived a nonlinear differential equation of the second order for $\beta(s)$. For a given periodic function $K(s)$, its solution uniquely defines $\beta(s)$ such that $\beta(s) = \beta(s + L)$, although in most practical cases the equation can only be solved numerically, usually with dedicated computer codes developed for this purpose. After $\beta(s)$ is found, the betatron phase ψ is obtained by a straightforward integration of Eq. (6.21). Note that, as follows from (6.21), for a periodic $\beta(s)$ the derivative ψ' is also periodic with the same period L, thus satisfying the requirement formulated after Eq. (6.16). In Fig. 6.2 we show the beta functions for the Advanced Light Source at Lawrence Berkeley National Laboratory (LBNL).

An important characteristic of the magnetic lattice of a ring is the betatron phase advance over its circumference C, $\Delta\psi = \int_0^C ds/\beta(s)$. The quantity $\Delta\psi/2\pi$ is called the *tune* ν (also denoted by Q in the European literature),

Fig. 6.2 The beta functions β_x and β_y for the Advanced Light Source at LBNL

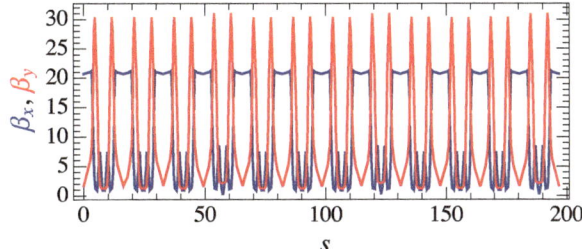

$$\nu = \frac{1}{2\pi} \int_0^C \frac{ds}{\beta(s)}. \tag{6.24}$$

As we will see in the following chapters, the tune plays an important role in beam dynamics.

The main advantage of introducing the two functions $\beta(s)$ and $\psi(s)$ is that, for a given magnetic lattice, they need to be calculated only once. Having found them, the general solution to the equation of motion (6.15) can be written as

$$x(s) = A\sqrt{\beta(s)}\cos[\psi(s) - \psi_0], \tag{6.25}$$

where A and ψ_0 are two arbitrary constants that depend on the initial conditions. Note that the phase term only needs to be adjusted by a constant offset for different initial conditions. Even without detailed knowledge of β and ψ, this equation gives important information about the structure of particle trajectories in the ring.

To conclude this section, we mention that although our analysis was motivated by circular accelerators, the same representation (6.16) of particle orbits is often used in linear accelerators. In the absence of the periodicity condition in such machines, to solve Eq. (6.23) one needs either to specify the initial β and its derivative β' at the entrance to the system or to impose equivalent boundary conditions.

Worked Examples

Problem 6.1 *The magnetic field $B_s(s)$ of a solenoid cannot be described with a single longitudinal component A_s of the vector potential. Show that this magnetic field can be represented with a vector potential that has two transverse components:*

$$A_x = -\frac{1}{2}B_s(s)y, \qquad A_y = \frac{1}{2}B_s(s)x, \qquad A_s = 0.$$

Solution: For the above vector potential, we can use Eqs. (5.4b)–(5.4d) setting $\rho \to \infty$ (there is no bending magnetic field inside the solenoid). We then find that the fields of a solenoid are

$$B_s = -\frac{\partial A_x}{\partial y} + \frac{\partial A_y}{\partial x} = B_s(s)\,,$$

$$B_x = -\frac{\partial A_y}{\partial s} + \frac{\partial A_s}{\partial y} = -\frac{x}{2}\frac{\partial B_s}{\partial s}\,,$$

$$B_y = -\frac{\partial A_s}{\partial x} + \frac{\partial A_x}{\partial x} = -\frac{y}{2}\frac{\partial B_s}{\partial s}\,.$$

The transverse fields are related to longitudinal derivatives of B_s and vanish inside the solenoid where B_s is constant. They play a role at the edges of the solenoid and are required to conserve magnetic flux ($\nabla \cdot \boldsymbol{B} = 0$).

Problem 6.2 *Using Eq. (5.17) and the Hamiltonian (6.10) show that*

$$\frac{dt}{ds} = \frac{1}{v}\left(1 + \frac{x}{\rho}\right)\,.$$

Explain the meaning of this relation. It follows from this equation that, for a relativistic particle, ds/dt can be larger than c. Does this constitute violation of the special theory of relativity which forbids motion of bodies faster than the speed of light?

Solution: We need to evaluate $dt/ds = -\partial K/\partial h$. Using $K = p_0 \mathcal{H}$, $\eta = (p - p_0)/p_0$, and $p = \sqrt{h^2/c^2 - m^2 c^2}$ for a relativistic particle, one obtains:

$$-\frac{\partial K}{\partial h} = -p_0 \frac{\partial \mathcal{H}}{\partial \eta}\frac{\partial \eta}{\partial p}\frac{\partial p}{\partial h} = -\frac{\partial \mathcal{H}}{\partial \eta}\frac{h}{c^2 p} = \frac{1}{v}\left(1 + \frac{x}{\rho}\right)\,,$$

where we used the relation $v = c^2 p/h$. It does not violate the special theory of relativity, since the change in the s coordinate differs from the actual space travelled by the particle. In particular, if for negative values of x the particle is travelling on a circle of shorter radius than ρ the rate of change of s can be larger than c.

Problem 6.3 *Find terms in the Hamiltonian $\mathcal{H}$ responsible for the skew quadrupole (the magnetic field given by Eq. (6.6)).*

Solution: A skew quadrupole has vector potential $A_s = G_{sq}xy$, adding an additional term to the Hamiltonian of

$$\mathcal{H}_{sq} = -\frac{e}{p_0}\left(1 + \frac{x}{\rho}\right)G_{sq}xy$$

$$\approx -\frac{e}{p_0}G_{sq}xy\,,$$

which makes the motion in x and y coupled.

Problem 6.4 *Using the vector potential for the solenoid*

$$A_x = -\frac{1}{2}B_s y , \qquad A_y = \frac{1}{2}B_s x ,$$

and starting from the Hamiltonian Eq. (5.18), find the contribution to K of the magnetic field of the solenoid. Assume that x and y, and hence A_x and A_y, are small and use the Taylor expansion in the Hamiltonian, keeping only linear and second order terms.

Solution: We set $A_s = 0$ and $\rho = \infty$ in the Hamiltonian (5.18),

$$K = -\left[\frac{1}{c^2}h^2 - (\Pi_x - eA_x)^2 - (\Pi_y - eA_y)^2 - m^2c^2\right]^{1/2} .$$

Defining $p^2 = h^2/c^2 - m^2c^2$ and expanding the square root for small A_x and A_y yields

$$K = -p\left[1 - \frac{1}{p^2}(\Pi_x^2 - 2e\Pi_x A_x + e^2 A_x^2 + \Pi_y^2 - 2e\Pi_y A_y + e^2 A_y^2)\right]^{1/2}$$

$$\approx -p + \frac{1}{2p_0}(\Pi_x^2 - 2e\Pi_x A_x + e^2 A_x^2 + \Pi_y^2 - 2e\Pi_y A_y + e^2 A_y^2) .$$

Substituting the specified values of the vector potential into this expression yields

$$K \approx -p + \frac{1}{2p_0}\left[\Pi_x^2 + \Pi_y^2 + \frac{1}{4}e^2 B_s^2(x^2 + y^2) + eB_s(\Pi_x y - \Pi_y x)\right] .$$

Note that the $e^2 B_s^2(x^2 + y^2)/8p_0$ term in the Hamiltonian provides focusing in both transverse planes, and the last term in K introduces a coupling between x and y.

There is a strong similarity between this expression and the result of Problem 1.2 for a coordinate system rotating in time, especially if we were to continue that earlier exercise to calculate the Hamiltonian. Although in the current exercise we consider a relativistic particle moving with small transverse angles and we are using s as the independent variable, this can be taken into account by replacing the mass with the nominal momentum and the rotation frequency by a rate of change with respect to the coordinate s. The rotating frame result will match the current example when the wavenumber of the rotation in s is set to $eB_s/2p_0$. This is half the rate at which a particle performs transverse oscillations inside the solenoid field.

Problem 6.5 *Find the solutions of Eq. (6.23) in free space where $K = 0$.*

Solution: Taking the case of $K = 0$ (drift space) we rewrite Eq. (6.23) as

$$\frac{1}{2}\beta'' = \frac{1}{\beta} + \frac{\beta'^2}{4\beta} . \tag{6.26}$$

Then differentiating both sides of the equation yields

$$
\frac{1}{2}\beta''' = -\frac{\beta'}{\beta^2} - \frac{\beta'^3}{4\beta^2} + \frac{\beta''\beta'}{2\beta}
$$
$$
= \frac{\beta'}{\beta^2}\left(-1 - \frac{1}{4}\beta'^2 + \frac{1}{2}\beta\beta''\right) = 0.
$$

The vanishing third derivative of β means that it is a quadratic polynomial of s,

$$
\beta(s) = As^2 + Bs + C,
$$

where A, B and C are constants. They however are not independent which can be observed by substituting this solution back into Eq. (6.26),

$$
AC = 1 + \frac{1}{4}B^2.
$$

A more convenient form for the beta function in free space that is often used in practice is obtained by introducing two parameters: $s_0 = -B/2A$ and $\beta_0 = 1/A$, which then requires $C = \beta_0 + s_0^2/\beta_0$. This allows us to rewrite the beta function as

$$
\beta(s) = \beta_0 + \frac{1}{\beta_0}(s - s_0)^2. \tag{6.27}
$$

The parameter s_0 is the location of the minimum of the beta function and β_0 is the value of the beta function at the minimum.

Problem 6.6 *Calculate a jump of the derivative of the beta function through a thin quadrupole. Such a quadrupole is defined by $K(s) = f^{-1}\delta(s - s_0)$, where f is called the focal length of the quadrupole.*

Solution: We can integrate Eq. (6.23) over a narrow window around s_0:

$$
\int_{s_0-}^{s_0+} ds\, \frac{1}{2}\beta\beta'' - \int_{s_0-}^{s_0+} ds\, \frac{1}{4}\beta'^2 + \int_{s_0-}^{s_0+} ds\, \frac{\beta^2}{f}\delta(s - s_0) = \int_{s_0-}^{s_0+} ds.
$$

For an infinitesimally narrow window, the right-hand side goes to 0. We also set the second term on the left to zero; because β is continuous, β' must be finite, and the integral in the limit of a narrow window must go to zero. The first term can be integrated by parts to find

$$
\frac{\beta_0}{2}\left(\beta'_{0+} - \beta'_{0-}\right) + f^{-1}\beta_0^2 = 0.
$$

So the jump in the first derivative of the β function is given by $\Delta\beta' = -2\beta/f$.

Fig. 6.3 The beta function versus longitudinal position in a FODO lattice, both scaled to the periodicity length l. This example is for $\mathcal{K}_0 = 2/l$

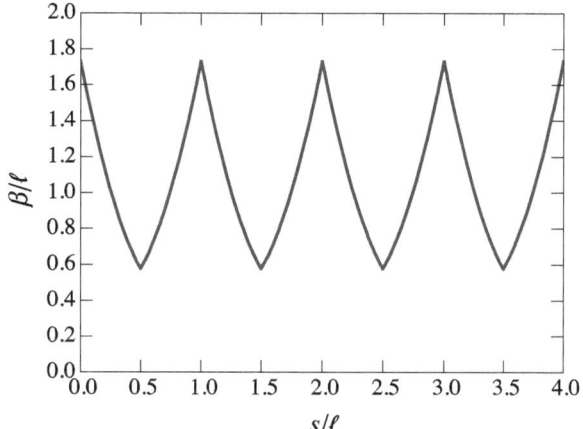

Problem 6.7 *A FODO lattice is a sequence of thin quadrupoles with alternating polarities:*

$$K_{\text{FODO}}(s) = \sum_{n=-\infty}^{\infty} \mathcal{K}_0 \delta(s - nl) - \mathcal{K}_0 \delta\left(s - \left[n + \frac{1}{2}\right]l\right),$$

where l is the period of the lattice. Solve Eq. (6.23) for the FODO lattice and find $\beta(s)$. For a given value of l find the maximum value of $\mathcal{K}_0$ for which the motion is stable.

Solution: The FODO lattice is periodic, with focusing quadrupole, drift, defocusing quadrupole, drift. Its beta function is also periodic with the period l and is symmetric about each quadrupole, see Fig. 6.3.

Due to the symmetry it is sufficient to find $\beta(s)$ in the first half of the cell, $0 < s < l/2$, where we can use Eq. (6.27) in which β_0 and s_0 are two unknowns.

From the previous problem we can identify $\mathcal{K}_0 = 1/f$, where f is the focal length, and calculate the jump of the beta function $\Delta\beta' = -2\beta(0)/f$ at $s = 0$, and the jump $\Delta\beta' = -2\beta(l/2)/f$ at $s = l/2$. From the symmetry of the beta function we conclude that the values of the derivatives of the beta function at the location of the quadrupoles are equal to the halves of the corresponding jumps:

$$\beta'(0^+) = -\frac{1}{f}\beta(0), \qquad \beta'\left(\frac{l}{2} - 0\right) = -\frac{1}{f}\beta\left(\frac{l}{2}\right).$$

Calculating $\beta'(s)$ from Eq. (6.27) and substituting it into these equations we obtain two equations for the two unknowns β_0 and s_0,

$$2\frac{s_0}{\beta_0} = \frac{1}{f}\left(\beta_0 + \frac{s_0^2}{\beta_0}\right), \qquad 2\frac{s_0 - l/2}{\beta_0} = \frac{1}{f}\left[\beta_0 + \frac{(s_0 - l/2)^2}{\beta_0}\right].$$

Solving these equations we find

$$s_0 = \frac{1}{4}(4f + l), \qquad \beta_0 = \frac{1}{4}\sqrt{16f^2 - l^2}.$$

This yields a constraint that $l < 4f$ so that β is real. This is as expected — if the drift is much longer than the focusing length, we would not expect the focusing to be sufficient. In terms of the magnetic strength this condition becomes $|\mathcal{K}_0| < 4/l$.

For the maximum and minimum values of the beta function we find

$$\beta_{\max} = \beta(0) = 2f\sqrt{\frac{4f + l}{4f - l}}, \qquad \beta_{\min} = \beta\left(\frac{l}{2}\right) = 2f\sqrt{\frac{4f - l}{4f + l}}.$$

Problem 6.8 *Consider two rings with circumferences C_1 and C_2. Assume that $C_1 = \lambda C_2$ and $K_2(s) = \lambda^2 K_1(\lambda s)$, and prove that $\beta_2(s) = \lambda^{-1}\beta_1(\lambda s)$.*

Solution: The second ring is related to the first by having its circumference and all lengths scaled by $1/\lambda$, and all focusing strengths around the ring scaled by λ^2. We start from the equation for β_2,

$$\frac{1}{2}\beta_2\beta_2'' - \frac{1}{4}\beta_2'^2 + K_2\beta_2^2 = 1.$$

If we take $\beta_1(\lambda s) = \lambda\beta_2(s)$, we find $\beta_2'(s) = (1/\lambda)\beta_1'(\lambda s)\lambda = \beta_1'(\lambda s)$ and $\beta_2''(s) = \lambda\beta_1''(\lambda s)$. For each derivative, we extract an extra factor of λ. Substituting in for β_1 above we find

$$\frac{1}{2}\beta_1\beta_1'' - \frac{1}{4}\beta_1'^2 + K_1\beta_1^2 = 1,$$

confirming that β_2 is a valid solution so long as β_1 is a solution for its ring.

Reference

1. W. Magnus, S. Winkler. *Hill's Equation* (Dover Publications, Mineola, 2004)

Chapter 7
Action-Angle Variables for Betatron Oscillations

In Chap. 3 we showed that choosing the action-angle canonical variables in a one-dimensional Hamiltonian system dramatically simplifies the dynamics: the action remains constant and the angle increases linearly with time. With minor modifications, the same transformation can be applied to the Hamiltonian (6.14) that describes betatron oscillations in an accelerator. This yields an invariant of the motion and is also a useful starting point for analyzing more complicated dynamics.

7.1 Action-Angle Variables

In Sect. 6.3, we proved that the general solution of the equations of motion for the Hamiltonian (6.14) can be written in the form (6.25),

$$x(s) = A\sqrt{\beta(s)} \cos \psi(s), \qquad (7.1)$$

(where ψ_0 is now included into ψ) which looks similar to the solution (4.1) of the linear harmonic oscillator, although the amplitude $A\sqrt{\beta(s)}$ varies with position, and the phase $\psi(s)$ is not necessarily a linear function of s. The canonical momentum P_x corresponding to $x(s)$ is equal to $x'(s)$ and is obtained by differentiating the above equation,

$$
\begin{aligned}
P_x(s) = x'(s) &= \frac{A}{\sqrt{\beta}} \cos \psi(s) \left(\frac{\beta'}{2} - \tan \psi(s) \right) \\
&= \frac{x}{\beta} \left(\frac{\beta'}{2} - \tan \psi(s) \right).
\end{aligned}
\qquad (7.2)
$$

© Springer International Publishing AG, part of Springer Nature 2018
G. Stupakov and G. Penn, *Classical Mechanics and Electromagnetism in Accelerator Physics*, Graduate Texts in Physics,
https://doi.org/10.1007/978-3-319-90188-6_7

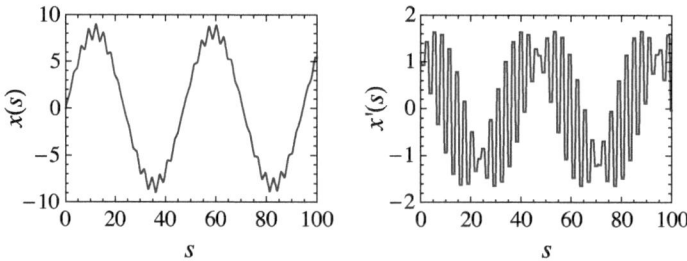

Fig. 7.1 Plots of x and x' versus s for a particular solution to Eq. (6.15) with initial conditions $x(0) = 0$ and $x'(0) = 1$

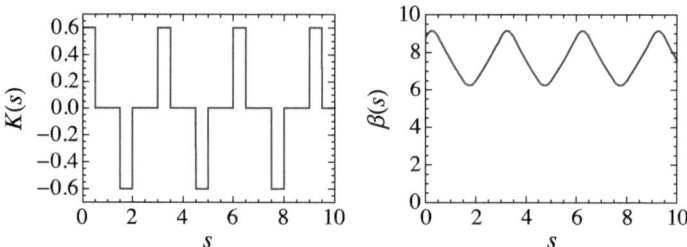

Fig. 7.2 Function $K(s)$ (left panel) and the corresponding β function (right panel)

These two dependences are illustrated in Fig. 7.1 for a particular choice of function $K(s)$ shown in the left panel of Fig. 7.2 — a sequence of rectangular pulses of opposite polarity with the period equal to 3. One can see that both functions show a complicated pattern consisting of an approximately sinusoidal oscillation with a superimposed jitter caused by the jumps in the function $K(s)$. We will now show how to "rectify" this complicated dynamics making a transformation to the action-angle variables ϕ and J.

We will use the same approach as in the case of the linear harmonic oscillator, i.e., assuming that $A = A(J)$ and replacing the phase in the representation of the solution (7.1) by the angular variable ϕ:

$$x(s) = A(J)\sqrt{\beta(s)} \cos \phi \qquad (7.3a)$$

$$P_x(s) = -\frac{x}{\beta(s)}(\alpha + \tan \phi) , \qquad (7.3b)$$

where we have introduced the notation

$$\alpha(s) = -\frac{1}{2}\beta'(s) . \qquad (7.4)$$

The next step is to calculate the generating function (3.6) of the first kind,

$$F_1(x, \phi, s) = \int P_x dx = -\frac{x^2}{2\beta} (\alpha + \tan \phi) .$$
(7.5)

With this generating function, we find the action

$$J = -\frac{\partial F_1}{\partial \phi} = \frac{x^2}{2\beta}(1 + \tan^2 \phi) ,$$
(7.6)

and expressing $\tan \phi$ from Eq. (7.3b),

$$\tan \phi = -\frac{\beta P_x}{x} - \alpha ,$$
(7.7)

we obtain J in terms of x and P_x:

$$J = \frac{1}{2\beta} \left[x^2 + (\beta P_x + \alpha x)^2 \right] .$$
(7.8)

Equations (7.7) and (7.8) give us the transformation from the old conjugate variables x and P_x to the new ones ϕ and J. The inverse transformation $(\phi, J) \rightarrow (x, P_x)$ can also be found. From Eq. (7.6) we have

$$x = \sqrt{2\beta J} \cos \phi .$$
(7.9)

Substituting this relation to Eq. (7.3b) we find P_x in terms of J and ϕ,

$$P_x = -\sqrt{\frac{2J}{\beta}} (\sin \phi + \alpha \cos \phi) .$$
(7.10)

To find the new Hamiltonian which we denote by $\hat{\mathcal{H}}(\phi, J)$, we need to take into account that the generating function depends on the time-like variable s and use Eq. (2.28b):

$$\hat{\mathcal{H}} = \mathcal{H} + \frac{\partial F_1}{\partial s}$$
$$= \frac{1}{2} P_x^2 + \frac{1}{2} K(s) x^2 + \frac{x^2}{4} \frac{\beta'' \beta - \beta'^2}{\beta^2} + \frac{x^2 \beta'}{2\beta^2} \tan \phi .$$
(7.11)

With the help of Eq. (6.23) we eliminate β'' from this equation and then substitute Eq. (7.7) for $\tan \phi$:

$$\hat{\mathcal{H}} = \frac{1}{2}P_x^2 + \frac{1}{2\beta^2}x^2 + \frac{\alpha^2}{2\beta^2}x^2 + \frac{\alpha}{\beta}P_x x$$

$$= \frac{J}{\beta}. \tag{7.12}$$

In the last step we have used the definition (7.8) of J in terms of the old variables. Since the new Hamiltonian is independent of ϕ the equation of motion for J is

$$J' = -\frac{\partial \hat{\mathcal{H}}}{\partial \phi} = 0, \tag{7.13}$$

which means that J is an integral of motion. The quantity $2J$ is called the Courant-Snyder invariant [1]. The Hamiltonian equation for ϕ reads

$$\phi' = \frac{\partial \hat{\mathcal{H}}}{\partial J} = \frac{1}{\beta(s)}. \tag{7.14}$$

Comparing this equation with Eq. (6.21) we see that the new coordinate ϕ is actually equal to the old betatron phase, $\phi = \psi + \phi_0$. This is not surprising, because we replaced the betatron phase in Eqs. (7.1) and (7.2) by ϕ to arrive at Eqs. (7.3a, 7.3b).

Note that Eq. (7.12) is not a universal expression relating the Hamiltonian to the action in action-angle coordinates. Equation (7.12) is a special, particularly simple form that is valid when the focusing is linear and the action is directly obtained from the particle's physical displacement.

7.2 Eliminating Phase Oscillations

As follows from Eq. (7.14), the phase coordinate ϕ monotonically grows with s, with a rate of change that oscillates around some average value due to the oscillations of the beta function as shown in Fig. 7.2b. We can do one more canonical transformation of the variables and "straighten out" these oscillations. This transformation replaces ϕ, J with new canonical variables ϕ_1 and J_1. It is carried out with the help of a generating function of the second type, $F_2(\phi, J_1, s)$, which is given by the following equation:

$$F_2(\phi, J_1, s) = J_1 \left(\frac{2\pi\nu s}{C} - \int_0^s \frac{ds'}{\beta(s')} \right) + \phi J_1, \tag{7.15}$$

where C is the circumference of the accelerator and the tune ν is given by Eq. (6.24). The new angle is given by the derivative of F_2 with respect to J_1,

$$\phi_1 = \frac{\partial F_2}{\partial J_1} = \phi + \frac{2\pi\nu s}{C} - \int_0^s \frac{ds'}{\beta(s')} = \phi + \frac{2\pi\nu s}{C} - \psi(s), \tag{7.16}$$

and the action is unchanged,

$$J = \frac{\partial F_2}{\partial \phi} = J_1 .$$ (7.17)

We denote the new Hamiltonian by $\hat{\mathcal{H}}_1$,

$$\hat{\mathcal{H}}_1 = \hat{\mathcal{H}} + \frac{\partial F_2}{\partial s} = \frac{2\pi\nu}{C} J_1 .$$ (7.18)

With this Hamiltonian, the equation of motion for ϕ_1 reads

$$\phi_1' = \frac{\partial \hat{\mathcal{H}}_1}{\partial J_1} = \frac{2\pi\nu}{C} ,$$ (7.19)

which means that ϕ_1 is a linear function of s with the slope given by $2\pi\nu/C$. We see that, indeed, we got rid of the oscillations exhibited by the phase ϕ and obtained a new phase that follows a straight line in s. The tune is identical in both sets of coordinates.

7.3 Phase Space Motion at a Given Location

As a particle travels in a circular accelerator, every revolution period it arrives at the same longitudinal position s. Let us consider the phase plane x, P_x at this location and plot the particle coordinates every time it passes through s. Because there is an integral of motion J, all these points are located on the curve $J = \mathrm{const}$. From expression (7.8) for J it follows that this curve is an ellipse whose size and orientation depend on the values of J, β, and α. A set of consecutive points x_i, $P_{x,i}$, $i = 1, 2\ldots$, are shown in Fig. 7.3, where for convenience x is normalized by the beta function at this location, $\beta(s)$. Particles with different values of J have geometrically similar ellipses enclosed inside each other.

From expression (7.8) it is easy to see that the ellipse turns into a circle if $\alpha = 0$ and again x is normalized by $\beta(s)$. In this case, the trajectory is very simple: on each revolution the representative point rotates by the betatron phase advance in the ring $\Delta\psi = 2\pi\nu$ in the clockwise direction.

A set of ellipses at another location in the ring will have a different shape which is defined by the local values of β and α. When one travels along the circumference of the accelerator, one sees a continuous transformation of these sets with the coordinate s. For a collection of particles in a bunch, this effect includes changes not only in the size of the beam but also in statistical correlations between x and P_x.

Worked Examples

Problem 7.1 *Using Eqs. (7.7) and (7.8), show by direct calculation of Poisson brackets that the transformation x, $P_x \rightarrow \phi$, J is canonical.*

Fig. 7.3 The phase space ellipse (solid curve) and a particle's positions at consecutive turns. Dashed lines show ellipses for particles with smaller and larger values of action J. The vertical axis is marked by x' which is equal to P_x

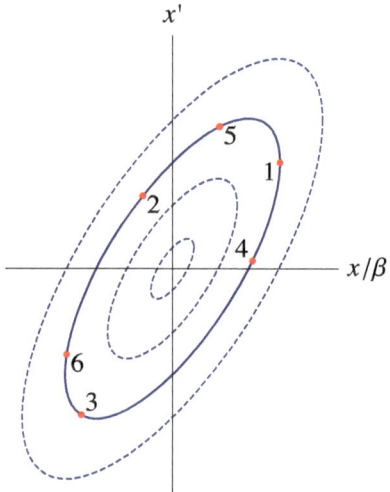

Solution: The only Poisson bracket that needs to be evaluated is $\{\phi, J\}_{x,P_x}$, because it is trivially true that $\{\phi, \phi\} = \{J, J\} = 0$. The coordinate transformation is defined by

$$\tan \phi = -\frac{\beta P_x}{x} - \alpha, \qquad J = \frac{1}{2\beta}\left[x^2 + (\beta P_x + \alpha x)^2\right].$$

We use the fact that $d(\tan \phi) = (1/\cos^2 \phi)d\phi$ to express

$$\frac{\partial \phi}{\partial x} = \cos^2 \phi \frac{\partial \tan \phi}{\partial x},$$

and similarly for $\partial \phi/\partial P_x$. This yields

$$\begin{aligned}
\{\phi, J\}_{x,P_x} &= \frac{\partial \phi}{\partial x}\frac{\partial J}{\partial P_x} - \frac{\partial \phi}{\partial P_x}\frac{\partial J}{\partial x} \\
&= (\cos^2 \phi)\frac{\beta P_x}{x^2}\frac{1}{2\beta}2(\beta P_x + \alpha x)\beta \\
&\quad - (\cos^2 \phi)\left(\frac{-\beta}{x}\right)\frac{1}{2\beta}[2x + 2(\beta P_x + \alpha x)\alpha] \\
&= \cos^2 \phi\left[1 + \frac{(\beta P_x + \alpha x)^2}{x^2}\right] = \cos^2 \phi\left(1 + \tan^2 \phi\right) = 1.
\end{aligned}$$

Thus, the value of the Poisson bracket is 1 as required for a canonical transformation.

Problem 7.2 *Find the major and minor half axes, and the tilt of the ellipse shown in Fig. 7.3 which is based on Eq. (7.8).*

Solution: In Eq. (7.8) we change variables to $\zeta = x/\beta$ and multiply both sides by $2/\beta$ to obtain:

$$\frac{2J}{\beta} = \zeta^2 + (P_x + \alpha\zeta)^2 = (1+\alpha^2)\zeta^2 + P_x^2 + 2\alpha P_x \zeta .$$

This is the formula for an ellipse, which, because of the $P_x \zeta$ cross term, is tilted. We look for a rotation to new variables (r, u) defined by

$$\zeta = r\cos\theta - u\sin\theta ,$$
$$P_x = r\sin\theta + u\cos\theta ,$$

such that we no longer have a cross term. Substituting in above, we find that the cross term is

$$2\alpha r u \left(\cos^2\theta - \sin^2\theta - \alpha\sin\theta\cos\theta\right) .$$

For $\alpha \neq 0$ (otherwise the ellipse is already a circle with radius $\sqrt{2J/\beta}$), setting this to zero requires $\alpha = 2\cos 2\theta / \sin 2\theta$ so

$$\tan 2\theta = \frac{2}{\alpha} .$$

This last expression can also be obtained by looking for an extremum in $\zeta^2 + (P_x)^2$ by calculating the variation dJ under the constraint that $\zeta d\zeta + P_x dP_x = 0$, and then solving for $dJ = 0$. This yields the equation

$$\left(\frac{P_x}{\zeta}\right)^2 + \alpha\frac{P_x}{\zeta} - 1 = 0 ,$$

which gives the same condition after setting $P_x/\zeta = \tan\theta$ and multiplying by $\cos^2\theta$.

The major and minor half axes are just the maximum and minimum value of the radius under fixed $2J/\beta$ and fixed α. As we already defined a pair of rotated coordinates for which the ellipse is upright, we can rewrite the equation for the ellipse in terms of r and u:

$$\frac{2J}{\beta} = r^2 \left(1 + \alpha^2\cos^2\theta + 2\alpha\sin\theta\cos\theta\right) + u^2 \left(1 + \alpha^2\sin^2\theta - 2\alpha\sin\theta\cos\theta\right)$$
$$= r^2 \cot^2\theta + u^2 \tan^2\theta ,$$

where the second equality comes from eliminating α using $\alpha = 2/\tan 2\theta$, which justifies dropping the ru cross term, and from trigonometric identities. The maximum and minimum values are found by setting either $u = 0$ or $r = 0$, giving the two values for the radius or half-axis in these scaled coordinates as $\sqrt{2J/\beta}\,|\tan\theta|$ and

$\sqrt{2J/\beta}\,|\cot\theta|$; for $\alpha > 0$ the $\tan\theta$ term will be the shorter, minor axis, and for $\alpha < 0$ it will be the longer, major axis.

Problem 7.3 *Prove that the transformation* $x, P_x \rightarrow \bar{x}, \bar{P}_x$ *with*

$$\bar{x} = \frac{1}{\sqrt{\beta}}x, \qquad \bar{P}_x = \frac{1}{\sqrt{\beta}}(\beta P_x + \alpha x),$$

is canonical. Prove that phase space orbits plotted in variables $\bar{x}, \bar{P}_x$ *are circles.*

Solution: For a linear transformation in one degree of freedom, in order to show it is symplectic and thus canonical it is sufficient to show that the determinant is 1. The transformation is equivalent to

$$\begin{pmatrix} \bar{x} \\ \bar{P}_x \end{pmatrix} = \begin{pmatrix} \dfrac{1}{\sqrt{\beta}} & 0 \\ \dfrac{\alpha}{\sqrt{\beta}} & \sqrt{\beta} \end{pmatrix} \begin{pmatrix} x \\ P_x \end{pmatrix},$$

and the determinant is clearly equal to 1.

This transformation is canonical regardless of what the quantities x and P_x represent. When we use the original parametrization for the particle position, $x = \sqrt{2\beta J}\cos\phi$, with the corresponding derivative being P_x, we find that

$$\bar{x} = \sqrt{2J}\cos\phi,$$

and

$$\bar{P}_x = \sqrt{2J}\sin\phi.$$

Note that $\bar{P}_x$ is β times the derivative of $\bar{x}$. Because the action J is constant, these expressions give the equation for a circle with radius $\sqrt{2J}$.

Reference

1. E.D. Courant, H.S. Snyder, Theory of the alternating-gradient synchrotron. Ann. Phys. **3**, 1–48 (1958)

Chapter 8
Magnetic Field and Energy Errors

Due to small magnetic field errors the magnetic field in any real machine differs from the ideal linear lattice studied in the previous chapters. Also, the particle energy in the beam deviates from the reference energy for which the ideal closed orbit is designed. It is important to understand what new effects the non-ideal magnetic fields introduce to the particle motion. In this chapter, we study how the dipole and quadrupole field errors as well as small energy deviations affect the orbits.

8.1 Closed Orbit Distortions

We first consider what happens with the reference orbit if the dipole magnetic field is not exactly equal to the design one.

Let $B_0(s)$ be the design vertical magnetic field and consider the magnetic field error to be also directed along $\hat{y}$ with the total field $\boldsymbol{B}(s) = \hat{y}[B_0(s) + \Delta B(s)]$. The deviation from the ideal field ΔB is small, $|\Delta B| \ll B_0$. Our curvilinear coordinate system x, y, s is associated with the ideal reference orbit of the magnetic field B_0, which is the line $x = y = 0$. The vector potential ΔA_s that describes the error field is given by Eq. (6.2) in which, because of the smallness of ΔB, we keep only the first order term, $\Delta A_s = -\Delta B(s)x$. This vector potential should be added to the Hamiltonian (6.12):

$$\mathcal{H}_x = \frac{1}{2}P_x^2 + \frac{1}{2}K(s)x^2 + \frac{e\Delta B(s)}{p_0}x, \qquad (8.1)$$

where $K = \rho^{-2} + eG/p_0$. From this Hamiltonian we find the following differential equation for x,

$$x'' + K(s)x = -\frac{e\Delta B(s)}{p_0}. \qquad (8.2)$$

© Springer International Publishing AG, part of Springer Nature 2018
G. Stupakov and G. Penn, *Classical Mechanics and Electromagnetism in Accelerator Physics*, Graduate Texts in Physics,
https://doi.org/10.1007/978-3-319-90188-6_8

This equation describes general betatron oscillations in the accelerator, but can also be used to find a new closed orbit in the perturbed magnetic field $B(s)$. Because the perturbation is small, this new orbit should be close to the old one; it is obtained as a periodic solution to Eq. (8.2). We denote this solution by $x_0(s)$; it is called the *closed orbit distortion*. By definition, closed orbits must satisfy the periodicity condition $x_0(s + C) = x_0(s)$, where C is the circumference of the ring.

To calculate $x_0(s)$ we first consider the case of a field perturbation localized at one point: $\Delta B(s) = \Delta B_0(s')\delta(s - s')$. Since the right-hand side of Eq. (8.2) is equal to zero everywhere except for the point $s = s'$, we seek a solution in the form of (6.25) with an unknown amplitude A and an initial phase ψ_0,

$$x_0(s) = A\sqrt{\beta(s)} \cos[\psi(s) - \psi_0]. \tag{8.3}$$

To be a periodic solution to (8.2), this function should satisfy two requirements. The first one is that $x_0(s)$ should be continuous at $s = s'$. This is achieved if we choose $\psi_0 = \psi(s') + \pi\nu$ and agree that $\psi(s)$ varies from $\psi(s')$ to $\psi(s') + 2\pi\nu$ when s varies from s' to $s' + C$. Indeed, when $\psi(s) = \psi(s')$, the argument of the cos function is equal to $-\pi\nu$, and $x_0(s') = A\sqrt{\beta(s')} \cos(-\pi\nu)$. After one turn around the ring, the argument of the cos function becomes equal to $\pi\nu$, and since cos is an even function, $x_0(s' + C) = x_0(s')$. Hence,

$$x_0(s) = A\sqrt{\beta(s)} \cos[\psi(s) - \psi(s') - \pi\nu]. \tag{8.4}$$

The second requirement is obtained when we integrate Eq. (8.2) over s from $s' - \epsilon$ to $s' + \epsilon$, where ϵ is an infinitesimally small number. The delta function in $\Delta B(s)$ gives a nonzero value on the right-hand side, and the integration of the left-hand side gives a jump in the first derivative of x_0 at s':

$$x_0'(s') - x_0'(s' + C) = -\frac{e\Delta B_0(s')}{p_0}. \tag{8.5}$$

Note that according to our convention of measuring s from s' to $s' + C$ (which means that s cannot be smaller than s'), for the derivative of x_0 before the delta function we use the notation $x_0'(s' + C)$ instead of $x_0'(s' - \epsilon)$. Substituting (8.4) into (8.5) we find

$$A = -\frac{\sqrt{\beta(s')}}{2\sin(\pi\nu)} \frac{e\Delta B_0(s')}{p_0}. \tag{8.6}$$

Having solved the problem for the delta-function perturbation, we can now extend this solution to the case of an arbitrary $\Delta B(s)$. For this, we represent $B(s)$ as a superposition of delta-function contributions using the following identity:

$$\Delta B(s) = \int_s^{s+C} ds' \Delta B(s')\delta(s - s'). \tag{8.7}$$

Fig. 8.1 An ideal (dashed blue line) and a distorted (solid orange line) orbit in a ring. The thick green line shows a betatron orbit with the offset $\xi(s)$ measured relative to the distorted orbit. Note that both the ideal and distorted orbits are closed lines, but the betatron oscillation is not

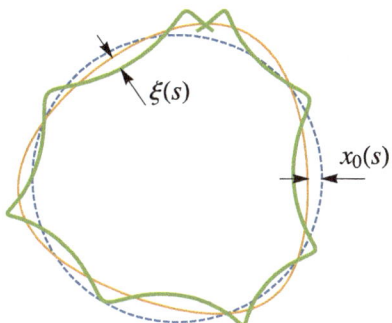

Because x_0 is linear in ΔB, we need to add contributions from all locations s', integrating the right-hand side of Eq. (8.4) over s':

$$x_0(s) = \frac{-e}{2p_0 \sin(\pi\nu)} \int_s^{s+C} ds' \, \Delta B(s') \sqrt{\beta(s)\beta(s')} \cos[\psi(s) - \psi(s') - \pi\nu]. \quad (8.8)$$

An important immediate consequence of this formula is that integer values for the tune ν are not allowed in a realistic magnetic lattice: such a lattice would be unstable with respect to small errors of the ideal magnetic field.

We found x_0 as a particular solution of the inhomogeneous differential equation (8.2). To find its general solution $x(s)$ we seek it as a sum

$$x(s) = x_0(s) + \xi(s). \quad (8.9)$$

Substituting this into (8.2) we see that $\xi(s)$ satisfies the homogeneous equation

$$\xi'' + K(s)\xi = 0, \quad (8.10)$$

which is the equation for the horizontal betatron oscillations. This makes the meaning of $\xi(s)$ clear — it is the betatron oscillation around the perturbed closed orbit $x_0(s)$. This is illustrated by Fig. 8.1.

The action-angle variables (7.8) derived in Sect. 7.1 for the Hamiltonian (6.14) can be naturally generalized for the Hamiltonian (8.1). The result of this generalization can be understood if we recall that in the expression for the action (7.8) the coordinate x and the momentum $P_x = x'$ are measured relative to the ideal closed orbit $x = 0$. With the perturbed magnetic field, both the distance and the angle has to be measured relative to the perturbed closed orbit $x_0(s)$. It is not surprising then that the new action is obtained from (7.8) by simply replacing $x \to x - x_0(s)$ and $P_x \to P_x - x_0'(s)$:

$$J(x, P_x, s) = \frac{1}{2\beta} \left\{ (x - x_0)^2 + \left[\beta(P_x - x_0') + \alpha(x - x_0) \right]^2 \right\}. \quad (8.11)$$

The derivation of this result is the subject of Problem 8.1.

As the perturbation of the fields becomes larger, higher-order terms that were neglected in this derivation become more important. At some point, it may even become impossible to find a closed orbit at all. Our derivation was carried out for the case when the perturbation of the magnetic field is vertical, i.e., in the same direction as the main magnetic field. The case of the horizontal perturbation, $\boldsymbol{B}(s) = \hat{\boldsymbol{y}} B_0(s) + \hat{\boldsymbol{x}} \Delta B(s)$, can be considered in a similar fashion: to lowest order the perturbation propagates into the Hamiltonian $\mathcal{H}_y$, Eq. (6.13), and the perturbed closed orbit gets a distortion $y_0(s)$ in the vertical direction.

8.2   Effect of Energy Deviation

Another effect that causes a distortion of the closed orbit is the deviation of the particle energy from the nominal one. We can directly adapt the results of the previous section to find the distortion for a particle with a relative energy that differs from the nominal one by η. From the Hamiltonian (6.10), we see that it has a term $-\eta x/\rho$ that couples η to the horizontal motion. Hence, instead of (6.12) we have

$$\mathcal{H}_x = \frac{1}{2} P_x^2 + \frac{1}{2} K(s) x^2 - \frac{\eta}{\rho} x . \tag{8.12}$$

This Hamiltonian can be formally obtained from (8.1) by the replacement

$$\Delta B \to -\frac{\eta p_0}{e\rho} . \tag{8.13}$$

Making this replacement in Eq. (8.8), we find the orbit distortion caused by the energy deviation η,

$$x_0(s) = D(s)\eta , \tag{8.14}$$

where the function D is

$$D(s) = \frac{1}{2 \sin(\pi\nu)} \int_s^{s+C} ds' \frac{\sqrt{\beta(s)\beta(s')}}{\rho(s')} \cos[\psi(s) - \psi(s') - \pi\nu] . \tag{8.15}$$

This function is called the *dispersion function* of the ring, and it too is a periodic function of s.

Using the expressions (8.11) and (8.14) we can also find the action variable for a particle with an energy deviation η,

$$J(x, P_x, \eta, s) = \frac{1}{2\beta} \left([x - \eta D(s)]^2 + \{\beta[P_x - \eta D'(s)] + \alpha[x - \eta D(s)]\}^2 \right) . \tag{8.16}$$

8.3 Quadrupole Errors

We will now address the issue of errors in the quadrupole component of the magnetic field in the ring. Let us assume that due to a perturbation of the quadrupole field the ideal focusing strength $K(s)$ is changed to $K(s) + \Delta K(s)$, where $|\Delta K| \ll |K|$. The perturbed Hamiltonian is obtained through the replacement $K(s) \to K(s) + \Delta K(s)$,

$$\mathcal{H} = \frac{1}{2} P_x^2 + \frac{1}{2} K(s) x^2 + \frac{1}{2} \epsilon \Delta K(s) x^2 , \tag{8.17}$$

where we have introduced a formal smallness parameter ϵ which will be set to unity at the end of the calculation. With the proper choice of function $K(s)$, this Hamiltonian can be applied to both horizontal (x) and vertical (y) coordinates. Since we know that the focusing function $K(s)$ determines the betatron oscillations in the system, it is clear that changing the focusing strength would result in the perturbation of the beta function and, hence, the tune of the ring.

To calculate these changes we first transform to the action-angle variables J and ϕ defined in Sect. 7.1. This transformation casts the first two terms of the Hamiltonian into J/β:

$$\frac{1}{2} P_x^2 + \frac{1}{2} K(s) x^2 \to \frac{J}{\beta(s)} . \tag{8.18}$$

In the last term of Eq. (8.17) we express x in terms of J and ϕ using Eq. (7.9):

$$\begin{aligned}
\mathcal{H} &= \frac{J}{\beta(s)} + \epsilon \Delta K(s) J \beta(s) \cos^2 \phi \\
&= J \left(\frac{1}{\beta(s)} + \frac{1}{2} \epsilon \Delta K(s) \beta(s) \right) + \frac{1}{2} \epsilon \Delta K(s) J \beta(s) \cos 2\phi ,
\end{aligned} \tag{8.19}$$

where we have split the perturbation term into an averaged part and one oscillating as $\cos 2\phi$. We will denote the last term in this equation by $\epsilon V(\phi, J, s)$:

$$V(\phi, J, s) = \frac{1}{2} \Delta K(s) J \beta(s) \cos 2\phi . \tag{8.20}$$

To solve for the Hamiltonian (8.19) we will use perturbation theory based on canonical transformations with the goal of eliminating the perturbation term (8.20) from the transformed Hamiltonian, thereby removing any dependence on the angular variable. If this goal is achieved and the transformed Hamiltonian becomes dependent only on the action, this action is a constant of motion and the equation of motion for the angular variable that is conjugate to the action can be easily integrated, as discussed in Sects. 7.1 and 7.2. As it turns out, there is no general method that completely eliminates the angular dependence on ϕ; however, this dependence can be made of higher order in ϵ. For our purposes, it is enough to obtain the transformed Hamiltonian with an accuracy of the order of ϵ^2.

Our first step is to make a canonical transformation to new variables, $(\phi, J) \rightarrow (\xi, I)$, where ξ and I are the new angle and action. We will use the generating function F_2 of the second type,

$$F_2(\phi, I, s) = \phi I + \epsilon G(\phi, I, s), \tag{8.21}$$

where the appropriate function G will be determined below. Using Eq. (2.29a) we obtain the following relations between the old and new variables:

$$\xi = \phi + \epsilon G_I(\phi, I, s), \qquad J = I + \epsilon G_\phi(\phi, I, s), \tag{8.22}$$

where we use the subscripts I and ϕ to denote differentiation with respect to the corresponding variables. Because G is multiplied by the small parameter ϵ, the difference between the old and the new variables is small, on the order of ϵ. We can solve these equations to the first order in ϵ:

$$\xi \approx \phi + \epsilon G_I(\phi, J, s), \qquad I \approx J - \epsilon G_\phi(\phi, J, s), \tag{8.23}$$

where terms of the order of ϵ^2 are neglected.

To find the new Hamiltonian, $\mathcal{H}_1(\xi, I, s)$, we use Eq. (2.29b) and then express ϕ and J in terms of I and ξ using (8.23) and neglecting terms of the second and higher order in ϵ,

$$\mathcal{H}_1 = I \left(\frac{1}{\beta} + \frac{1}{2}\epsilon\Delta K\beta \right) + \epsilon V + \frac{1}{\beta}\epsilon G_\phi + \epsilon G_s + O(\epsilon^2). \tag{8.24}$$

We can cancel out the angle-dependent part of the perturbation V in the new Hamiltonian by choosing G in such a way that the ϕ-dependent terms in (8.24), which are all linear in ϵ, cancel:

$$V + \frac{1}{\beta}G_\phi + G_s = 0. \tag{8.25}$$

Note that because the old and new variables differ by small terms of the order of ϵ, we can write V as if it were a function of the variables ϕ, I, s to match the form of the function G from the generating function. It does not matter that $\mathcal{H}_1$ has to be expressed as a function of I, ξ, s because we are ensuring that to first order in ϵ the dependence on ϕ will vanish.

One needs to find a solution to (8.25) that is periodic in s with the period equal to the ring circumference C. This solution is given by the expression

$$G(\phi, I, s) = -\frac{I}{4\sin(2\pi\nu)} \int_s^{s+C} ds' \Delta K(s')\beta(s') \sin 2[\phi - \psi(s) + \psi(s') - \pi\nu]. \tag{8.26}$$

Indeed, this function is periodic in s, because the integrand does not change when both s' and s change to $s' + C$ and $s + C$, respectively. It also satisfies Eq. (8.25) as is directly established by differentiation of the right-hand side of (8.26). We see that to avoid a singularity in the formula we need to require that ν is not equal to integer or half-integer values.

With the function G defined above, the new Hamiltonian becomes

$$\mathcal{H}_1 = I \left(\frac{1}{\beta} + \frac{1}{2} \epsilon \Delta K \beta \right) , \tag{8.27}$$

where we have neglected higher order terms. Because $\mathcal{H}_1$ does not depend on the angle ξ, the action I is an integral of motion. Equation (8.23) now gives the transformation from the old to new variables to the lowest order in ϵ. For the new action we have

$$I = J + \frac{\epsilon J}{2 \sin(2\pi\nu)} \int_s^{s+C} ds' \, \Delta K(s') \beta(s') \cos 2[\phi - \psi(s) + \psi(s') - \pi\nu] . \tag{8.28}$$

Changing the strength of the focusing in the lattice by $\epsilon \Delta K$ also modifies the beta function in the ring. Let us denote the new beta function and corresponding betatron phase by β_1 and ϕ_1, respectively. It is tempting to use the variable ξ given by the first equation in Eq. (8.22) for the new betatron phase ϕ_1, but this is not justified. We have already seen from Eq. (7.17) that the same action can be paired with different conjugate angle coordinates, and by comparing the expressions (7.14) and (7.19) we note that the derivative of the phase terms are different. In the same example, we found that the form of the Hamiltonian also changed, so we also cannot simply assume $\mathcal{H}_1 = I/\beta_1$. The proper definition of the beta function and betatron phase comes from Eqs. (7.8)–(7.10). These expressions should hold equally in the new action-angle coordinates by taking $J \to I$, $\beta \to \beta_1$ and $\phi \to \phi_1$:

$$x = \sqrt{2\beta_1 I} \cos \phi_1 ,$$

$$P_x = -\sqrt{\frac{2I}{\beta_1}} \, (\sin \phi_1 + \alpha_1 \cos \phi_1) ,$$

$$I = \frac{1}{2\beta_1} \left[x^2 + (\beta_1 P_x + \alpha_1 x)^2 \right] , \tag{8.29}$$

where we again define $\alpha_1 = -\beta_1'/2$.

Even though both β_1 and ϕ_1 are unknown, we can overcome this by using the fact that they are only functions of s, and that the above expressions have to hold true for all combinations of x and P_x. Thus we may limit our attention to the case $\phi = \pi/2$. This requires $x = 0$, so ϕ_1 is constrained to be $\pi/2$ as well and

$$J = \frac{1}{2} \beta P_x^2 , \qquad I = \frac{1}{2} \beta_1 P_x^2 , \tag{8.30}$$

which implies

$$\beta_1 = \frac{\beta}{J} I|_{\phi=\pi/2}. \tag{8.31}$$

We now use Eq. (8.28), our solution for the new action to first order in the perturbed fields. For brevity, we write this expression as $I = J + \epsilon \Delta J$, and similarly we will write the new beta function to first order in ϵ as $\beta_1 = \beta + \epsilon \Delta \beta$, where $\epsilon \Delta \beta$ is the difference between the new and the old beta functions. Then

$$\Delta\beta(s) = \frac{\beta(s)}{J} \Delta J|_{\phi=\pi/2}$$

$$= \frac{\beta(s)}{2\sin(2\pi\nu)} \int_s^{s+C} ds' \Delta K(s')\beta(s') \cos 2[\pi/2 - \psi(s) + \psi(s') - \pi\nu]$$

$$= -\frac{\beta(s)}{2\sin(2\pi\nu)} \int_s^{s+C} ds' \Delta K(s')\beta(s') \cos 2[-\psi(s) + \psi(s') - \pi\nu], \tag{8.32}$$

using $\cos(\pi + \theta) = -\cos\theta$.

An important conclusion that follows from the above equation is that one should avoid integer or half-integer values of the tune — they are unstable with respect to errors in the focusing strength of the lattice. Having found the correction to the beta function, we can find the correction to the tune by using Eq. (6.24).

Worked Examples

Problem 8.1 *The action-angle variables defined by Eqs. (7.8)–(7.10) have to be modified in the case of dipole field errors. Starting from the Hamiltonian (8.1) transform to the action-angle variables using the following generating function:*

$$F_1(x, \phi, s) = \frac{[x - x_0(s)]^2}{2\beta(s)} \left(\frac{\beta'(s)}{2} - \tan\phi \right) + x x_0'(s).$$

Show that in this case

$$J(x, P_x, s) = \frac{1}{2\beta(s)} \left([x - x_0(s)]^2 + \{\beta(s)[P_x - x_0'(s)] + \alpha(s)[x - x_0(s)]\}^2 \right),$$

and obtain the Hamiltonian (7.12).

Solution: For the given generating function of the first type the new, action-angle variables satisfy

$$J = -\frac{\partial F_1}{\partial \phi} = \frac{[x - x_0(s)]^2}{2\beta(s)} \sec^2\phi,$$

and

$$P_x = -\frac{\partial F_1}{\partial x} = \frac{x - x_0(s)}{\beta(s)} \left(\frac{\beta'(s)}{2} - \tan\phi \right) + x_0'(s) \,. \tag{8.33}$$

We want to express J in terms of x and P_x. Solving for $\tan\phi$ from our expression for P_x we find (in the following we leave out most of the s-dependences):

$$\tan\phi = \frac{\beta'}{2} - \frac{\beta(P_x - x_0')}{x - x_0} = -\frac{\beta(P_x - x_0') + \alpha(x - x_0)}{x - x_0} \,,$$

and using $\sec^2 = 1 + \tan^2$ the result is

$$J = \frac{1}{2\beta} \left\{ (x - x_0)^2 + [\beta(P_x - x_0') + \alpha(x - x_0)]^2 \right\} \,.$$

Finally, our new Hamiltonian is related to the old one by

$$\hat{\mathcal{H}} = H + \frac{\partial F_1}{\partial s} = \frac{1}{2} P_x^2 + \frac{1}{2} K(s)x^2 + \frac{e \Delta B(s)}{p} x + \frac{(x - x_0)^2}{2\beta^2} \frac{1}{2} \beta \beta''$$
$$- (P_x - x_0') \left[x_0' - \frac{\alpha}{\beta}(x - x_0) \right] + x x_0'' \,,$$

where we used $\alpha = -\beta'/2$ and the fact that

$$\frac{\beta'}{2} - \tan\phi = \frac{\beta(P_x - x_0')}{x - x_0'} \,,$$

that follows from Eq. (8.33). Finally, we use Eq. (6.23) for β'' and the fact that x_0 is a solution of Eq. (8.2) to obtain:

$$\hat{\mathcal{H}} = \frac{1}{2}(P_x - x_0')^2 + \frac{1 + \alpha^2}{2\beta^2}(x - x_0)^2 + \frac{\alpha}{\beta}(x - x_0)(P_x - x_0') + \frac{1}{2}(x_0')^2 - \frac{1}{2} K x_0^2$$
$$= \frac{J}{\beta} + \frac{1}{2}(x_0')^2 - \frac{1}{2} K x_0^2 \,.$$

The last two terms are functions of s only and do not affect the dynamics. If we had chosen the last term in F_1 to be $(x - x_0)x_0'$, then the residual term would be simply $x_0 e \Delta B / 2p$.

Problem 8.2 *For a ring, how does the sensitivity of the beam orbit in the horizontal plane to the size of the error magnetic field $\Delta B_y(s)$ scale with the parameters of the ring?*

Solution: A crude estimate of the magnitude of the closed orbit distortion from Eq. (8.8) is

$$|x_0| \sim \frac{e}{p_0 \sin(\pi\nu)} \Delta B \bar{\beta} C \,,$$

where $\bar{\beta}$ is the typical magnitude of the beta function in the ring. From the definition of the bending radius,

$$|x_0| \sim \frac{1}{\sin(\pi\nu)} \frac{\Delta B}{B} \bar{\beta} \frac{C}{\rho} \,,$$

and usually the circumference is roughly $2\pi\rho$. We see from this equation that the effect of magnetic field errors is less in machines with small beta functions, that is with stronger focusing. The effect of errors can be greatly enhanced when the tune is close to an integer.

Problem 8.3 *Verify by direct calculation that G given by Eq. (8.26) satisfies Eq. (8.25).*

Solution: We start with $G(\phi, I, s)$, and keep in mind that all partial derivatives involve holding two of the arguments fixed while varying the third one. To calculate $G_\phi = \partial G/\partial\phi$, we note that ϕ only appears in the sin term at the end, and the derivative changes this to $2 \cos 2[\phi - \psi(s) + \psi(s') - \pi\nu]$. To evaluate $G_s = \partial G/\partial s$, we note that in addition to $\psi(s)$ both limits of integration also depend on s. We can calculate these terms separately, then add them. The $\psi(s)$ derivative changes the sin term at the end to $-(2d\psi/ds) \cos 2[\phi - \psi(s) + \psi(s') - \pi\nu]$. Because $d\psi/ds = 1/\beta(s)$, this exactly cancels $(1/\beta)G_\phi$. The remainder comes from evaluating the integrand at $s' = s + C$ and $s' = s$, which combine to form

$$-\frac{I}{4 \sin(2\pi\nu)} \Delta K(s)\beta(s)\big\{ \sin 2[\phi - \psi(s) + \psi(s + C) - \pi\nu]$$

$$- \sin 2[\phi - \psi(s) + \psi(s) - \pi\nu]\big\} \,,$$

where we use the fact that ΔK and β are periodic in s. The phases are not periodic, however; $\psi(s + C) = \psi(s) + 2\pi\nu$ by definition of the tune. This gives

$$\frac{1}{\beta}G_\phi + G_s = -\frac{I}{4 \sin(2\pi\nu)} \Delta K(s)\beta(s) [\sin 2(\phi + \pi\nu) - \sin 2(\phi - \pi\nu)]$$

$$= -\frac{I}{4 \sin(2\pi\nu)} \Delta K(s)\beta(s) 2 \cos(2\phi) \sin(2\pi\nu) = -\frac{1}{2} I \Delta K(s)\beta(s) \cos(2\phi) \,.$$

This is equal and opposite to V up to order ϵ, and because these terms are all small to begin with the difference will be neglected as being $O(\epsilon^2)$.

Fig. 8.2 Transformation of
the integration domain in the
double integral

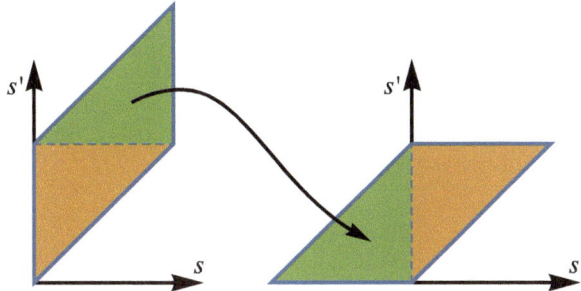

Problem 8.4 *Show that due to the perturbation of the quadrupole strength $\Delta K(s)$
the tune change is given by the following equation:*

$$\Delta\nu = \frac{1}{4\pi}\int_0^C ds\,\Delta K(s)\beta(s)\,. \tag{8.34}$$

Solution: From the definition of the tune we find

$$\Delta\nu = \frac{1}{2\pi}\int_0^C ds\,\Delta\left(\frac{1}{\beta(s)}\right) \simeq -\frac{1}{2\pi}\int_0^C ds\,\frac{\Delta\beta(s)}{\beta^2(s)}$$

$$= \frac{1}{4\pi\sin(2\pi\nu)}\int_0^C \frac{ds}{\beta(s)}\int_s^{s+C} ds'\,\Delta K(s')\beta(s')\cos 2\left[-\psi(s)+\psi(s')-\pi\nu\right],$$

where we used Eq. (8.32). Note that because $d\psi(s)/ds = 1/\beta(s)$, the quantity

$$\frac{1}{\beta(s)}\cos 2\left[-\psi(s)+\psi(s')-\pi\nu\right] = -\frac{1}{2}\frac{d}{ds}\sin 2\left[-\psi(s)+\psi(s')-\pi\nu\right]$$

is a total derivative, which indicates that a change in the order of integration might
be helpful. The only problem is that the limits of the internal integral have an explicit
dependence on s. We first split the double integral into two regions in the s, s' plane as

$$\int_0^C ds\int_s^{s+C} ds' = \int_0^C ds\int_s^C ds' + \int_0^C ds\int_C^{s+C} ds'\,.$$

The integration area in the second integral on the right-hand side is then downshifted
by the ring circumference C in both variables, as shown Fig. 8.2, to yield

$$\int_0^C ds\int_C^{s+C} ds' \;\rightarrow\; \int_{-C}^0 ds\int_0^s ds'\,.$$

This transformation does not change the result because $\Delta K(s')$ and $\beta(s')$ are periodic functions with the period C, and the combination $\psi(s) - \psi(s')$ does not change when we simultaneously replace s' with $s' - C$ and s with $s - C$ (both phases have offsets of $2\pi\nu$ that cancel each other out). It is then straightforward to see that the resulting two integrals are equal to one double integral with the reversed order of integration,

$$\int_0^C ds \int_s^C ds' + \int_{-C}^0 ds \int_0^s ds' = \int_0^C ds' \int_{s'-C}^{s'} ds .$$

The internal integral can now be easily integrated,

$$\Delta\nu = \frac{1}{4\pi \sin(2\pi\nu)} \int_0^C ds' \Delta K(s')\beta(s') \left(-\frac{1}{2}\right) \left\{ \sin 2 \left[-\psi(s) + \psi(s') - \pi\nu\right] \right|_{s=s'-C}^{s'}$$

$$= \frac{1}{4\pi \sin(2\pi\nu)} \int_0^C ds' \Delta K(s')\beta(s') \left(-\frac{1}{2}\right) [\sin(-2\pi\nu) - \sin(2\pi\nu)] .$$

We can write this more simply as Eq. (8.34).

Because the beta function correction is proportional to $1/\sin(2\pi\nu)$ which in theory could be sensitive to small changes in ν, the result is often expressed in terms of the change in $\cos(2\pi\nu)$; this is proportional to $\sin(2\pi\nu)\Delta\beta$ and may be a better approximation. The resulting expression is

$$\Delta \cos(2\pi\nu) = -\frac{1}{2} \sin(2\pi\nu) \int_0^C ds \Delta K(s)\beta(s) .$$

Problem 8.5 *Calculate the beta beat and the tune change for a localized perturbation of the lattice:* $\Delta K = \Delta K_0 \delta(s - s_0)$.

Solution: We use the formula (8.34) from Problem 8.4 for calculating the tune change from an arbitrary quadrupole error. Substituting in $\Delta K = \Delta K_0 \delta(s - s_0)$ yields

$$\Delta\nu = \frac{1}{4\pi} \Delta K_0 \beta(s_0) .$$

We calculate the beta beat for the same example by using Eq. (8.32). The betatron function becomes

$$\beta_1(s) = \beta(s) - \frac{\beta(s)}{2\sin(2\pi\nu)} \Delta K_0(s_0)\beta(s_0) \cos 2 \left[-\psi(s) + \psi(s_0) - \pi\nu\right] .$$

The remaining dependence on lattice location s shows that the beta beating term oscillates twice as fast as the betatron motion itself.

Chapter 9
Nonlinear Resonance and Resonance Overlapping

Higher-order components in the magnetic field in a ring introduce nonlinear terms into the Hamiltonian and generate nonlinear resonances. This can lead to complicated motion for particles with large amplitudes of betatron oscillations. We derive the resonant structure in the phase space due to a sextupole magnet when the fractional part of the tune is close to $\pm\frac{1}{3}$. For a Hamiltonian system with many resonances, they can interact with each other and lead to stochastic orbits in phase space. To understand this effect, we study a model called the *standard map*, that illustrates qualitative features of what can occur in a Hamiltonian system with many resonances. The impact on dynamics is similar whether originating from effects as diverse as nonlinear magnetic fields, RF cavities, space-charge forces among the charged particles in a bunch, or interactions between bunches.

9.1 The Third-Order Resonance

We will now study the effect of sextupoles on betatron oscillations. The sextupole vector potential is given by Eq. (6.9). We will limit our analysis to one-dimensional betatron oscillations in the x direction, set $y = 0$, and use $A_s = -S(s)x^3/6$ for the vector potential. Recalling that, in the lowest order, the vector potential enters the Hamiltonian in the combination $-eA_s/p_0$ (see Eq. (5.23)) we need to add it to the generic Hamiltonian (6.14),

$$\mathcal{H} = \frac{1}{2}P_x^2 + \frac{1}{2}K(s)x^2 + \frac{1}{6}\mathcal{S}(s)x^3\,, \tag{9.1}$$

where $\mathcal{S} = eS/p_0$. In what follows, we will assume that the last term on the right-hand side is small compared to the first two terms and treat it as a perturbation.

We first make a transformation to the action-angle variables J_1 and ϕ_1 defined in Sect. 7.2 (in what follows we will drop the subscript 1 to simplify the notation).

© Springer International Publishing AG, part of Springer Nature 2018   107
G. Stupakov and G. Penn, *Classical Mechanics and Electromagnetism in Accelerator Physics*, Graduate Texts in Physics,
https://doi.org/10.1007/978-3-319-90188-6_9

This transformation converts the first two terms of the Hamiltonian (9.1) into a linear function of J:

$$\frac{1}{2}P_x^2 + \frac{1}{2}K(s)x^2 \rightarrow \frac{2\pi\nu}{C}J. \tag{9.2}$$

Transforming the last, nonlinear term we obtain:

$$\mathcal{H} = \frac{2\pi\nu}{C}J + \frac{\sqrt{2}}{3}J^{3/2}\mathcal{S}(s)\beta^{3/2}(s)\cos^3\left[\phi - \frac{2\pi\nu s}{C} + \psi(s)\right]. \tag{9.3}$$

It is convenient to change the scale of the independent variable s replacing it by the angle $\theta = 2\pi s/C$, so that θ increases by 2π every revolution in the ring. It is easy to check that the new Hamiltonian which incorporates the rescaling is obtained from (9.3) through multiplication by $(C/2\pi)$,

$$\mathcal{H} = \nu J + V(\phi, J, \theta), \tag{9.4}$$

where the perturbation V is

$$V(\phi, J, \theta) = \frac{\sqrt{2}}{3}\frac{C}{2\pi}J^{3/2}\mathcal{S}(\theta)\beta^{3/2}(\theta)\cos^3[\phi - \nu\theta + \psi(\theta)] \tag{9.5}$$

$$= \frac{C}{12\pi\sqrt{2}}J^{3/2}\mathcal{S}(\theta)\beta^{3/2}(\theta)\{\cos 3[\phi - \nu\theta + \psi(\theta)] + 3\cos[\phi - \nu\theta + \psi(\theta)]\}.$$

The new Hamiltonian is periodic in θ with period 2π. The equations of motion for the action-angle variables are:

$$\frac{\partial J}{\partial \theta} = -\frac{\partial \mathcal{H}}{\partial \phi} = -\frac{\partial V}{\partial \phi},$$
$$\frac{\partial \phi}{\partial \theta} = \frac{\partial \mathcal{H}}{\partial J} = \nu + \frac{\partial V}{\partial J}. \tag{9.6}$$

Let us now analyze the relative role of the two terms on the right-hand side of (9.5). Note that if we neglect the perturbation V in the Hamiltonian (9.4), the phase evolves as a linear function of θ, $\phi = \nu\theta + \phi_0$. After one turn, the combination $\nu\theta - \psi(\theta)$ does not change, and hence the argument $\phi - \nu\theta + \psi(\theta)$ in (9.5) changes by $2\pi\nu$. If the fractional part of ν is close to one-third or two-thirds, $\nu \approx n \pm 1/3$, where n is an integer, $\cos 3[\phi - \nu\theta + \psi(\theta)]$ returns to approximately to same value after θ changes by 2π, and the effect of this part of the perturbation accumulates with each subsequent period leading to relatively large excursions in J on the orbit. On the contrary, $\cos[\phi - \nu\theta + \psi(\theta)]$ has a phase jumping by $\approx \pm 2\pi/3$ after each turn, and due to continuous change of the sign of the cos function the effect of this term averages out almost to zero over many revolutions in the ring. This term would be resonant for the tune close to an integer but, as we know, the integer values of the tune are already unstable.

In the rest of this section we will focus on the most interesting case when $\nu \approx n \pm 1/3$ and drop the $\cos[\phi - \nu\theta + \psi(\theta)]$ term in the perturbation,

$$V(\phi, J, \theta) = \frac{C}{12\pi\sqrt{2}} J^{3/2} \mathcal{S}(\theta) \beta^{3/2}(\theta) \cos 3[\phi - \nu\theta + \psi(\theta)]. \tag{9.7}$$

To further simplify our analysis let us consider a ring with one sextupole magnet of length much shorter than the ring circumference C.[1] Without loss of generality we can assume that the magnet is located at $\theta = 0$. For such a magnet, $\mathcal{S}(\theta)$ can be approximated by a periodic delta function,

$$\mathcal{S}(\theta) = \mathcal{S}_0 \tilde{\delta}(\theta), \tag{9.8}$$

where $\tilde{\delta}(\theta) = \sum_{k=-\infty}^{\infty} \delta(\theta + 2\pi k)$. The requirement of the periodicity of $\mathcal{S}(\theta)$ follows from the fact that two values of θ that differ by 2π correspond to the same position in the ring. The term V can now be written as

$$\begin{aligned} V(\phi, J, \theta) &= \frac{C}{12\pi\sqrt{2}} J^{3/2} \mathcal{S}_0 \beta^{3/2}(\theta) \tilde{\delta}(\theta) \cos 3[\phi - \nu\theta + \psi(\theta)] \\ &= \frac{1}{3} R J^{3/2} \tilde{\delta}(\theta) \cos 3[\phi - \nu\theta + \psi(\theta)], \end{aligned} \tag{9.9}$$

where we have introduced the notation $R = C\mathcal{S}_0 \beta_0^{3/2}/(4\pi\sqrt{2})$, with $\beta_0 = \beta(0)$. The equations of motion (9.6) take the form

$$\begin{aligned} \frac{\partial J}{\partial \theta} &= R J^{3/2} \tilde{\delta}(\theta) \sin 3[\phi - \nu\theta + \psi(\theta)], \\ \frac{\partial \phi}{\partial \theta} &= \nu + \frac{1}{2} R J^{1/2} \tilde{\delta}(\theta) \cos 3[\phi - \nu\theta + \psi(\theta)]. \end{aligned} \tag{9.10}$$

Let us consider how J and ϕ evolve over one turn in the ring, when θ changes from 0 to 2π, starting from $\theta = -0$, that is right before the delta function kick. Without loss of generality, we can assume that $\psi(0) = 0$. We first need to integrate these equations through the delta-function kick, that is from $\theta = -0$ to $\theta = +0$, for arbitrary initial values J_1 and ϕ_1. For this one integration, the equations are simplified:

$$\begin{aligned} \frac{\partial J}{\partial \theta} &= J^{3/2} \delta(\theta) \sin 3\phi, \\ \frac{\partial \phi}{\partial \theta} &= \frac{1}{2} J^{1/2} \delta(\theta) \cos 3\phi, \end{aligned} \tag{9.11}$$

---

[1] To be more precise, the length of the magnet should be much shorter than the betatron wavelength, which typically is a fraction of the circumference C.

where we rescaled the action introducing $\mathcal{J} = R^2 J$, set $\theta = 0$ everywhere except in the argument of the delta function, and discarded the constant ν term from the second part of Eq. (9.10) in comparison to the delta function. The initial conditions for these equations are $\mathcal{J} = \mathcal{J}_1 = R^2 J_1$ and $\phi = \phi_1$.

Going back to our starting point of Eq. (9.1), we see that when crossing $\theta = 0$ (and thus $s = 0$) there is no direct contribution from the sextupole field to the evolution of x, based on $dx/ds = \partial \mathcal{H}/\partial P_x$. From the equation $dP_x/ds = -\partial \mathcal{H}/\partial x$, it follows that a short magnet only gives a transverse kick of magnitude $\Delta P_x = -C S_0 x^2/4\pi$. Unfortunately, when we switch to action-angle coordinates we obtain two sets of coupled equations for ϕ and $\mathcal{J}$ which are more challenging to solve, and once we discard the $\cos \phi$ term from the perturbation V it is no longer consistent to insist on using the same value for ΔP_x nor requiring $\Delta x = 0$. So while our equations (9.11) are physically meaningful when the tune is close to an integer $\pm 1/3$, we have more work to do in order to evaluate the effect of the sextupole.

One step we can take analytically is find an integral of motion across the sextupole using the following formal transformation which takes advantage of the fact that the delta function $\delta(\theta)$ is a derivative of the step function $h(\theta)$,

$$\frac{dh}{d\theta} = \delta(\theta) , \tag{9.12}$$

where $h(\theta)$ is equal to 1 for $\theta > 0$ and zero otherwise. Replacing the delta function by this derivative and then noting that

$$\frac{\partial}{\partial \theta} = \frac{dh}{d\theta} \frac{\partial}{\partial h} = \delta(\theta) \frac{\partial}{\partial h} , \tag{9.13}$$

we replace Eq. (9.11) with

$$\frac{\partial \mathcal{J}}{\partial h} = \mathcal{J}^{3/2} \sin 3\phi ,$$
$$\frac{\partial \phi}{\partial h} = \frac{1}{2} \mathcal{J}^{1/2} \cos 3\phi , \tag{9.14}$$

where the independent variable h now changes from 0 to 1 when θ traverses the delta-function. It is straightforward to check that Eq. (9.14) represents Hamilton's equations of motion for the following Hamiltonian:

$$\mathcal{H}(\phi, \mathcal{J}) = \frac{1}{3} \mathcal{J}^{3/2} \cos 3\phi , \tag{9.15}$$

Note that the Hamiltonian would be proportional to x^3 if we had not neglected the $\cos \phi$ term. Since the Hamiltonian does not depend on the independent variable h, it is conserved, and its trajectories can be easily found from the equation $\mathcal{H}(\phi, \mathcal{J}) = $ const. They are shown in Fig. 9.1. Remarkably, due to the rescaling of variables,

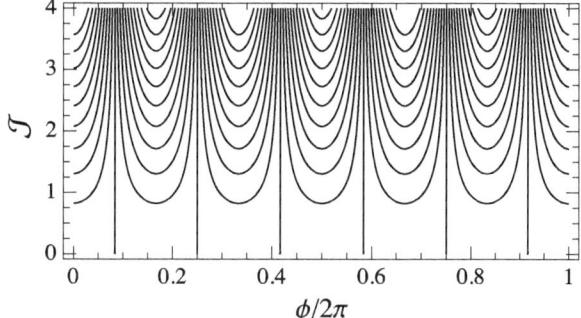

Fig. 9.1 Trajectories in the phase space $\phi - \mathcal{J}$ for the Hamiltonian (9.15). Note the symmetry along the ϕ axis with the period $\pi/3$

our new Hamiltonian (9.15) is universal — it does not contain any original parameters of the problem, such as the strength of the sextupole or the circumference of the ring.

We have now recovered a method to calculate the action-angle variables at the end of the sextupole, which we will denote as $\tilde{J}$ and $\tilde{\phi}$. By numerically integrating Eq. (9.14) from $h = 0$ to $h = 1$, we can generate the map that moves our particle through the sextupole magnet according to our approximate Hamiltonian: $\tilde{\mathcal{J}} = f(\phi_1, \mathcal{J}_1)$ and $\tilde{\phi} = g(\phi_1, \mathcal{J}_1)$. To get an entire full-turn map we then need to evaluate the equations of motion from the exit from the sextupole, $\theta = +0$, through the whole ring, to the next entrance to the magnet, $\theta = 2\pi - 0$. We will denote the action-angle variables at $\theta = 2\pi - 0$ as $\mathcal{J}_2$ and ϕ_2. The action remains unchanged until the sextupole is again crossed, and the phase linearly grows until by the sextupole entrance it has increased by $2\pi\nu$, so we obtain:

$$\mathcal{J}_2 = f(\phi_1, J_1), \qquad \phi_2 = g(\phi_1, \mathcal{J}_1) + 2\pi\nu. \tag{9.16}$$

It is now clear that every revolution in the ring repeats the transformation (9.16), and the values $\mathcal{J}_n$, ϕ_n at the start of the n-th revolution are expressed through the values at the previous one,

$$\mathcal{J}_n = f(\phi_{n-1}, \mathcal{J}_{n-1}), \qquad \phi_n = g(\phi_{n-1}, \mathcal{J}_{n-1}) + 2\pi\nu. \tag{9.17}$$

To illustrate the dynamics of this map, we iterated it 300 times starting from different initial conditions and numerically integrating Eq. (9.14) on each step. Each pair ϕ_n, $\mathcal{J}_n$ was converted to the original canonical variables x and P_x using Eqs. (7.9) and (7.10) and the relation $J = \mathcal{J}/R^2$:

$$xR/\sqrt{\beta_0} = \sqrt{2\mathcal{J}}\cos\phi, \qquad P_x R\sqrt{\beta_0} = -\sqrt{2\mathcal{J}}\sin\phi, \tag{9.18}$$

where for simplicity we have assumed $\alpha = 0$. The result of this simulation when the fractional part $[\nu]$ of the tune is equal to $\frac{1}{3} + 0.1$ is shown in Fig. 9.2. The horizontal and vertical axes are the normalized dimensionless coordinate $xR/\sqrt{\beta_0}$ and the normalized angle $P_x R\sqrt{\beta_0}$. We see that the orbits with large amplitudes of

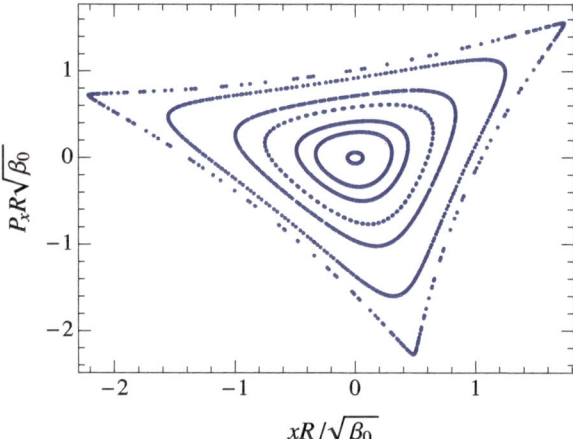

Fig. 9.2 Orbits in phase space for the case $[\nu] - 1/3 = 0.1$. Particles starting from outside of the largest, triangular-shaped orbit quickly leave the system

betatron oscillations get a triangular shape reflecting the three-fold symmetry of the sextupole field (see Fig. 6.1b). The largest orbit shown in Fig. 9.2 is a separatrix — orbits that start outside of it move away from the axis and leave the system in several iterations.

An example of experimentally measured third-order resonance orbits at the IUCF cooler ring (at the Indiana University Bloomington campus) can be found in Ref. [1].

9.2 Standard Model and Resonance Overlapping

As we saw in the previous section, the effect of a sextupole on betatron oscillations can be reduced to a map, Eq. (9.17), which demonstrates particle confinement near the axis and particle losses outside of the separatrix. Many other nonlinear beam dynamics phenomena can also be formulated in terms of Hamiltonian maps. In this section, we will consider one such map often called the *standard*, or Chirikov-Taylor, map. The remarkable feature of this map is that it demonstrates a transition from regular to chaotic motion in a non-integrable, time-dependent Hamiltonian system with only one degree of freedom.

The standard map describes the evolution in time of a system with the following Hamiltonian:

$$H(\theta, I, t) = \frac{1}{2}I^2 + K\tilde{\delta}(t) \cos \theta, \tag{9.19}$$

where K is a parameter, $\tilde{\delta}(t) = \sum_{n=-\infty}^{\infty} \delta(t+n)$ is the periodic δ function that describes kicks repeating with the unit period (note that this definition of $\tilde{\delta}(t)$ is different from the one defined in Sect. 9.1 where the period was equal to 2π). Here I can be considered as an action, and θ as an angle variable; they are both dimensionless.

The equations of motion for I and θ are

$$\dot{I} = -\frac{\partial H}{\partial \theta} = K\tilde{\delta}(t)\sin\theta\,, \qquad \dot{\theta} = \frac{\partial H}{\partial I} = I\,. \tag{9.20}$$

If I_n and θ_n are the values at $t = n - 0$ (before the delta-function kick), then integrating the first of Eq. (9.20) from $t = n - 0$ to $t = n + 0$ (through the delta-function kick) gives $I_{n+1} = I_n + K\sin\theta_n$, which is then conserved over the rest of the unit interval from $t = n + 0$ to $t = (n + 1) - 0$ (where there are no kicks). Integrating the second equation in (9.20) along this part of the interval and remembering that the action here is already equal to I_{n+1} gives $\theta_{n+1} = \theta_n + I_{n+1}$. Combining these two equations we obtain

$$I_{n+1} = I_n + K\sin\theta_n\,,$$
$$\theta_{n+1} = \theta_n + I_{n+1}\,. \tag{9.21}$$

These equations transform the action-angle variables from their values at time $t = n$ to time $t = n + 1$. This transformation is called the *standard map*.[2]

The periodic delta-function used in Eq. (9.19) can be expanded as a Fourier series,

$$\tilde{\delta}(t) = 1 + 2\sum_{n=1}^{\infty} \cos(2\pi nt)\,. \tag{9.22}$$

Substituting this representation into the Hamiltonian (9.19) we can rewrite the latter in the following form:

$$\begin{aligned}
H(\theta, I, t) &= \frac{1}{2}I^2 + K\cos(\theta) + 2K\cos(\theta)\sum_{n=1}^{\infty}\cos(2\pi nt) \\
&= \frac{1}{2}I^2 + K\cos(\theta) + K\sum_{\substack{n=-\infty \\ n\neq 0}}^{\infty}\cos(\theta - 2\pi nt)\,,
\end{aligned} \tag{9.23}$$

where we have used the relation $2\cos(\theta)\cos(2\pi nt) = \cos(\theta - 2\pi nt) + \cos(\theta + 2\pi nt)$. The two first terms on the right-hand side comprise the Hamiltonian of the pendulum[3] and the sum over terms with $n \neq 0$ is a time-dependent periodic driver with the frequencies equal to $2\pi n$.

We can get some insight into the structure of the phase space by selecting only one term n in the infinite sum over all n (similar to what we did analyzing the Hamiltonian (9.4)):

---

[2]One can also find in the literature a definition of the standard map which differs from Eq. (9.21) by numerical factors.

[3]This can be seen if one sets $ml^2 \to 1$ in the pendulum Hamiltonian (1.32) and associates K with $-\omega_0^2$. In case of positive K, one also needs to shift the origin of the angle, $\theta \to \theta + \pi$.

Fig. 9.3 Sketch illustrating the superposition of the phase portraits for various resonances of the standard map. The width of each resonance is $4\sqrt{|K|}$

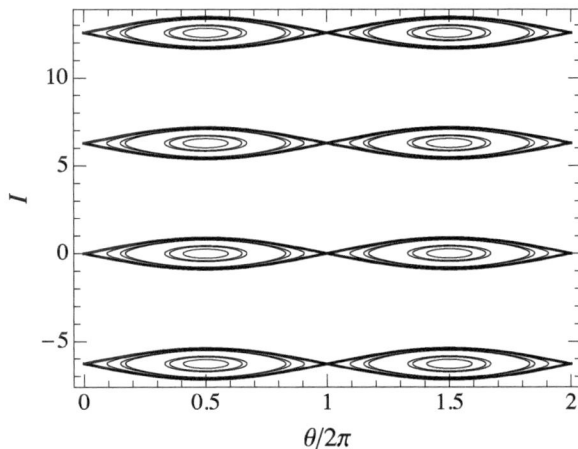

$$\mathcal{H}(\theta, I, t) = \frac{1}{2}I^2 + K \cos(\theta - 2\pi n t). \qquad (9.24)$$

Making the canonical transformation $\theta, I \rightarrow \phi, J$ where

$$J = I - 2\pi n, \qquad \phi = \theta - 2\pi n t, \qquad (9.25)$$

we eliminate the time variable and find the new Hamiltonian,

$$\mathcal{H}'(\phi, J) = \frac{1}{2}J^2 + K \cos\phi, \qquad (9.26)$$

which is again the Hamiltonian of a pendulum. The structure of the phase space for this Hamiltonian is shown in Fig. 4.4; the separatrix that encompasses trajectories with a limited angular variation is defined by the equation $\mathcal{H}'(\phi, J) = |K|$. From this equation we find that the maximum deviation of J on the separatrix is $J = \pm 2\sqrt{|K|}$. For the original action variable I this translates into the deviation relative to the value $2\pi n$, $I = 2\pi n \pm 2\sqrt{|K|}$.

Trying to understand the overall structure of the phase space of the original Hamiltonian, we can naively superimpose the phase portraits for the Hamiltonians (9.26) with different values of n. This makes a cartoon shown in Fig. 9.3. Such superposition makes sense only if $|K| \ll 1$, when the resonances for different values of n are well separated and, in the first approximation, do not interact with each other.

Computer simulations of the standard map show that, indeed, as long as the distance between the islands is much larger than the width of the separatrix, then to a good approximation resonances with different values of n can be considered separately. However, increasing $|K|$ leads to an *overlapping of the resonances* and the motion becomes much more complicated. Because the distance between the resonances is 2π and the width is $4\sqrt{|K|}$, formally the overlapping occurs for $|K| > \pi^2/4$, but given the qualitative nature of our argument one should not expect a drastic transition at this exact value of K. Indeed, simulations show that when $|K|$ increases, there is a gradual transformation of the flow shown in Fig. 9.3 into a regime in which the laminar orbits are destroyed and the motion becomes stochastic. This is illustrated by Fig. 9.4. Qualitatively, the transition from the laminar to stochastic motion occurs at

$$K \sim 1 . \tag{9.27}$$

When $|K|$ becomes much larger than one, regular orbits are destroyed and the particle motion becomes chaotic. In this limit, after each kick the particle loses memory of its previous phase, and the consecutive phases θ_n can be considered to be uncorrelated. As a result, the subsequent values of action I_n can be described as a random walk, and over many steps a statistical description of the process as a diffusion along the I axis becomes appropriate.

We can easily estimate the rate of diffusion for the action. From Eq. (9.21) we calculate the change of the action in one step, $\Delta I_n = K \sin \theta_n$, and taking the square of ΔI_n and averaging it over the random phase θ_n we obtain

$$\langle \Delta I_n^2 \rangle = \frac{1}{2} K^2 . \tag{9.28}$$

In a random walk, the average squares $\langle \Delta I_n^2 \rangle$ add up, and after N steps the average square of the accumulated action I_N is

$$\langle I_N^2 \rangle \approx \langle \Delta I_n^2 \rangle N \approx \frac{1}{2} K^2 N . \tag{9.29}$$

The linear growth of $\langle I_N^2 \rangle$ with the number of steps is a characteristic feature of the diffusion process.

Figure 9.5 shows the result of a numerical simulation of the map (9.21) for a particular value $K = 8.41$. The black line shows one sample orbit that demonstrates wild fluctuations of I_N^2 on top of a systematic growth with N. The red line is $\langle I_N^2 \rangle$ averaged over an ensemble of 400 orbits starting from different initial conditions — it agrees well with the analytical formula (9.29).[4]

9.3 Dynamic Aperture in Accelerators

We have seen in this chapter and the previous one that the severity of field errors and nonlinearities depends strongly on the tune. Nonlinear fields (both external and self-

---

[4]A more detailed study [2] reveals that Eq. (9.29) is only the leading term in the formula for $\langle I_N^2 \rangle$ in the limit $K \gg 1$. The correction to (9.29) is noticeable when $K \lesssim 10$, however it happens to vanish for the particular value of $K = 8.41$. This explains the very good agreement between the simulation and theory in Fig. 9.5.

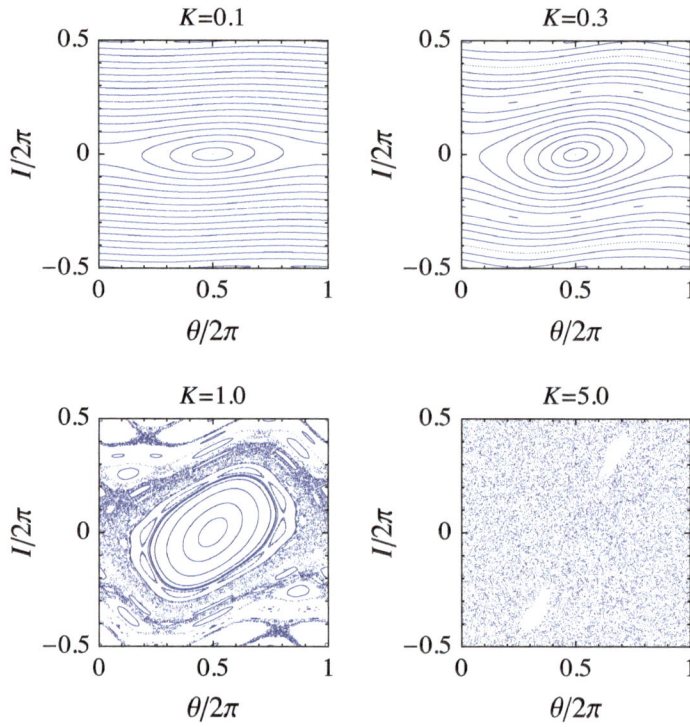

Fig. 9.4 The result of computer simulations for the standard map. Shown is a part of the phase space bounded in the vertical direction by the inequality $-\pi < I < \pi$, for four different values of the parameter K, $K = 0.1, 0.3, 1, 5$ from left to right and from top to bottom. The last two pictures show an increase and then an almost complete domination of the stochastic component of the motion in a large part of the phase space

Fig. 9.5 The calculated ensemble average of $\langle I_N^2 \rangle$ versus the number of iteration N for $K = 8.41$ (red line). The straight blue line is the theoretical prediction of Eq. (9.29). The black line shows one particular orbit that starts from $I = \pi$, $\theta = \pi/10$

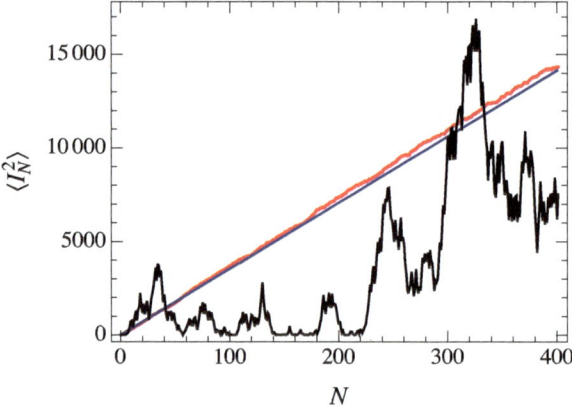

forces) have the effect of varying the tune corresponding to different particle orbits, leading to a *tune spread*. This means that even if a particle following the reference orbit is not near a resonance, other particles at higher amplitudes may be strongly perturbed by the nonlinear fields. We have not discussed coupling between different degrees of freedom, but that leads to more opportunities for resonances to appear (for example, if the vertical and horizontal tunes differ by 1/2). In a typical situation, nonlinear fields make the phase space at some distance from the reference orbit prone to stochastic motion, leading to a random walk of the particle until it is lost.

At best there can only be a limited region near the reference orbit where particles are properly confined. This region in phase space is called the *dynamic aperture* of the machine. It is computed with the help of accelerator codes by launching particles at various locations away from the reference orbit and tracking their motion. Rather than having a sharp boundary, the dynamic aperture is usually surrounded by an intermediate zone where the rate of diffusion which particles experience gets gradually worse. A related concept that focuses on the short-term tune of particles within a bunch is the analysis of *frequency maps* [3].

A modern circular accelerator has many magnets that play various roles in confining the beam in the ring. Even as industry and researchers learn to reduce errors in the manufacture and installation of magnets, more aggressive designs and improvements in other areas tend to make nonlinearities a major constraint on the operation of accelerators, limiting the total charge contained in storage rings and the luminosity of colliders.

Worked Examples

Problem 9.1 *Prove that the standard map defines a canonical transformation* $(I_n, \theta_n) \rightarrow (I_{n+1}, \theta_{n+1})$.

Solution: We can see that $\{I_{n+1}, I_{n+1}\} = 0$ and $\{\theta_{n+1}, \theta_{n+1}\} = 0$ trivially, since $\{f, f\}$ always vanishes for the Poisson bracket. Then with

$$I_{n+1} = I_n + K \sin \theta_n ,$$
$$\theta_{n+1} = \theta_n + I_n + K \sin \theta_n ,$$

we find

$$\{\theta_{n+1}, I_{n+1}\} = \frac{\partial \theta_{n+1}}{\partial \theta_n} \frac{\partial I_{n+1}}{\partial I_n} - \frac{\partial \theta_{n+1}}{\partial I_n} \frac{\partial I_{n+1}}{\partial \theta_n}$$
$$= (1 + K \cos \theta_n) - K \cos \theta_n = 1 .$$

Therefore the map is canonical.

Problem 9.2 *Prove the following property of the standard map: for two trajectories starting from the same initial value θ_0 but with different values $I_0^{(1)}$ and $I_0^{(2)}$, such that $I_0^{(2)} - I_0^{(1)} = 2\pi m$, where m is an integer, the difference $I_n^{(2)} - I_n^{(1)}$ remains equal to $2\pi m$ for all values of n.*

Solution: We will do a proof by induction. First we assume it holds true for some value of n: $I_n^{(2)} - I_n^{(1)} = 2\pi m$. While the orbits start from the same value of angle θ_0, we assume that on step n their angles differ by an integer number of 2π, $\theta_n^{(2)} = \theta_n^{(1)} + 2\pi p$, where p can be any integer. Then

$$
\begin{aligned}
I_{n+1}^{(2)} - I_{n+1}^{(1)} &= (I_n^{(2)} - I_n^{(1)}) + K(\sin\theta_n^{(2)} - \sin\theta_n^{(1)}) \\
&= 2\pi m \, ,
\end{aligned}
$$

$$
\begin{aligned}
\theta_{n+1}^{(1)} - \theta_{n+1}^{(2)} &= (\theta_n^{(2)} - \theta_n^{(1)}) + (I_n^{(2)} - I_n^{(1)}) + K(\sin\theta_n^{(2)} - \sin\theta_n^{(1)}) \\
&= 2\pi(m + p) \, .
\end{aligned}
$$

We are given that it holds for $n = 0$, so it must hold for all n.

Problem 9.3 *Prove that Eq. (9.25) defines a canonical transformation, find the corresponding generating function F_2, and obtain the Hamiltonian (9.26).*

Solution: First we show the transformation is canonical:

$$
\{\phi, J\} = \frac{\partial J}{\partial I}\frac{\partial \phi}{\partial \theta} - \frac{\partial J}{\partial \theta}\frac{\partial \phi}{\partial I} = 1 \, .
$$

Then we can find the generating function $F_2(\theta, J, t)$ by integrating the expressions for ϕ and I in terms of the partial derivatives of F_2:

$$
\phi = \frac{\partial F_2}{\partial J} = \theta - 2\pi nt \quad \Longrightarrow \quad F_2 = J(\theta - 2\pi nt) + g(\theta, t) \, ,
$$

$$
I = \frac{\partial F_2}{\partial \theta} = J + 2\pi n \quad \Longrightarrow \quad F_2 = \theta(J + 2\pi n) + f(J, t) \, ,
$$

where g and f are arbitrary functions of their arguments. Combining these two equations for F_2 we conclude that

$$
F_2 = \theta J + 2\pi n\theta - 2\pi nJt + h(t) \, ,
$$

where h is an arbitrary function of time. The quantity $h(t)$ can be ignored because it does not couple to the dynamic variables. Finally, we can find the new Hamiltonian from

$$
\begin{aligned}
H' &= H + \frac{\partial F_2}{\partial t} \\
&= \frac{I^2}{2} + K\cos(\theta - 2\pi nt) - 2\pi nJ \\
&= \frac{(J + 2\pi n)^2}{2} + K\cos(\phi) - 2\pi nJ \\
&= \frac{J^2}{2} + K\cos(\phi) + 2\pi^2 n^2 \, ,
\end{aligned}
$$

as given. Again, the constant term or any purely time-dependent term does not affect the dynamics and can be ignored.

Problem 9.4 *For Eq. (9.21), the main fixed points (ignoring periodicity) when $K >$ 0 are an (unstable) saddle point at $\theta = 0$, $I = 0$ and, for small K, a stable fixed point at $\theta = \pi$, $I = 0$. Find the value of $K > 0$ above which the motion around $\theta = \pi$, $I = 0$ also becomes unstable.*

Solution: Because we are expanding around $\theta = \pi$, it is convenient to define a shifted variable $\delta = \theta - \pi$. Then the map becomes $I_{n+1} = I_n - K \sin \delta_n$, $\delta_{n+1} = \delta_n + I_{n+1}$. Expanding for $\delta, I \ll 1$ and only keeping the linear term, $\sin \delta \approx \delta$, we can write the linear single-turn map in matrix form as

$$\begin{pmatrix} \delta_{n+1} \\ I_{n+1} \end{pmatrix} = \begin{pmatrix} 1 - K & 1 \\ -K & 1 \end{pmatrix} \begin{pmatrix} \delta_n \\ I_n \end{pmatrix} .$$

This matrix, obtained by linearizing the single-turn map, is identical to the Jacobian matrix for this map as described in Chap. 3, but specifically at the fixed point (note that the determinant is 1). We can evaluate the local stability by looking for eigenvalues of the matrix:

$$0 = \begin{vmatrix} 1 - K - \lambda & 1 \\ -K & 1 - \lambda \end{vmatrix} = \lambda^2 + \lambda(K - 2) + 1 .$$

Solving this equation we find

$$\lambda = \frac{1}{2} \left(2 - K \pm \sqrt{K^2 - 4K} \right) .$$

Whenever there is some $|\lambda| > 1$, the motion is unstable because there is some combination of small offsets which will grow exponentially each turn. A simple analysis shows that $|\lambda| > 1$ for $K > 4$. The loss of the stable point at $\theta = \pi$ and $I = 0$ can be clearly seen in Fig. 9.4 comparing the phase space at $K = 1$ and $K = 5$.

References

1. D.D. Caussyn, M. Ball, B. Brabson, J. Collins, S.A. Curtis, V. Derenchuck, D. DuPlantis, G. East, M. Ellison, T. Ellison, D. Friesel, B. Hamilton, W.P. Jones, W. Lamble, S.Y. Lee, D. Li, M.G. Minty, T. Sloan, G. Xu, A.W. Chao, K.Y. Ng, S. Tepikian, Experimental studies of nonlinear beam dynamics. Phys. Rev. A **46**(12), 7942–7952 (1992)
2. A.B. Rechester, R.B. White, Calculation of turbulent diffusion for the Chirikov-Taylor model. Phys. Rev. Lett. **44**, 1586–1589 (1980)
3. D. Robin, C. Steier, J. Laskar, L. Nadolski, Global dynamics of the advanced light source revealed through experimental frequency map analysis. Phys. Rev. Lett. **85**, 558 (2000)

Chapter 10
The Kinetic Equation

In the preceding chapters we focused our attention on the motion of a single particle. In this chapter, we will introduce the concept of the distribution function and describe the formalism of the kinetic equation for treating large ensembles of particles in a beam. While this chapter focuses on deterministic Hamiltonian motion, kinetic equations in general can also include stochastic motion and damping.

10.1 The Distribution Function in Phase Space and the Kinetic Equation

We begin from a simple case of one degree of freedom when each particle is characterized by two canonically conjugate variables q and p. A large ensemble of particles that represents a beam shall contain various pairs of values of q and p and constitutes a "cloud" in the phase space as illustrated in Fig. 10.1. With time, each particle will move along its own orbit in the physical space, and the corresponding phase point travels along a trajectory in the phase space. The "cloud" gradually changes shape. The particle motion is governed by external fields, as well as interactions between the particles. In this chapter, however, we neglect the interaction effects, and assume that each particle moves independently due to external electromagnetic fields only.

Consider an infinitesimally small region $dq \times dp$ in the phase plane with the center located at p, q, as shown in Fig. 10.2, and let the number of particles at time t in this phase space element be given by dN. This mathematically infinitesimal phase element should be considered to be physically large enough to include many particles, so that $dN \gg 1$. We define the distribution function of the beam $f(q, p, t)$ so that

$$dN(t) = f(q, p, t)dp\,dq\,. \tag{10.1}$$

© Springer International Publishing AG, part of Springer Nature 2018
G. Stupakov and G. Penn, *Classical Mechanics and Electromagnetism in Accelerator Physics*, Graduate Texts in Physics,
https://doi.org/10.1007/978-3-319-90188-6_10

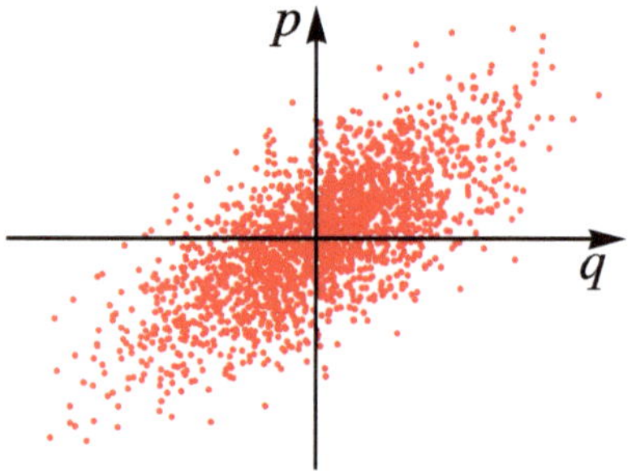

Fig. 10.1 Illustration of the phase space of an ensemble of particles with the position of each particle indicated by a red point

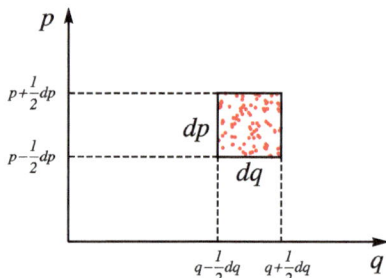

Fig. 10.2 A characterization of the distribution function f. The red dots show particles inside an infinitesimally small rectangle with the sides dq and dp

We can say that the distribution function gives the *density of particles in the phase space*.

As was emphasized above, the phase points are moving along trajectories, and the distribution function evolves with time. Our goal is to derive an equation that governs this evolution.

At time $t + dt$ the number of particles in the region $dq \times dp$ will change because of the flow of particles through the four boundaries of the rectangle. Due to the motion in the q direction, the number of particles that flow in through the left boundary is

$$f \left(q - \frac{1}{2} dq, p, t \right) \times dp \times \dot{q} \left(q - \frac{1}{2} dq, p, t \right) dt \, . \tag{10.2}$$

In this equation, $\dot{q} dt$ is the distance from which the flow brings new particles into the rectangle during time dt, and we take the values of both f and $\dot{q}$ in the middle of the left side of the rectangle. Similarly, the number of particles that flow out through the right boundary is

$$f \left(q + \frac{1}{2} dq, p, t \right) \times dp \times \dot{q} \left(q + \frac{1}{2} dq, p, t \right) dt \, . \tag{10.3}$$

Using the same logic, we calculate the number of particles which flow in through the lower horizontal boundary,

$$f\left(q, p - \frac{1}{2}dp, t\right) \times dq \times \dot{p}\left(q, p - \frac{1}{2}dp, t\right) dt, \qquad (10.4)$$

and the number of particles that flow out through the upper horizontal boundary,[1]

$$f\left(q, p + \frac{1}{2}dp, t\right) \times dq \times \dot{p}\left(q, p + \frac{1}{2}dp, t\right) dt. \qquad (10.5)$$

We are now ready to calculate the change of the number of particles in the phase volume $dq \times dp$. On one hand, this number is due to the change of the distribution function during the time interval dt, $dN(t + dt) - dN(t) = [f(q, p, t + dt) - f(q, p, t)]dp\,dq$. On the other hand, it is equal to the sum of the four contributions calculated above. Equating these two expressions we obtain,

$$[f(q, p, t + dt) - f(q, p, t)]dp\,dq$$
$$= f\left(q - \frac{1}{2}dq, p, t\right)\dot{q}\left(q - \frac{1}{2}dq, p, t\right) dp\,dt - f\left(q + \frac{1}{2}dq, p, t\right)\dot{q}\left(q + \frac{1}{2}dq, p, t\right) dp\,dt$$
$$+ f\left(q, p - \frac{1}{2}dp, t\right)\dot{p}\left(q, p - \frac{1}{2}dp, t\right) dq\,dt - f\left(q, p + \frac{1}{2}dp, t\right)\dot{p}\left(q, p + \frac{1}{2}dp, t\right) dq\,dt.$$
$$(10.6)$$

Expanding both sides of this equation in the Taylor series and keeping only linear terms in dp, dq, dt, and then dividing both sides by $dp\,dq\,dt$, we arrive at the following result:

$$\frac{\partial f(q, p, t)}{\partial t} + \frac{\partial}{\partial q}[\dot{q}(q, p, t)f(q, p, t)] + \frac{\partial}{\partial p}[\dot{p}(q, p, t)f(q, p, t)] = 0.$$
$$(10.7)$$

This is the *continuity* equation for the function f — it is a mathematical expression of the fact that the particles in the phase space are not created and do not disappear; they are being transported from one place to another along smooth paths. Integrating this equation over the whole phase space and taking into account that f vanishes at infinity when $|q|, |p| \to \infty$, gives $dN/dt = 0$, where $N = \int f\,dq\,dp$ is the total number of particles in the system. As expected, the total number of particles is conserved.

In the derivation above, we did not use the Hamiltonian nature of the phase flow. We will now show that, for a Hamiltonian system, the medium represented by the distribution function f is *incompressible*. This follows immediately from Liouville's

---

[1]Our classification of "flow in" and "flow out" tacitly assumes that the corresponding velocities $\dot{q}$ and $\dot{p}$ are positive. For negative values of these velocities, the values of the "flow in" and "flow out" will be negative, corresponding to an actual outflow and inflow respectively.

theorem (see Sect. 3.4) that states that the volume $dp\,dq$ of a space phase element does not change in Hamiltonian motion. Since f is proportional to the number of particles in this volume, and this number is conserved, f too is conserved, but only within a *moving* phase space volume element. The density at a given point q, p of the phase space does, however, change because a fluid element located at this point at a given time will be replaced by a new element at a later time.

Mathematically, the fact of incompressibility is reflected in the following transformation of the continuity equation (10.7). From the Hamiltonian equations for $\dot{q}$ and $\dot{p}$ it follows that

$$\frac{\partial}{\partial q}\dot{q} = \frac{\partial}{\partial q}\frac{\partial H}{\partial p}, \qquad \frac{\partial}{\partial p}\dot{p} = -\frac{\partial}{\partial p}\frac{\partial H}{\partial q}, \tag{10.8}$$

and hence $\partial \dot{q}/\partial q + \partial \dot{p}/\partial p = 0$. Using this relation, we can rewrite Eq. (10.7) as follows:

$$\frac{\partial f}{\partial t} + \frac{\partial H}{\partial p}\frac{\partial f}{\partial q} - \frac{\partial H}{\partial q}\frac{\partial f}{\partial p} = 0, \tag{10.9}$$

where, for brevity, we have dropped the arguments of f and H. In accelerator and plasma physics this version of the kinetic equation is often called the *Vlasov* equation. Specifically, there are no scattering or damping terms. It provides an extremely powerful tool in accelerator physics giving a detailed description of the beam dynamics.

Note that we can also use the formalism of the Poisson bracket to write the Vlasov equation as

$$\frac{\partial f}{\partial t} = \{H, f\}. \tag{10.10}$$

In this form, the Vlasov equation is also valid for n degrees of freedom, with canonical variables q_i and p_i, $i = 1, 2, \ldots, n$, and the distribution function f defined as a density in $2n$-dimensional phase space and depending on all these variables, $f(q_1, \ldots, q_n, p_1, \ldots, p_n, t)$. Equivalently, the Vlasov equation can be written as

$$\frac{\partial f}{\partial t} = \sum_{i=1}^{n}\left(\frac{\partial H}{\partial q_i}\frac{\partial f}{\partial p_i} - \frac{\partial H}{\partial p_i}\frac{\partial f}{\partial q_i}\right). \tag{10.11}$$

To conclude this section, we note that in some cases it is more convenient to normalize f by the number of particles N; in this case, the integral of f over the phase space is equal to one. With such a normalization, $f(q, p, t)dq\,dp$ can be understood as a probability to find a particle in the phase volume $dq\,dp$ in the vicinity of the phase point q, p.

Fig. 10.3 A trajectory in the extended phase space connecting two neighboring points

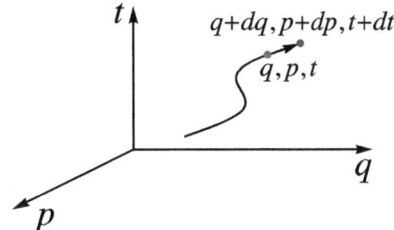

10.2   Integration of the Vlasov Equation Along Trajectories

In this section we will prove by direct calculation that any distribution function which satisfies the Vlasov equation (10.11) remains constant in each "fluid" element of the phase space as it moves along a particle trajectory. This property of the distribution function immediately follows from the Liouville theorem; because the phase volume of a small fluid element is conserved, the value of the distribution function equal to the ratio of the number of particles in this element to its volume does not change. In addition to confirming this important property, we will derive from it a powerful method for solving the Vlasov equation. For simplicity we will limit our consideration to Hamiltonian systems with one degree of freedom.

Let us consider a trajectory in the *extended* phase space that in addition to q and p axes includes the time axis t, as shown in Fig. 10.3. We want to calculate the difference of f at two close points, at time t and $t + dt$, on this single trajectory. We have

$$df = f(q + dq, p + dp, t + dt) - f(q, p, t)$$
$$= \frac{\partial f}{\partial t} dt + \frac{\partial f}{\partial q} dq + \frac{\partial f}{\partial p} dp . \tag{10.12}$$

Because the two points are on the same trajectory, $dq = \dot{q} dt = (\partial H / \partial p) \, dt$ and $dp = \dot{p} dt = -(\partial H / \partial q) \, dt$. Substituting these relations into (10.12) we find that

$$df = \frac{\partial f}{\partial t} dt - \frac{\partial H}{\partial q} \frac{\partial f}{\partial p} dt + \frac{\partial H}{\partial p} \frac{\partial f}{\partial q} dt = 0 , \tag{10.13}$$

or

$$\frac{df}{dt} = 0 . \tag{10.14}$$

On the last step we invoked Eq. (10.9). Equation (10.14) is the mathematical expression of the fact that f remains constant along a trajectory. This derivative, which describes changes along a trajectory, is referred to as the convective derivative, and can be written as

$$\frac{d}{dt}\bigg|_{\text{conv}} \equiv \frac{\partial}{\partial t} + \dot{q}\frac{\partial}{\partial q} + \dot{p}\frac{\partial}{\partial p}.$$ (10.15)

We have encountered this derivative in Sect. 1.5.

Knowing that f is constant along trajectories, we can find solutions to the Vlasov equation if the phase space orbits are known. Let $q(q_0, p_0, t)$ and $p(q_0, p_0, t)$ be solutions of the Hamiltonian equations with initial values q_0 and p_0 at $t = 0$, and $F(q_0, p_0)$ be the initial distribution function at $t = 0$. To find the value of f at q, p at time t we need to trace back the trajectory that passes through q, p at t, and find the initial values q_0, p_0 where it starts at $t = 0$. Hence, we need to invert the relations

$$q = q(q_0, p_0, t), \qquad p = p(q_0, p_0, t),$$ (10.16)

and find q_0, p_0 in terms of q, p: $q_0 = q_0(q, p, t)$ and $p_0 = p_0(q, p, t)$. The value of f at q, p at time t is then equal to the value of F at q_0, p_0:

$$f(q, p, t) = F(q_0(q, p, t), p_0(q, p, t)).$$ (10.17)

For simple trajectories, the inversion of (10.16) can be done analytically, and Eq. (10.17) then defines f for an arbitrary initial function F.

As an illustration of the method, let us consider an ensemble of linear oscillators with frequency ω, whose motion is described by the Hamiltonian

$$H(x, p) = \frac{p^2}{2} + \omega^2\frac{x^2}{2}.$$ (10.18)

The distribution function $f(x, p, t)$ for these oscillators satisfies the Vlasov equation,

$$\frac{\partial f}{\partial t} + p\frac{\partial f}{\partial x} - \omega^2 x\frac{\partial f}{\partial p} = 0.$$ (10.19)

Solving the Hamiltonian equations, it is easy to find the trajectory which has initial value x_0 and p_0 at $t = 0$,

$$x = x_0\cos(\omega t) + \frac{p_0}{\omega}\sin(\omega t),$$
$$p = -\omega x_0\sin(\omega t) + p_0\cos(\omega t).$$ (10.20)

Inverting these equations, we find

$$x_0 = x\cos(\omega t) - \frac{p}{\omega}\sin(\omega t),$$
$$p_0 = \omega x\sin(\omega t) + p\cos(\omega t).$$ (10.21)

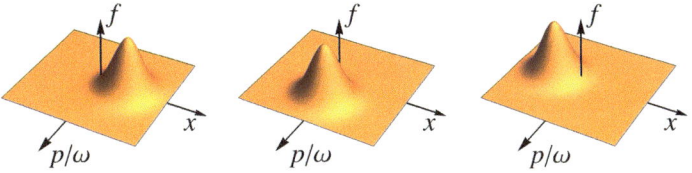

Fig. 10.4 Plots of the distribution function f at three distinct moments in time: $\omega t = 0, \omega t = \pi/2$ and $\omega t = \pi$. A bell-shaped distribution function is initially shifted along the x-axis

If $F(x, p)$ is the initial distribution function at $t = 0$, then, according to Eq. (10.17) we have

$$f(x, p, t) = F\left(x \cos(\omega t) - \frac{p}{\omega}\sin(\omega t), \omega x \sin(\omega t) + p \cos(\omega t)\right). \qquad (10.22)$$

This solution describes a rotation of the distribution function in the phase space which is illustrated by Fig. 10.4. The figure shows three consecutive positions of a bell-shaped function f with an initial offset in the x direction.

10.3 Action-Angle Variables in the Vlasov Equation

The Vlasov equation (10.9) has the same form independent of the choice of the canonical variables q and p. For a particular problem, a judicious choice of these variables can greatly simplify finding a solution to the Vlasov equation. In this section, we will demonstrate the advantages of using in the Vlasov equation the action-angle variables ϕ, J introduced for 1D systems in Sect. 3.2.

Consider a 1D system with the action-angle variables ϕ, J and a time-independent Hamiltonian $H(J)$. For a distribution function f that depends on ϕ, J, and t, $f = f(\phi, J, t)$, the Vlasov equation reads,

$$\frac{\partial f}{\partial t} + \frac{\partial H}{\partial J}\frac{\partial f}{\partial \phi} - \frac{\partial H}{\partial \phi}\frac{\partial f}{\partial J} = \frac{\partial f}{\partial t} + \frac{\partial H}{\partial J}\frac{\partial f}{\partial \phi} = 0, \qquad (10.23)$$

where we have used the fact that H does not depend on ϕ. Noting that according to Eqs. (3.20) and (3.23) the derivative $\partial H/\partial J$ is equal to the revolution frequency $\omega(J)$ along the orbit with action J, we rewrite (10.23) as

$$\frac{\partial f}{\partial t} + \omega(J)\frac{\partial f}{\partial \phi} = 0. \qquad (10.24)$$

This equation is satisfied by an arbitrary function f of the following form:

$$f(\phi, J, t) = F(\phi - \omega(J)t, J), \qquad (10.25)$$

which is easily verified by substituting f into (10.24). This result has a simple geometrical meaning: the values of the distribution function on the orbit with a given action J rotate along this orbit with the angular frequency $\omega(J)$. In general, this is a differential rotation — different layers of the phase space rotate with different frequencies.

Using Eq. (10.25) we can find a general form of a steady-state distribution function that does not depend on time. Because $\partial f/\partial t = -\omega(J)\partial F/\partial \phi$, from $\partial f/\partial t = 0$ is follows that F does not depend on ϕ. We come to the conclusion that any function f that depends only on J is a steady-state solution to the Vlasov equation.

The particular form of the function $f(J)$ for a beam in an accelerator cannot be found from Eq. (10.24) alone, and in practice is often determined by either initial conditions (how the beam was generated or injected into an accelerator) or some slow diffusion or collision processes in the ring. In many cases, a negative exponential dependence of f versus J is a good approximation to the equilibrium beam state,

$$
f = \text{const} \, e^{-J/\epsilon_0} = \text{const} \exp\left(-\frac{1}{2\beta\epsilon_0}\left[x^2 + (\beta P_x + \alpha x)^2\right]\right), \qquad (10.26)
$$

where we have used the expression (7.8) for J in a linear magnetic lattice. The quantity ϵ_0 is called the beam *emittance*. It is an important characteristic of the beam quality.

10.4   Phase Mixing

From Eq. (10.25) we can draw some important conclusions about the evolution of the distribution function in the limit $t \to \infty$. Because ϕ is an angular variable, two values of ϕ that differ by 2π correspond to the same point in phase space. Hence F is a periodic function of ϕ with period 2π and can be expanded into the Fourier series

$$
F(\phi, J) = \sum_{n=-\infty}^{\infty} F_n(J)e^{in\phi}, \qquad (10.27)
$$

where

$$
F_n(J) = \frac{1}{2\pi} \int_0^{2\pi} F(\phi, J)e^{-in\phi}\, d\phi. \qquad (10.28)
$$

Using this representation of F we can rewrite Eq. (10.25) as

$$
f(\phi, J, t) = \sum_{n=-\infty}^{\infty} F_n(J)e^{in[\phi-\omega(J)t]}. \qquad (10.29)
$$

In the limit $t \to \infty$ all terms in this sum, except for $n = 0$, become rapidly oscillating functions of the action J due to the factor $e^{-in\omega(J)t}$. When calculating any integral of f over the phase space, the contribution of these terms averages out to almost zero and becomes negligible. Hence in this limit we only have to keep the $n = 0$ term:

$$f(t, \phi, J) \to F_0(J) \equiv \frac{1}{2\pi} \int_0^{2\pi} f(0, \phi, J) \, d\phi. \qquad (10.30)$$

This is simply the average over the angle coordinate of the initial distribution function f. This derivation naturally explains why the steady-state distribution function depends only on action — the fact established in the previous section.[2]

The mechanism that is responsible for the evolution of the distribution function to a steady state through rapid oscillations of the phase factor $e^{-in\omega(J)t}$ is called *phase mixing*. We can make a rough estimate of the time needed to approach equilibrium in this scenario. If we use $\Delta\omega$ to characterize the frequency spread in the system due to the function $\omega(J)$ and the distribution of particles found in the beam, the phase variation $n\omega(J)t$ at time t can be estimated as $n\Delta\omega t$, and the phases on different orbits start to diverge at $t \gtrsim \pi/n\Delta\omega$. The longest time needed to mix the phases corresponds to the $n = 1$ term, giving an estimate $t \gtrsim \pi/\Delta\omega$. Hence, the distribution function reaches the steady state at times $t \gg \pi/\Delta\omega$.

10.5 Damping and Stochastic Motion

In previous chapters we have discussed how the amplitude of motion of a single particle can decrease due to damping, or take a random walk as a result of stochastic motion. Here, we briefly describe how these effects are incorporated in the formalism of the kinetic equation.

The most straightforward way to see the impact of damping is to return to the continuity equation and recalculate the convective derivative, this time with corrections to the Hamiltonian dynamics from Eq. (3.38) due to non-conservative forces:

$$
\begin{aligned}
0 &= \frac{\partial f}{\partial t} + \sum_i \left[\frac{\partial}{\partial q_i} (\dot{q}_i f) + \frac{\partial}{\partial p_i} (\dot{p}_i f) \right] \\
&= \frac{\partial f}{\partial t} + \sum_i \left[\dot{q}_i \frac{\partial f}{\partial q_i} + \dot{p}_i \frac{\partial f}{\partial p_i} + f \frac{\partial \dot{q}_i}{\partial q_i} + f \frac{\partial \dot{p}_i}{\partial p_i} \right] \\
&= \frac{df}{dt}\bigg|_{\text{conv}} + f \sum_i \frac{\partial F_i}{\partial p_i},
\end{aligned} \qquad (10.31)
$$

---

[2] A linear oscillator in which ω is constant and does not depend on J is an exception: it does not exhibit phase mixing.

thus we find that

$$\left.\frac{df}{dt}\right|_{\text{conv}} = -f \sum_i \frac{\partial F_i}{\partial p_i} . \tag{10.32}$$

Here we have used the expression for the convective derivative of Eq. (10.15). This result is consistent with the more general result Eq. (3.39) given the additional constraint on the distribution function from the continuity equation. Using Eq. (3.46) we can connect the evolution of the distribution function with the time derivative of the determinant of the Jacobian matrix M of the dynamic flow,

$$\left.\frac{1}{f}\frac{df}{dt}\right|_{\text{conv}} = -\frac{1}{\det M}\frac{d \det M}{dt} . \tag{10.33}$$

Integrating this equation over time, we find that $f(t)$ evaluated along a particle trajectory scales in time as the inverse of the determinant of the matrix $M(t)$: $f(q_i, p_i, t)/f(q_i, p_i, 0) = 1/[\det M(t)]$. This is consistent with the notion that the phase space density increases only when trajectories converge in phase space due to non-Hamiltonian dynamics.

For one degree of freedom and $F = -\gamma \dot{x} = -\gamma p$, the right-hand side of Eq. (10.32) becomes simply γf, and

$$f(x(t), p(t), t) = f(x(0), p(0), 0)\, e^{\gamma t} . \tag{10.34}$$

Random kicks with a small correlation time can also be incorporated into the formalism of the distribution function in a natural way if the coordinates are chosen so that only the momenta are directly impacted by the kicks. Because these kicks lead to a random walk of individual particles, this appears in the distribution function as a diffusion process when a large ensemble is considered. Considering a single degree of freedom and very short time scales, so that the dynamics have a negligible impact, uncorrelated random kicks with typical magnitude Δp and a typical time Δt between kicks lead to a random walk with $\langle [p(t) - p_0]^2 \rangle = (t - t_0)\langle (\Delta p)^2 / \Delta t \rangle$, where $p(t_0) = p_0$ (cf. Eq. (4.15)). This process, convolved with the initial distribution, leads to a spreading out of the distribution function. Statistically, it can also be described as the result of a differential operator

$$\frac{\partial f}{\partial t} = D_{\text{s}} \frac{\partial^2 f}{\partial p^2} , \tag{10.35}$$

where $D_{\text{s}} = \langle (\Delta p)^2 / \Delta t \rangle$. Because this expression of the dynamics depends on infinitesimal time scales, it is only necessary to add back the full dynamics by replacing $\partial f / \partial t$ with the convective derivative — for completeness we include the correction from frictional forces:

$$\frac{df}{dt}\Big|_{\text{conv}} = -f\frac{\partial F}{\partial p} + D_{\text{s}}\frac{\partial^2 f}{\partial p^2}\,. \tag{10.36}$$

As seen in Sect. 4.3, the impact of this differential operator will be mixed with that of the particle dynamics to yield a spread in both momentum and position, especially when the frequency of motion is fast compared to the impact of the scattering. For a one-dimensional system and the simple form of damping $F = -\gamma p$, we can expand this to find the partial time derivative:

$$\begin{aligned}
\frac{\partial f}{\partial t} &= -\frac{\partial H}{\partial p}\frac{\partial f}{\partial x} + \frac{\partial H}{\partial q}\frac{\partial f}{\partial p} + \gamma p\frac{\partial f}{\partial p} + \gamma f + D_{\text{s}}\frac{\partial^2 f}{\partial p^2} \\
&= -\{f, H\} + \frac{\partial}{\partial p}\left(\gamma p f\right) + D_{\text{s}}\frac{\partial^2 f}{\partial p^2}\,.
\end{aligned} \tag{10.37}$$

The second term in the final expression, related to damping, combines the effect of the distribution function having a convective derivative, as found above, with the fact that the flow in phase space is itself no longer fully defined by the Poisson bracket, as seen in Eq. (3.39). In some literature, Eq. (10.37) is called the Vlasov-Fokker-Planck equation.

Worked Examples

Problem 10.1 *Write the Vlasov equation for a beam distribution $f(x, P_x, s)$ in terms of variables x and P_x. Use the Hamiltonian from Eq. (6.14).*

Solution: The Vlasov equation is given by

$$-\frac{\partial f}{\partial s} + \frac{\partial H}{\partial x}\frac{\partial f}{\partial P_x} - \frac{\partial H}{\partial P_x}\frac{\partial f}{\partial x} = 0\,.$$

Using our Hamiltonian (6.14) for an accelerator, $H_0 = P_x^2/2 + K(s)x^2/2$, we find

$$-\frac{\partial f}{\partial s} + K(s)x\frac{\partial f}{\partial P_x} - P_x\frac{\partial f}{\partial x} = 0\,. \tag{10.38}$$

Problem 10.2 *Give a direct proof that the function (10.26) satisfies the Vlasov equation.*

Solution: The partial derivatives of the distribution function (10.26) can be written as

$$\frac{1}{f}\frac{\partial f}{\partial x} = -\frac{1}{\epsilon_0}\left(\frac{1+\alpha^2}{\beta}x + \alpha P_x\right),$$

$$\frac{1}{f}\frac{\partial f}{\partial P_x} = -\frac{1}{\epsilon_0}(\beta P_x + \alpha x),$$

$$\frac{1}{f}\frac{\partial f}{\partial s} = -\frac{1}{\epsilon_0}\left[-\frac{(1+\alpha^2)\beta'}{2\beta^2}x^2 + \frac{\alpha\alpha'}{\beta}x^2 + \frac{\beta'}{2}P_x^2 + \alpha'x P_x\right].$$

The partial derivative in s only acts on β and $\alpha = -\beta'/2$. From the previous problem, the Vlasov equation (10.38) has 3 terms proportional to x^2, P_x^2, and $x P_x$. The P_x^2 term is $(-\alpha - \beta'/2)P_x^2 = 0$ by the definition of α. The rest of the equation is then:

$$0 = x^2\left[\alpha K + \frac{(1+\alpha^2)\beta'}{2\beta^2} - \frac{\alpha\alpha'}{\beta}\right] + x P_x\left(\beta K - \frac{1+\alpha^2}{\beta} - \alpha'\right)$$

$$= \frac{x}{\beta^2}(\beta P_x + \alpha x)\left(K\beta^2 - 1 - \alpha^2 - \beta\alpha'\right).$$

The last expression vanishes because that is the equation which defines the beta function, see Eq. (6.23).

Problem 10.3 *Show how the average of the squares of x and P_x are expressed through ϵ_0 and β for the distribution function of Eq. (10.26).*

Solution: To find an average value of some quantity, we need to calculate the following integrals:

$$\langle\ldots\rangle = \frac{\int(\ldots)f\,dx\,dP_x}{\int f\,dx\,dP_x}.$$

We begin by calculating the normalization, $\int f\,dx\,dP_x$. It is simplest to do the integral over P_x first:

$$\int f\,dx\,dP_x = \int dx\,e^{-x^2/2\beta\epsilon_0}\int dP_x\,\exp\left[-\frac{1}{2\beta\epsilon_0}(\beta P_x + \alpha x)^2\right]$$

$$= \int dx\,e^{-x^2/2\beta\epsilon_0}\sqrt{\frac{2\pi\epsilon_0}{\beta}} = \sqrt{2\pi\epsilon_0\beta}\sqrt{\frac{2\pi\epsilon_0}{\beta}} = 2\pi\epsilon_0,$$

where we used the fact that we could change P_x to $u = P_x + \alpha x/\beta$ to simplify the exponent and still have an integral over all u. To calculate the mean of x^2, the integral over P_x is unaffected:

$$\langle x^2 \rangle = \frac{1}{2\pi\epsilon_0} \int dx \, x^2 e^{-x^2/2\beta\epsilon_0} \int dP_x \, \exp\left[-\frac{1}{2\beta\epsilon_0}(\beta P_x + \alpha x)^2\right]$$

$$= \frac{1}{2\pi\epsilon_0} \sqrt{2\pi} \, (\epsilon_0\beta)^{3/2} \sqrt{\frac{2\pi\epsilon_0}{\beta}} = \beta\epsilon_0 \,.$$

We can again use the trick $u = P_x + \alpha x/\beta$ to calculate the moment

$$\left\langle (\beta P_x + \alpha x)^2 \right\rangle = \frac{1}{2\pi\epsilon_0} \int dx \, e^{-x^2/2\beta\epsilon_0} \int dP_x \, (\beta P_x + \alpha x)^2 \exp\left[-\frac{1}{2\beta\epsilon_0}(\beta P_x + \alpha x)^2\right]$$

$$= \frac{1}{2\pi\epsilon_0} \sqrt{2\pi\epsilon_0\beta}\sqrt{2\pi\epsilon_0\beta} = \epsilon_0\beta \,.$$

Finally, we can see from symmetry that the average of $x(\beta P_x + \alpha x)$ has to vanish because again we can switch to the variable u whose integral is 0 by symmetry. As a consequence, $\langle (\beta x P_x + \alpha x^2) \rangle = 0$. Combining these expressions yields

$$\langle x^2 \rangle = \beta\epsilon_0 \,, \qquad \langle x P_x \rangle = -\alpha\epsilon_0 \,, \qquad \langle P_x^2 \rangle = \frac{1 + \alpha^2}{\beta} \epsilon_0 \,.$$

Part II
Electricity and Magnetism

Chapter 11
Self Field of a Relativistic Beam

The electromagnetic field generated by a high-current, relativistic beam of charged particles plays an important role in beam dynamics. On the one hand, the force exerted on the beam by this field can lead to beam instabilities and a deterioration of its properties in the process of beam generation, acceleration and transport. On the other hand, this field induces currents and charges in the beam environment that can be used for diagnostic purposes. Hence calculation of this field and the forces associated with it constitutes an essential part of beam physics for accelerators.

In this chapter, we analyze the electromagnetic field of a relativistic beam moving with a constant velocity in free space. In the following chapter we will look at the impact of material surfaces around the beam path. The authors recommend books by Jackson [1], by Landau and Lifshitz [2], and by Landau and Lifshitz and Pitaevskii [3] for an overview of the electromagnetic field and its interactions with matter.

11.1 Relativistic Field of a Particle Moving with Constant Velocity

A beam consists of many charged particles, so it makes sense to begin with a review of the electromagnetic field of a point charge q moving with a relativistic velocity v in free space. The most straightforward way to derive this field is to use the Lorentz transformation starting from the particle rest frame where there is only the static Coulomb field,

$$\mathbf{E}' = \frac{1}{4\pi\epsilon_0} \frac{q\mathbf{r}'}{r'^3} \, , \tag{11.1}$$

with $\mathbf{r}'$ defined as the vector drawn from the position of the particle to the observation point and $\mathbf{E}'$ being the electric field at the observation point. The magnetic field $\mathbf{B}'$ is zero in this frame. Here and below the prime denotes variables in the rest frame.

© Springer International Publishing AG, part of Springer Nature 2018
G. Stupakov and G. Penn, *Classical Mechanics and Electromagnetism
in Accelerator Physics*, Graduate Texts in Physics,
https://doi.org/10.1007/978-3-319-90188-6_11

To find the electric and magnetic fields in the lab frame we will use the Lorentz transformation for the fields, Eq. (B.17). We assume that the particle is moving along the z-axis with its trajectory given by $x = y = 0$ and $z = vt$, so that the origin in the rest frame ($x' = y' = z' = 0$) coincides with the particle. For cases like this where the magnetic field vanishes in the rest frame we have

$$E_x = \gamma E_x' , \qquad E_y = \gamma E_y' , \qquad E_z = E_z' . \tag{11.2}$$

We also need to transform the vector $\mathbf{r}'$ in Eq. (11.1) into the lab frame using Eq. (B.2). For the length of this vector we obtain $r' = \sqrt{x^2 + y^2 + \gamma^2 (z - vt)^2}$, which gives for the electric field $\mathbf{E}$ in the lab frame expressed in Cartesian coordinates:

$$
\begin{aligned}
E_x &= \frac{1}{4\pi\epsilon_0} \frac{q\gamma x}{[x^2 + y^2 + \gamma^2 (z - vt)^2]^{3/2}} , \\
E_y &= \frac{1}{4\pi\epsilon_0} \frac{q\gamma y}{[x^2 + y^2 + \gamma^2 (z - vt)^2]^{3/2}} , \\
E_z &= \frac{1}{4\pi\epsilon_0} \frac{q\gamma (z - vt)}{[x^2 + y^2 + \gamma^2 (z - vt)^2]^{3/2}} .
\end{aligned}
\tag{11.3}
$$

These three equations can be combined into a vectorial one:

$$\mathbf{E} = \frac{1}{4\pi\epsilon_0} \frac{q\mathbf{r}}{\gamma^2 \mathcal{R}^3} . \tag{11.4}$$

Here the vector $\mathbf{r}$ is drawn from the current position of the particle to the observation point, $\mathbf{r} = (x, y, z - vt)$, and $\mathcal{R}$ denotes the following expression:

$$\mathcal{R} = \sqrt{(z - vt)^2 + \gamma^{-2}(x^2 + y^2)} . \tag{11.5}$$

As follows from Eq. (B.17), the moving charge also carries a magnetic field,

$$\mathbf{B} = \frac{1}{c^2} v \times \mathbf{E} . \tag{11.6}$$

The magnetic field points in the azimuthal direction, with the magnetic field lines encircling the z-axis.

The above result can also be obtained through the Lorentz transformation of the potentials. Indeed, in the particle rest frame we have the Coulomb electrostatic potential ϕ' and zero vector potential $\mathbf{A}'$,

$$\phi' = \frac{1}{4\pi\epsilon_0} \frac{q}{r'} , \qquad \mathbf{A}' = 0 . \tag{11.7}$$

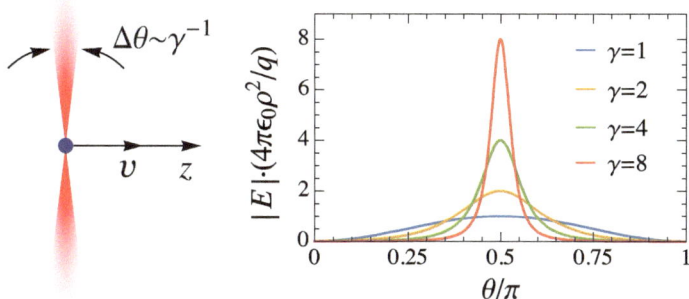

Fig. 11.1 A sketch of the regions of intense electromagnetic fields around a particle moving along a trajectory $z = vt$ (left), and a plot of the magnitude of the electric field E (right, measured in units $q/4\pi\epsilon_0\rho^2$) for a given transverse offset $\rho = \sqrt{x^2 + y^2}$ and as a function of the polar angle $\theta = \arctan[\rho/(z - vt)]$ relative to the particle and its direction of motion, for several values of γ. The width of each curve scales as $1/\gamma$. Both the electric and magnetic fields of a relativistic particle are mainly localized around the plane transverse to the velocity

Using the Lorentz transformation (B.18) we find their values in the lab frame,

$$\phi = \gamma\phi', \qquad \mathbf{A} = \frac{1}{c}\boldsymbol{\beta}\phi. \tag{11.8}$$

Expressing r' in terms of the coordinates in the lab frame, $r' = \gamma\mathcal{R}$, gives

$$\phi = \frac{1}{4\pi\epsilon_0}\frac{q}{\mathcal{R}}, \qquad \mathbf{A} = \frac{Z_0}{4\pi}\boldsymbol{\beta}\frac{q}{\mathcal{R}}, \tag{11.9}$$

where $Z_0 = 1/(\epsilon_0 c) = \mu_0 c \simeq 377$ Ohm. This is often called the wave impedance of free space because it is equal to the ratio of the field amplitudes $E/H = \mu_0 E/B$ for an electromagnetic wave traveling in a vacuum. Using the expressions for the fields through the potentials, it is now straightforward to show that Eq. (11.9) gives the fields (11.4) and (11.6).

The electromagnetic field of a relativistic particle moving with velocity v close to the speed of light, or equivalently $\gamma \gg 1$, has some remarkable properties that follow from Eq. (11.4). For a given distance r from the path of the charge, this field reaches a maximum in the mid-plane $z = vt$, where it scales as $E \propto \gamma/r^2$ and is γ times larger than the field of a charge at rest. The field remains strong within an angular distance $\sim 1/\gamma$ from this plane, see Fig. 11.1. On the other hand, the field on axis ($x = y = 0$) is suppressed by a factor of γ^2, $E \sim 1/r^2\gamma^2$. The absolute value of the magnetic field is almost equal to that of the electric field,[1] $cB \approx E$.

---

[1] We actually compare the magnitude of E with cB which have the same dimensions in the SI system of units that we use in this book. In the Gaussian system of units, the electric and magnetic fields have the same dimensions and their magnitude can be compared directly without extra factors.

In the case of very large values of γ, to which we refer as the *ultra-relativistic* limit, we can take the limit $\gamma \to \infty$ and consider the electromagnetic field as localized in an infinitely thin "pancake" region around the transverse plane $z = vt$. Because in this plane the field is directed along the vector drawn from the current position of the charge to the observation point, we can write $\mathbf{E} = A\boldsymbol{\rho}\,\delta(z - ct)$ where $\boldsymbol{\rho} = \hat{\mathbf{x}}x + \hat{\mathbf{y}}y$ and A is a factor which is determined by the requirement that the areas under the curves $E_x(z)$ and $E_y(z)$ agree with the ones given by Eq. (11.4) in the limit $\gamma \to \infty$. The calculation of these integrals is carried out in Problem 11.3; they give the following expressions for the fields:

$$\mathbf{E} = \frac{1}{4\pi\epsilon_0} \frac{2q\boldsymbol{\rho}}{\rho^2}\delta(z - ct)\,, \qquad \mathbf{B} = \frac{1}{c}\hat{\mathbf{z}} \times \mathbf{E}\,. \tag{11.10}$$

11.2 Interaction of Moving Charges in Free Space

Charged particles in beams move in the same general direction and almost on parallel paths. To get a feeling of the strength of the electromagnetic interaction inside the bunch we consider two such particles and calculate the force of their interaction.

Consider a *source* particle of charge q that is moving with velocity v along the z-axis, $z = vt$, and a *test* particle of the same charge moving on a parallel path. The test particle lags behind the leading one at a distance l with an offset a: $x = a$, $y = 0$, $z = vt - l$. We denote by $\mathbf{F}$ the force with which the source particle acts on the test one. Using Eq. (11.4), we find for the longitudinal component F_z,

$$F_z = qE_z = -\frac{1}{4\pi\epsilon_0} \frac{q^2 l}{\gamma^2(l^2 + a^2/\gamma^2)^{3/2}}\,. \tag{11.11}$$

The transverse component of the force is directed along x and is equal to

$$F_x = q(E_x - vB_y) = \frac{1}{4\pi\epsilon_0} \frac{q^2 a}{\gamma^4(l^2 + a^2/\gamma^2)^{3/2}}\,. \tag{11.12}$$

Consider the limit $\gamma \gg 1$ and assume that the offset x is fixed. In the region $l \gtrsim |a|/\gamma$ we can neglect the terms a^2/γ^2 in the denominators of (11.11) and (11.12) and obtain the scalings $F_z \propto 1/l^2\gamma^2$ and $F_x \propto |a|/l^3\gamma^4$ which means that the interaction forces rapidly decrease as γ increases. The reason why the transverse force so rapidly decreases with γ is due to the fact that in the relativistic limit the electric and magnetic forces are almost equal to each other and are in opposite directions. The scalings are somewhat different inside the region $l \lesssim |a|/\gamma$, however for a given a the length of this region decreases as $1/\gamma$, and hence the probability to find particles in that region is relatively small.

From this simple analysis, we see an indication that in the relativistic limit the direct interaction between particles in a bunch gets suppressed as the particle energy increases. In the next section, we will confirm this observation by direct calculation of the fields inside and outside of a relativistic bunch.

11.3   Field of a Relativistic Bunch of Particles

We first calculate the transverse radial component of the electric field outside of the bunch at a distance $\rho = \sqrt{x^2 + y^2}$ from the axis, again assuming that the bunch is moving along the z-axis. We will be interested in distances ρ that greatly exceed the transverse size of the beam, $\rho \gg \sigma_\perp$. Here, we can completely neglect the transverse size $\sigma_\perp$, and consider the beam as an infinitely thin line charge. The magnetic fields follow from Eq. (11.6).

To simplify the notation, we assume that the fields are calculated at time $t = 0$ and drop the variable t from our equations. The one-dimensional longitudinal distribution function of the bunch at this time is denoted by $\lambda(z)$ and is normalized to unity, $\int_{-\infty}^{\infty} \lambda(z)dz = 1$. The time dependence can be easily recovered in the final results by simply replacing $\lambda(z) \to \lambda(z - vt)$.

We carry out calculations in the lab frame. Each infinitesimally small element of the beam generates the electric field given by Eq. (11.4). From this equation we find that the radial component dE_ρ created by charge dq' located at coordinate z' is

$$dE_\rho(z, z', \rho) = \frac{1}{4\pi\epsilon_0} \frac{\rho \, dq'}{\gamma^2[(z - z')^2 + \rho^2/\gamma^2]^{3/2}} , \qquad (11.13)$$

where z and ρ are the coordinates of the observation point. To find the field of the whole bunch we note that the charge dq' within dz' is equal to $Q\lambda(z')dz'$, with Q the total charge of the bunch. The total field is then obtained by integration of the elementary contributions dE_ρ:

$$
\begin{aligned}
E_\rho(z, \rho) &= \int dE_\rho(z, z', \rho) \\
&= \frac{Q\rho}{4\pi\epsilon_0\gamma^2} \int_{-\infty}^{\infty} \frac{\lambda(z')dz'}{[(z - z')^2 + \rho^2/\gamma^2]^{3/2}} .
\end{aligned}
\qquad (11.14)
$$

This integral can be simplified in two limiting cases. The denominator of the integrand has a sharp peak of width $\Delta z \sim \rho/\gamma$ around $z' = z$. At distances $\rho \ll \gamma\sigma_z$ from the bunch, the width of the peak is smaller than the width σ_z of the distribution function and we can replace the inverse denominator by the delta function:

$$\frac{1}{[(z - z')^2 + \rho^2/\gamma^2]^{3/2}} \to \frac{2\gamma^2}{\rho^2}\delta(z - z') . \qquad (11.15)$$

The factor in front of the delta function on the right-hand side follows from the requirement that the area under both functions should be equal, using the mathematical identity

$$\int_{-\infty}^{\infty} \frac{dz'}{[(z-z')^2 + a^2]^{3/2}} = \frac{2}{a^2} \, . \tag{11.16}$$

The approximation (11.15) is equivalent to using Eqs. (11.10) instead of (11.4). Substituting Eq. (11.15) into (11.14) we obtain

$$E_\rho(z, \rho) = \frac{1}{4\pi\epsilon_0} \frac{2Q\lambda(z)}{\rho} \, . \tag{11.17}$$

We see that the Lorentz factor γ does not enter this formula — in agreement with what one would expect remembering that Eqs. (11.10) are valid in the limit $\gamma \to \infty$.

Note that in the above calculation we assumed $\rho \ll \gamma\sigma_z$ and $\sigma_\perp \ll \rho$, which means $\sigma_\perp \ll \gamma\sigma_z$. In the beam frame of reference, its length is γ times larger than in the lab frame, $\sigma'_z = \gamma\sigma_z$, while $\sigma'_\perp = \sigma_\perp$ (see also Sect. 11.4). We see that in the beam frame the length of the bunch is much larger than its transverse size, $\sigma'_z \gg \sigma'_\perp$, and hence this regime corresponds to a *long-thin* approximation for the bunch. When the bunch does not satisfy the long-thin approximation, then the above calculation does not apply. It is important to remember that for $\gamma \gg 1$ the long-thin approximation can hold even when $\sigma_z \approx \sigma_\perp$ in the lab frame.

In the opposite limit, $\rho \gg \gamma\sigma_z$, the distribution function $\lambda(z)$ in Eq. (11.14) can be considered as a relatively narrow function and replaced by the delta function $\delta(z)$, which gives for the bunch field an expression identical to the field of a point charge,

$$E_\rho(z, \rho) = \frac{1}{4\pi\epsilon_0} \frac{Q\gamma\rho}{(\gamma^2 z^2 + \rho^2)^{3/2}} \, . \tag{11.18}$$

This result should not be surprising — at a large distance the field cannot resolve the details of the charge distribution in the bunch and is determined only by the total charge Q.

In the intermediate region, $\rho \sim \gamma\sigma_z$, the field transitions from Eq. (11.17) to (11.18). This transition is illustrated in Fig. 11.2 which shows a numerically computed radial electric field (11.14) for a Gaussian distribution function $\lambda(z) = (1/\sqrt{2\pi}\sigma_z)e^{-z^2/2\sigma_z^2}$ for several intermediate values of the parameter $\rho/\sigma_z\gamma$, still assuming $\sigma_\perp \ll \rho$, $\sigma_z\gamma$.

We now calculate the longitudinal component of the electric field inside the bunch. It is instructive to try again to use the model of an infinitely thin beam as we did in the calculation of the radial electric field above. In this line-charge model, an elementary charge dq' located at coordinate z' creates the following field dE_z at the observation point z on the axis:

$$dE_z(z, z') = \frac{dq'}{4\pi\epsilon_0\gamma^2} \frac{z - z'}{|z - z'|^3} \, . \tag{11.19}$$

Fig. 11.2 Transverse electric field of a relativistic bunch with Gaussian distribution for various values of the parameter $\rho/\sigma_z\gamma$. This parameter takes the values of 0.1, 0.5, 1 and 3 with larger values corresponding to broader curves. The field is normalized by $(4\pi\epsilon_0)^{-1}Q/\gamma\sigma_z^2$ to yield a universal set of curves that only depend on $\rho/\gamma\sigma_z$

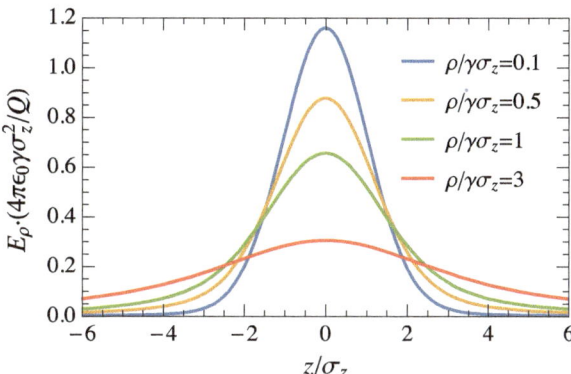

To find the total field, we need to integrate this expression as we did in the derivation of (11.14),

$$E_\parallel(z) = \int dE_z(z, z') = \frac{Q}{4\pi\epsilon_0\gamma^2} \int dz'\lambda(z') \frac{z - z'}{|z - z'|^3} \,. \tag{11.20}$$

Unfortunately, this integral diverges at $z' \to z$. The divergence indicates that in this particular problem, one cannot assume $\sigma_\perp \to 0$, and has to take into account the finite transverse size of the beam.

With the understanding of the importance of the radial charge distribution, let us calculate the longitudinal electric field in a specific model where the beam radius is a, and the charge is uniformly distributed over each cross section from $\rho = 0$ to $\rho = a$. We now slice the beam into infinitesimal disks of thickness dz' and charge $dq' = Q\lambda(z')dz'$. The longitudinal electric field generated by this slice on the axis at point z is

$$dE_z(z, z') = -\frac{1}{4\pi\epsilon_0} \frac{2dq'}{a^2}(z - z') \left(\frac{1}{\sqrt{a^2/\gamma^2 + (z - z')^2}} - \frac{1}{|z - z'|} \right). \tag{11.21}$$

We delegate the derivation of this expression to Problem 11.4. The longitudinal electric field on the axis of the bunch is obtained by integrating over all slices,

$$E_z(z) = \int_{-\infty}^{\infty} dE_z(z, z') \tag{11.22}$$

$$= -\frac{Q}{4\pi\epsilon_0} \frac{2}{a^2} \int_{-\infty}^{\infty} dz'\lambda(z')(z - z') \left(\frac{1}{\sqrt{a^2/\gamma^2 + (z - z')^2}} - \frac{1}{|z - z'|} \right).$$

Now the integral converges at $z' = z$ and can be calculated numerically. For the Gaussian distribution function $\lambda(z)$ the result of the calculation is shown in Fig. 11.3. One can show that in the limit $a/\gamma\sigma_z \ll 1$ a crude estimate for E_z is given by the

Fig. 11.3 Longitudinal electric field of a relativistic bunch with Gaussian distribution function for various values of the parameter $a/\gamma\sigma_z$. This parameter takes the values 0.1, 0.01, and 0.001 with smaller values corresponding to higher fields. The field is normalized by $(2\pi\epsilon_0)^{-1}Q/\gamma^2\sigma_z^2$

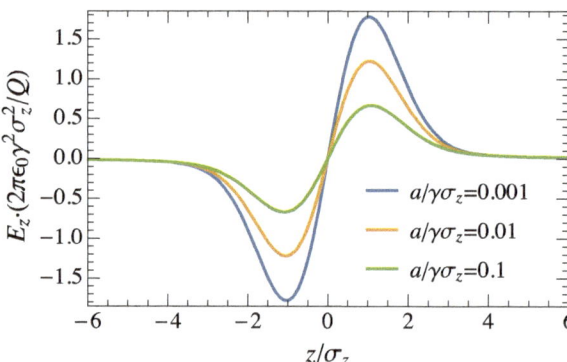

following expression:

$$E_z \sim \frac{1}{4\pi\epsilon_0} \frac{Q}{\gamma^2\sigma_z^2} \log \frac{\gamma\sigma_z}{a} . \tag{11.23}$$

This expression diverges in the limit of an infinitely thin beam ($a \to 0$) which explains our failure to integrate Eq. (11.20). Note that due to the factor $1/\gamma^2$ in this expression the effect of the longitudinal electric field for relativistic beams is usually small. It is often referred to as the *space charge effect*.

11.4   Electric Field of a 3D Gaussian Distribution

In accelerator physics, a bunch of charged particles is often represented as having a Gaussian distribution function in all three directions so that the charge density ρ of a bunch moving with velocity v along the z-axis is given by the following expression:

$$\rho(x, y, z, t) = \frac{Q}{(2\pi)^{3/2}\sigma_x\sigma_y\sigma_z} e^{-x^2/2\sigma_x^2-y^2/2\sigma_y^2-(z-vt)^2/2\sigma_z^2}, \tag{11.24}$$

where Q is the total charge of the bunch and σ_x, σ_y, and σ_z are the rms bunch lengths in the corresponding directions. Calculation of the electromagnetic field of such a bunch is more complicated than in the simplified models considered in the previous section. Here we will outline the main steps in this calculation.

First, we will transform into the beam frame where the beam is at rest. The charge distribution in this frame, $\rho'(x', y', z')$, does not depend on time and is obtained from (11.24) with the help of the Lorentz transformation for coordinates. Expressing x, y and $z - vt$ through x', y' and z' we find that the exponential factor in (11.24) becomes

$$e^{-x'^2/2\sigma_x^2 - y'^2/2\sigma_y^2 - z'^2/2\sigma_z^2\gamma^2} , \tag{11.25}$$

from which we conclude that the rms bunch length along z' is γ times larger than σ_z — a relativistic bunch is much longer in the rest frame than in the lab frame. The transverse dimensions of the bunch, however, are the same. From the total charge conservation in the Lorentz transformation, it is now clear that

$$\rho'(x', y', z') = \frac{Q}{(2\pi)^{3/2}\sigma_x\sigma_y\sigma_z'} e^{-x'^2/2\sigma_x^2 - y'^2/2\sigma_y^2 - z'^2/2\sigma_z^2} , \tag{11.26}$$

where $\sigma_z' = \gamma\sigma_z$.

The electrostatic potential ϕ' of the beam at rest is given by Coulomb's law,

$$\phi'(x', y', z') = \frac{1}{4\pi\epsilon_0} \int \frac{\rho(\xi, \psi, \zeta)d\xi d\psi d\zeta}{[(x'-\xi)^2 + (y'-\psi)^2 + (z'-\zeta)^2]^{1/2}} . \tag{11.27}$$

Unfortunately, with the distribution function given by Eq. (11.26) this integral cannot be done analytically. There is however a useful trick that considerably simplifies the calculation. It uses the following mathematical identity,

$$\frac{1}{R} = \sqrt{\frac{2}{\pi}} \int_0^\infty e^{-\lambda^2 R^2/2}d\lambda . \tag{11.28}$$

Setting $R = [(x'-\xi)^2 + (y'-\psi)^2 + (z'-\zeta)^2]^{1/2}$ and replacing $1/R$ in the integrand of Eq. (11.27) by (11.28) we first arrive at a four-dimensional integral

$$\phi' = \frac{1}{4\pi\epsilon_0}\sqrt{\frac{2}{\pi}} \int_0^\infty d\lambda \int e^{-\lambda^2[(x'-\xi)^2 + (y'-\psi)^2 + (z'-\zeta)^2]/2}\rho(\xi, \psi, \zeta)d\xi d\psi d\zeta . \tag{11.29}$$

With the Gaussian distribution (11.26) the integration over ξ, ψ and ζ can now be easily carried out, e.g.,

$$\int_{-\infty}^\infty e^{-\frac{1}{2}\lambda^2(x'-\xi)^2 - \xi'^2/2\sigma_x^2} d\xi = \frac{\sqrt{2\pi}}{\sqrt{\lambda^2 + \sigma_x^{-2}}} \exp\left[-\frac{x'^2\lambda^2}{2(\lambda^2\sigma_x^2 + 1)}\right] , \tag{11.30}$$

which gives for the potential

$$\phi' = \frac{1}{4\pi\epsilon_0}\sqrt{\frac{2}{\pi}}\frac{Q}{\sigma_x\sigma_y\sigma_z'} \int_0^\infty \frac{d\lambda}{\left[(\lambda^2 + \sigma_x^{-2})(\lambda^2 + \sigma_y^{-2})(\lambda^2 + \sigma_z'^{-2})\right]^{1/2}}$$
$$\times \exp\left[-\frac{x'^2\lambda^2}{2(\lambda^2\sigma_x^2 + 1)} - \frac{y'^2\lambda^2}{2(\lambda^2\sigma_y^2 + 1)} - \frac{z'^2\lambda^2}{2(\lambda^2\sigma_z^2 + 1)}\right] . \tag{11.31}$$

This integral involves only one integration and is much easier to evaluate numerically then the original expression (11.27).

Having found the potential in the beam frame, it is now easy to transform it to the lab frame using the Lorentz transformation. First we need to express σ_z' in (11.31) through the bunch length in the beam frame, $\sigma_z' = \gamma\sigma_z$. Second, from (B.18) we see that the potential in the lab frame is γ times larger than in the beam frame (note that $A_z' = 0$ in the beam frame). Third, we need to transform the coordinates x', y', z' in (11.31) to the lab frame. The resulting expression is:

$$\phi(x, y, z, t) = \frac{1}{4\pi\epsilon_0} \sqrt{\frac{2}{\pi}} \frac{Q}{\sigma_x\sigma_y\sigma_z} \int_0^\infty \frac{d\lambda}{\left[(\lambda^2 + \sigma_x^{-2})(\lambda^2 + \sigma_y^{-2})(\lambda^2 + \gamma^{-2}\sigma_z^{-2})\right]^{1/2}}$$
$$\times \exp\left[-\frac{x^2\lambda^2}{2(\lambda^2\sigma_x^2 + 1)} - \frac{y^2\lambda^2}{2(\lambda^2\sigma_y^2 + 1)} - \frac{(z - vt)^2\lambda^2}{2(\lambda^2\sigma_z^2 + \gamma^{-2})}\right]. \quad (11.32)$$

According to (B.18), in addition to the electrostatic potential, in the lab frame there is also a vector potential A_z responsible for the magnetic field of the moving bunch. It is equal to $A_z = v\phi/c^2$ with ϕ given by (11.32).

Examples of using Eq. (11.31) for the calculation of fields are given in Problems 11.6 and 11.7.

Worked Examples

Problem 11.1 *Verify by direct calculation that Eqs. (A.5) applied to the potentials* (11.9) *give the fields* (11.4) *and* (11.6).

Solution: With $\mathcal{R} = \sqrt{(z - vt)^2 + (x^2 + y^2)/\gamma^2}$, we differentiate directly to find

$$\begin{aligned}
\mathbf{E} &= \frac{q}{4\pi\epsilon_0} \frac{\hat{\mathbf{z}}(z - vt) + (\hat{\mathbf{x}}x + \hat{\mathbf{y}}y)/\gamma^2}{\mathcal{R}^3} + \left(\frac{Z_0\beta q}{4\pi}\right) \frac{-v\hat{\mathbf{z}}(z - vt)}{\mathcal{R}^3} \\
&= \frac{q}{4\pi\epsilon_0} \frac{\hat{\mathbf{z}}(1 - \beta^2)(z - vt) + (\hat{\mathbf{x}}x + \hat{\mathbf{y}}y)/\gamma^2}{\mathcal{R}^3} \\
&= \frac{q}{4\pi\epsilon_0} \frac{\mathbf{r}}{\gamma^2\mathcal{R}^3},
\end{aligned}$$

using $Z_0 = 1/\epsilon_0 c$ and $\mathbf{r} = (x, y, z - vt)$. Similarly,

$$\begin{aligned}
\mathbf{B} &= \frac{Z_0\beta q}{4\pi} \frac{(\hat{\mathbf{x}}y - \hat{\mathbf{y}}x)/\gamma^2}{\mathcal{R}^3} = \frac{qv}{4\pi\epsilon_0 c^2} \frac{\hat{\mathbf{x}}y - \hat{\mathbf{y}}x}{\gamma^2\mathcal{R}^3} \\
&= \frac{1}{c^2} v \times \mathbf{E}.
\end{aligned}$$

Problem 11.2 *For a fixed transverse offset ρ, calculate the dependence of the magnitude of E in Eq. (11.3) versus polar angle θ, where θ is the angle between the vector r, pointing from particle to the observation point, and the z-axis. What form does this take in the regime $\gamma \gg 1$?*

Solution: Setting for simplicity $t = 0$, with $\rho = (x, y, 0)$, we have

$$E_\rho = \frac{1}{4\pi\epsilon_0} \frac{q\gamma\rho}{\left(\rho^2 + \gamma^2 z^2\right)^{3/2}} = \frac{1}{4\pi\epsilon_0} \frac{q\gamma}{\rho^2 \left(1 + \gamma^2 \cot^2 \theta\right)^{3/2}} ,$$

$$E_z = \frac{1}{4\pi\epsilon_0} \frac{q\gamma z}{\left(\rho^2 + \gamma^2 z^2\right)^{3/2}} = \frac{1}{4\pi\epsilon_0} \frac{q\gamma \cot \theta}{\rho^2 \left(1 + \gamma^2 \cot^2 \theta\right)^{3/2}} ,$$

where $\tan \theta = \rho/z$. The magnitude is then

$$E = \sqrt{E_\rho^2 + E_z^2} = \frac{q}{4\pi\epsilon_0\rho^2} \frac{\gamma \sin^2 \theta}{\left(\sin^2 \theta + \gamma^2 \cos^2 \theta\right)^{3/2}} .$$

For $\gamma \ll 1$, this becomes very small unless $\cos \theta \ll 1$, so we approximate $\sin \theta \simeq 1$ and express $\theta = \pi/2 + \delta\theta$, yielding

$$E \simeq \frac{q\gamma}{4\pi\epsilon_0\rho^2} \left[1 + \gamma^2(\delta\theta)^2\right]^{-3/2} .$$

The fields are localized around $|\theta - \pi/2| \lesssim 1/\gamma$ and the peak magnitude scales with γ.

Problem 11.3 *Using Eq. (11.4) show that in the limit $\gamma \to \infty$ the following relations hold:*

$$\int_{-\infty}^{\infty} E_x dz = \frac{1}{4\pi\epsilon_0} \frac{2qx}{\rho^2} , \qquad \int_{-\infty}^{\infty} E_y dz = \frac{1}{4\pi\epsilon_0} \frac{2qy}{\rho^2} .$$

Solution: Starting from Eq. (11.4) we evaluate

$$\int_{-\infty}^{\infty} E_x dz = \frac{qx}{4\pi\epsilon_0\gamma^2} \int_{-\infty}^{\infty} \frac{dz}{\left[(z - vt)^2 + (x^2 + y^2)/\gamma^2\right]^{3/2}}$$

$$= \frac{qx}{4\pi\epsilon_0} \frac{z - vt}{(x^2 + y^2) \left[(z - vt)^2 + (x^2 + y^2)/\gamma^2\right]^{1/2}} \Bigg|_{z=-\infty}^{\infty}$$

$$= \frac{qx}{2\pi\epsilon_0(x^2 + y^2)}$$

$$= \frac{qx}{2\pi\epsilon_0\rho^2} .$$

The formula for E_y follows identically.

Problem 11.4 *Derive Eq. (11.21) for dE_z and analyze it considering the limits $|z - z'| \ll a/\gamma$ and $|z - z'| \gg a/\gamma$, using cylindrical coordinates.*

Solution: Consider a disc of radius a located at position z'. A thin ring of the disk within the radii ρ and $\rho + d\rho$ has a charge $dq = q/(\pi a^2) \times 2\pi\rho d\rho$, where q is the charge of the disk. The ring creates the longitudinal field on axis at coordinate z equal to

$$dE_z = \frac{dq}{4\pi\epsilon_0\gamma^2} \frac{z - z'}{\left[(z - z')^2 + \rho^2/\gamma^2\right]^{3/2}} = \frac{2q\rho d\rho}{4\pi\epsilon_0\gamma^2 a^2} \frac{z - z'}{\left[(z - z')^2 + \rho^2/\gamma^2\right]^{3/2}} .$$

To obtain the total field we need to integrate this expression from $\rho = 0$ to $\rho = a$:

$$E_z = \frac{2q}{a^2} \frac{1}{4\pi\epsilon_0\gamma^2} \int_0^a \rho d\rho \frac{z - z'}{\left[(z - z')^2 + \rho^2/\gamma^2\right]^{3/2}}$$

$$= \frac{2q}{a^2} \frac{1}{4\pi\epsilon_0} \frac{-(z - z')}{\left[(z - z')^2 + \rho^2/\gamma^2\right]^{1/2}} \Bigg|_{\rho=0}^{a}$$

$$= \frac{2q}{a^2} \frac{(z - z')}{4\pi\epsilon_0} \left(\frac{-1}{\left[(z - z')^2 + a^2/\gamma^2\right]^{1/2}} + \frac{1}{|z - z'|} \right) .$$

When $|z - z'| \ll a/\gamma$ and the disc looks like an infinite sheet to the test charge, the second term dominates, and the field reduces to

$$E_z = \frac{2q}{a^2} \frac{\text{sgn}(z - z')}{4\pi\epsilon_0} .$$

The result is independent of z and γ, as we expect from an infinite sheet of charge, and depends on the sign of $z - z'$. In the opposite limit when the disc looks like a point charge, $|z - z'| \gg a/\gamma$, we can approximate

$$E_z = -\frac{2q}{a^2} \frac{(z - z')}{4\pi\epsilon_0} \left[\frac{1}{|z - z'|} \left(1 - \frac{a^2}{2\gamma^2(z - z')^2} \right) - \frac{1}{|z - z'|} \right]$$

$$= \frac{2q}{a^2} \frac{\text{sgn}(z - z')}{4\pi\epsilon_0} \frac{a^2}{2\gamma^2(z - z')^2} ,$$

and we recover the $1/z^2$ field we expect from a point charge.

Problem 11.5 *A bunch of electrons in a future linear collider will have a charge of about 1 nC, bunch length $\sigma_z \approx 200$ μm, and will be accelerated in the linac from 5 GeV to 250 GeV over the length of $L = 10$ km. Estimate the energy spread in the beam induced by the space charge, assuming the bunch radius of 50 μm.*

Solution: With $a/\gamma\sigma_z = 50/(200\gamma) \ll 1$, we can estimate the longitudinal field as having the form of Eq. (11.23),

$$E_z \sim \frac{1}{4\pi\epsilon_0} \frac{Q}{\gamma^2 \sigma_z^2} \log \frac{\gamma \sigma_z}{a} \ .$$

We assume linear acceleration, $\gamma = \gamma_0 + (\gamma_f - \gamma_0)z/L$, with $\gamma_0 \approx 10^4$ and $\gamma_f \approx 5 \times 10^5$, and an accelerating gradient of 25 MeV/m ($L \approx 10$ km). We can approximate $\log(\gamma \sigma_z/a) = \log(\gamma_f \sigma_z/a) = 14$ as a constant, since the log term does not change much when γ varies from γ_0 to γ_f. The energy spread from some particles seeing this maximum electric field and others not being accelerated at all would be roughly

$$\begin{aligned}
W &= e \int_0^L E_z dz \\
&\approx -\frac{1}{4\pi\epsilon_0} \frac{eQ}{\sigma_z^2} \log\left(\frac{\gamma_f \sigma_z}{a}\right) \frac{L/(\gamma_f - \gamma_0)}{\gamma_0 + (\gamma_f - \gamma_0)z/L}\Bigg|_{z=0}^{L} \\
&\approx 6.5 \text{ keV} \ .
\end{aligned}$$

Problem 11.6 *Derive an expression for the field $E_z(z)$ on the beam axis for a Gaussian bunch using the result of Sect. 11.4. Assume $\sigma_x = \sigma_y$.*

Solution: We will carry out calculation of the longitudinal field in the beam frame using Eq. (11.31) for the potential in which we set $\sigma_z' = \gamma \sigma_z$ and $\sigma_x = \sigma_y = \sigma_\perp$,

$$\phi'(\rho', z') = \frac{1}{4\pi\epsilon_0} \sqrt{\frac{2}{\pi}} \frac{Q}{\sigma_\perp^2 \gamma \sigma_z} \int_0^\infty d\lambda \frac{e^{-\frac{\rho'^2 \lambda^2}{2(\lambda^2 \sigma_\perp^2 + 1)}} e^{-\frac{z'^2 \lambda^2}{2(\lambda^2 \gamma^2 \sigma_z^2 + 1)}}}{(\lambda^2 + \sigma_\perp^{-2})\sqrt{\lambda^2 + \gamma^{-2}\sigma_z^{-2}}} \ .$$

From $E_z' = -\partial\phi'/\partial z'$ we then find on axis ($\rho = 0$)

$$E_z' = \frac{1}{4\pi\epsilon_0} \sqrt{\frac{2}{\pi}} \frac{Q}{\sigma_\perp^2 \sigma_z \gamma} \int_0^\infty d\lambda \frac{z'\lambda^2}{\gamma^2 \sigma_z^2} \frac{e^{-\frac{z'^2 \lambda^2}{2(\lambda^2 \gamma^2 \sigma_z^2 + 1)}}}{(\lambda^2 + \sigma_\perp^{-2})(\lambda^2 + \gamma^{-2}\sigma_z^{-2})^{3/2}} \ .$$

The Lorentz transformation does not change the longitudinal field, $E_z = E_z'$. Substituting $z' = \gamma(z - \beta ct)$ we get the E_z field in the lab frame,

$$E_z = \frac{1}{4\pi\epsilon_0} \sqrt{\frac{2}{\pi}} \frac{Q}{\sigma_\perp^2 \sigma_z^3 \gamma^2}(z - \beta ct) \int_0^\infty d\lambda\, \lambda^2 \frac{e^{-\frac{\gamma^2(z-\beta ct)^2 \lambda^2}{2(\lambda^2 \gamma^2 \sigma_z^2 + 1)}}}{(\lambda^2 + \sigma_\perp^{-2})(\lambda^2 + \gamma^{-2}\sigma_z^{-2})^{3/2}} \ .$$

We could also have transformed the potentials into the lab frame to find the E_z field, but then we would have to consider a nonzero vector potential as well.

Problem 11.7 *Show that at large distances from the center, Eq.* (11.31) *reduces to*

$$\phi' = \frac{Q}{4\pi\epsilon_0\sqrt{x'^2 + y'^2 + z'^2}} \ .$$

Solution: We would like to evaluate Eq. (11.31),

$$\phi' = \frac{1}{4\pi\epsilon_0}\sqrt{\frac{2}{\pi}}\frac{Q}{\sigma_x\sigma_y\sigma_z'}\int_0^\infty d\lambda \frac{e^{-\frac{\lambda^2 x'^2}{2(\lambda^2\sigma_x^2+1)}}\, e^{-\frac{\lambda^2 y'^2}{2(\lambda^2\sigma_y^2+1)}}\, e^{-\frac{\lambda^2 z'^2}{2(\lambda^2\sigma_z^2+1)}}}{\sqrt{\lambda^2 + \sigma_x^{-2}}\sqrt{\lambda^2 + \sigma_y^{-2}}\sqrt{\lambda^2 + \sigma_z'^{-2}}} \ ,$$

for the case of $x \gg \sigma_x$ (and likewise for y, z). In this case, the exponentials are non-vanishing only when $\lambda^2 x^2 \lesssim 1 \implies \lambda^2\sigma_x^2 \ll 1$, so in the denominator we can approximate $\sqrt{\lambda^2 + \sigma_x^{-2}} \approx 1/\sigma_x$, and in the exponential function we can neglect $\lambda^2\sigma_x^2$ in comparison with 1. The full integral becomes

$$\phi' = \frac{1}{4\pi\epsilon_0}\sqrt{\frac{2}{\pi}}Q\int_0^\infty d\lambda e^{-\lambda^2(x'^2+y'^2+z'^2)/2}$$

$$= \frac{1}{4\pi\epsilon_0}\frac{Q}{\sqrt{x'^2 + y'^2 + z'^2}} \ ,$$

where we used $\int_0^\infty e^{-ax^2/2}\, dx = \sqrt{\pi/2a}$ to do the integral.

References

1. J.D. Jackson, *Classical Electrodynamics*, 3rd edn. (Wiley, New York, 1999)
2. L.D. Landau, E.M. Lifshitz, *The Classical Theory of Fields*, volume 2 of *Course of Theoretical Physics*, 4th edn. (Elsevier, Butterworth-Heinemann, Burlington MA, 1980). (translated from Russian)
3. L.D. Landau, E.M. Lifshitz, L.P. Pitaevskii, *Electrodynamics of Continuous Media*, volume 8 of *Course of Theoretical Physics*, 2nd edn. (Elsevier, Butterworth-Heinemann, Burlington, MA, 1984). (translated from Russian)

Chapter 12
Effect of Environment
on Electromagnetic Field of a Beam

In the previous chapter we calculated the electromagnetic field of a beam moving with constant velocity in free space. In reality, beams propagate inside a vacuum chamber, and one has to take into account the effect of the material walls of the chamber on the field. In this chapter, we study several important effects that characterize the interaction with metal walls. First, we discuss the skin effect that defines the penetration of the electromagnetic field into the metal. We then discuss an approximation that treats the wall as a perfectly conducting medium. Finally, we calculate the longitudinal field inside a round metal pipe excited by a relativistic point charge.

12.1 Skin Effect and the Leontovich Boundary Condition

Metals are good conductors and an electromagnetic field that rapidly varies with time only penetrates into a thin surface layer of the metal — this is the so-called *skin effect*. If the skin depth is much smaller than the characteristic dimension of the problem under consideration, the metal surface can be represented by an effective boundary condition for the Maxwell equations. In this section we will derive this boundary condition and, in Sect. 12.3, we will apply it to the calculation of the electromagnetic field excited by a relativistic charge moving along the axis of a cylindrical metal pipe.

Consider a metal that has a constant conductivity σ and magnetic permeability μ. The equations that describe the electromagnetic field inside the metal are the Maxwell equations (A.1) in which we neglect the displacement current $\partial \boldsymbol{D}/\partial t$ in comparison with the current density $\boldsymbol{j}$:

$$\nabla \times \boldsymbol{H} = \boldsymbol{j}, \qquad \nabla \cdot \boldsymbol{B} = 0, \qquad \nabla \times \boldsymbol{E} + \frac{\partial \boldsymbol{B}}{\partial t} = 0, \qquad (12.1)$$

where $\boldsymbol{B} = \mu \boldsymbol{H}$. Dropping the term $\partial \boldsymbol{D}/\partial t$ from the Maxwell equations is justified for metals with high conductivity σ where the current density $\boldsymbol{j}$ inside the metal,

© Springer International Publishing AG, part of Springer Nature 2018
G. Stupakov and G. Penn, *Classical Mechanics and Electromagnetism in Accelerator Physics*, Graduate Texts in Physics,
https://doi.org/10.1007/978-3-319-90188-6_12

Fig. 12.1 Geometry used
for deriving the boundary
condition on the surface of a
metal

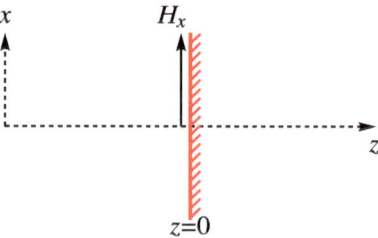

$$j = \sigma E , \qquad (12.2)$$

is dominant. Combining Eqs. (12.1) and (12.2) we obtain an equation for the magnetic
induction B,

$$
\begin{aligned}
\frac{\partial B}{\partial t} &= -\nabla \times E \\
&= -\sigma^{-1}\nabla \times j \\
&= -\sigma^{-1}\nabla \times \nabla \times H \\
&= \sigma^{-1}[\nabla^2 H - \nabla(\nabla \cdot H)] \\
&= (\sigma\mu)^{-1}\nabla^2 B ,
\end{aligned}
\qquad (12.3)
$$

where we have used the relation $\nabla \cdot H = \mu^{-1}\nabla \cdot B = 0$. We have derived the *diffusion* equation for the magnetic field B in the metal.[1]

We will now apply this equation to the case of an electromagnetic field that
penetrates into a metal occupying the half space $z > 0$, as shown in Fig. 12.1. Let
us assume that the field has the time dependence $E, H \propto e^{-i\omega t}$. We choose the
coordinate x so that the tangential component of the magnetic field on the metal
surface is directed along x, $H_x = H_0 e^{-i\omega t}$. Due to the continuity of the tangential
components of H, H_x is the same on both sides of the metal boundary, at $z = +0$
and $z = -0$. We seek a solution inside the metal in the form $H_x = h(z)e^{-i\omega t}$. The
x-component of Eq. (12.3) then reduces to

$$\frac{d^2 h}{dz^2} + i\mu\sigma\omega h = 0 , \qquad (12.4)$$

which has two solutions, $h = H_0 e^{ikz}$, where $k = \pm\sqrt{i\omega\mu\sigma}$. Only one of these solutions, with $Im\, k > 0$, should be selected as it corresponds to the magnetic field
vanishing far from the boundary, at $z \to \infty$. The complex wavenumber k can be
written as

$$k = \sqrt{i\omega\mu\sigma} = [\text{sgn}(\omega) + i]\sqrt{\frac{\mu\sigma|\omega|}{2}} . \qquad (12.5)$$

---

[1]In the Gaussian system of units, the diffusion coefficient is $c^2/4\pi\sigma\mu$, and the conductivity has
dimensions of inverse seconds.

In this equation we allow ω to take both positive and negative values. The *skin depth* δ, defined as the quantity

$$\delta(\omega) = \sqrt{\frac{2}{\mu\sigma|\omega|}}\,, \tag{12.6}$$

has dimension of length and it characterizes how deeply the electromagnetic field penetrates into the metal. If the magnetic properties of the metal can be neglected, then $\mu = \mu_0$, and the skin depth can be written as

$$\delta(\omega) = \sqrt{\frac{2c}{Z_0\sigma|\omega|}}\,, \tag{12.7}$$

where $Z_0 = \mu_0 c \simeq 377$ Ohm.

The electric field inside the metal has only a y component; it can be found from Eq. (12.2) in which the current is expressed through the magnetic field using the first equation in (12.1),

$$E_y = \frac{j_y}{\sigma} = \frac{1}{\sigma}\frac{dH_x}{dz} = \frac{ik}{\sigma}H_x\,. \tag{12.8}$$

We can now understand the mechanism that prevents the penetration of the magnetic field deep inside the metal. The time varying magnetic field in the metal, through the third equation in (12.1), generates a tangential electric field that in turn drives a large current in the skin layer. The magnetic field induced by this current is directed opposite to the magnetic field outside of the metal and shields it from penetrating deeper than several skin depths.

The relation (12.8) can also be written in the vector notation:

$$\boldsymbol{E}_t = \zeta\boldsymbol{H}_t \times \boldsymbol{n}\,, \tag{12.9}$$

where $\boldsymbol{n}$ is the unit vector normal to the surface and directed into the metal, $\boldsymbol{E}_t$ and $\boldsymbol{H}_t$ are the components of the fields oriented tangential to the surface, and the *surface impedance* ζ is given by

$$\zeta(\omega) = -\frac{ik}{\sigma} = [1 - i\,\text{sgn}(\omega)]\sqrt{\frac{Z_0|\omega|}{2c\sigma}}\,. \tag{12.10}$$

Because both $\boldsymbol{E}_t$ and $\boldsymbol{H}_t$ are continuous on the surface of the metal, Eq. (12.9) also applies to the vacuum components of the fields on the wall. With this understanding, Eq. (12.9) becomes a boundary condition for Maxwell's equations on the metal wall and is often called the *Leontovich boundary condition*. We need to emphasize that Eq. (12.9) is valid in the frequency domain, as indicated by the frequency dependence of the parameter ζ. To be able to use it in the time domain, one has to make the Fourier transformation from ω to t.

We have derived the boundary condition (12.9) for a flat metal surface. One can also apply it to curved surfaces if the radius of curvature is much larger than the skin depth δ: in this case one can locally approximate the surface by a plane when solving the diffusion equation (12.3). Because Eq. (12.9) was derived assuming that the metal extends to infinity, it can only be used if the thickness of the metal is much larger than the skin depth.

12.2 Perfectly Conducting Boundary Conditions

It follows from Eq. (12.10) that in the limit $\sigma \to \infty$ the surface impedance vanishes, $\zeta \to 0$, and the boundary condition (12.9) reduces to the requirement for the tangential electric field to be equal to zero,

$$E_t = 0\,. \tag{12.11}$$

We refer to this boundary condition as the *perfect conductivity* limit.

It follows from Eq. (12.11) that the *time dependent normal* component of the magnetic field, B_n, also vanishes on the metal wall. To prove this, we consider a small piece of the wall which can locally be approximated by a flat surface. We then introduce a Cartesian coordinate system with the origin on the surface, with the axes x and y located in the plane of the surface and the axis z normal to the surface. According to Eq. (12.11), $E_x(x, y, z = 0) = E_y(x, y, z = 0) = 0$. It follows then from the Maxwell equations that

$$\left.\frac{\partial B_z}{\partial t}\right|_{z=0} = -\left(\frac{\partial E_y}{\partial x} - \frac{\partial E_x}{\partial y}\right)\bigg|_{z=0} = 0\,. \tag{12.12}$$

Hence, unless there is a static magnetic field in the system, $B_z = 0$, or using the more general notation B_n for the component perpendicular to the metal surface,

$$B_n = 0\,. \tag{12.13}$$

In applications, one often encounters problems where one needs to calculate the electromagnetic field of a beam of relativistic particles traveling along the axis of a straight cylindrical pipe of a given cross section. In some cases it may be easier to do calculations in the beam frame and then transform it to the laboratory frame using the Lorentz transformations. Of course, the Maxwell equations are the same in both frames, but in general the boundary condition on the wall is expressed differently in the beam frame compared to the boundary condition in the lab frame. A remarkable property of the perfectly conducting boundary condition (12.11) is that it is the same in both frames. Indeed, using the prime to denote the field components in the beam frame, and from the Lorentz transformations for the fields (B.17), we can write the tangential components of the electric field in the beam frame as:

$$E'_{tz} = E_{tz}, \qquad E'_{t\perp} = \gamma\left[E_{t\perp} + v(\hat{z} \times B)_t\right], \qquad (12.14)$$

where we have assumed that the velocity is directed along the z axis, $\boldsymbol{v} = v\hat{z}$, and used the subscript $_{t\perp}$ to indicate the component of the field that is both tangential to the metal surface and perpendicular to the velocity. It is now easy to see that the components of the vector product $\hat{z} \times B$ that are tangential to the metal wall involve only the magnetic field component normal to the wall, $(\hat{z} \times B)_t = (\hat{z} \times B_n)_t$. Using the boundary conditions in the laboratory frame, Eqs. (12.11) and (12.13), we conclude that $E'_{tz} = 0$ and $E'_{t\perp} = 0$ and hence the whole tangential component of the electric field in the beam frame vanishes, $E'_t = 0$. We thus proved that the perfectly conducting boundary condition holds in any reference frame moving with a constant velocity directed parallel to the surface of the metal.

12.3 Round Pipe with Resistive Walls

As an example of the applications of the Leontovich boundary condition, in this section we will calculate how a round, conducting metal pipe influences the force between two charges, q_1 and q_2, moving along the axis of a pipe of radius a with relativistic velocity v, as shown in Fig. 12.2. We assume that the metal walls of the pipe have conductivity σ.

An important simplification in this problem comes from the fact that when the velocity v is close to the speed of light we can take the limit $v \to c$ and set $v = c$ in Maxwell's equations. In this limit, the electromagnetic field of charge q_2 does not propagate ahead of it and hence charge q_2 does not exert any force on charge q_1. On the other hand, charge q_1 generates a longitudinal electric field E_z behind it that does act on charge q_2. Our goal now is to find the field E_z at the location of charge q_2.

Assuming that the conductivity of the pipe is sufficiently large, we will calculate the fields using perturbation theory in which the solution is obtained as a Taylor expansion in the small parameter σ^{-1}. In the lowest approximation, we consider the pipe as being a perfect conductor. In this case, the electromagnetic field of charge

Fig. 12.2 Point charges q_1 and q_2 moving in a round pipe of radius a

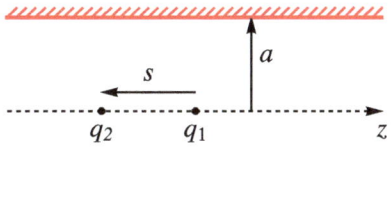

q_1 is the same as the free space solution given by Eq. (11.10), where it is assumed that at time t the charge is located at $z = ct$. Indeed, it follows from Eq. (11.10) that the electric field is perpendicular to the wall of the pipe as required by the boundary condition (12.11) in the limit of perfect conductivity. For what follows, we will only need the magnetic field B_θ,

$$B_\theta(\rho, z, t) = \frac{1}{4\pi\epsilon_0} \frac{2q_1}{c\rho} \delta(z - ct),$$ (12.15)

where ρ is the radial coordinate in the cylindrical coordinate system. Note that this magnetic field also satisfies the boundary condition (12.13).

Using the mathematical identity

$$\delta(z - ct) = \frac{1}{2\pi c} \int_{-\infty}^{\infty} d\omega e^{-i\omega(t-z/c)},$$ (12.16)

we Fourier transform B_θ and represent it as

$$B_\theta(\rho, z, t) = \int_{-\infty}^{\infty} d\omega \tilde{B}_\theta(\rho) e^{-i\omega t + i\omega z/c},$$ (12.17)

where

$$\tilde{B}_\theta(\rho) = \frac{1}{4\pi\epsilon_0 c^2} \frac{q_1}{\pi\rho} = \frac{\mu_0 q_1}{4\pi^2 \rho}.$$ (12.18)

We can now make the next step in the perturbation approach and take into account the finite conductivity σ. In this approximation, the tangential electric field on the metal wall is not zero. To find the Fourier component $\tilde{E}_z$ of the longitudinal electric field on the surface of the metal we will substitute into the Leontovich boundary condition (12.9) the magnetic field (12.18) computed in the limit $\sigma \to \infty$,

$$\tilde{E}_z|_{\rho=a} = -\zeta \frac{\tilde{B}_\theta(a)}{\mu_0} = -[1 - i\,\text{sgn}(\omega)]\sqrt{\frac{Z_0|\omega|}{2c\sigma}} \frac{q_1}{4\pi^2 a}.$$ (12.19)

Having found the electric field on the wall, we now need an equation that determines the radial dependence of E_z. This equation is provided by the z-component of the wave equation (A.3) which in the cylindrical coordinate system has the form

$$\frac{1}{c^2} \frac{\partial^2 E_z}{\partial t^2} - \frac{\partial^2 E_z}{\partial z^2} - \frac{1}{\rho} \frac{\partial}{\partial\rho}\rho\frac{\partial E_z}{\partial\rho} = 0.$$ (12.20)

It is clear that together with the magnetic field (12.15), the electric field depends on z and t in the same combination $z - ct$. With this dependence, the first two terms in Eq. (12.20) cancel out leaving only the radial term, which we now write for the Fourier transform $\tilde{E}_z$,

$$\frac{1}{\rho}\frac{\partial}{\partial\rho}\rho\frac{\partial\tilde{E}_z}{\partial\rho} = 0. \tag{12.21}$$

This equation has a general solution $\tilde{E}_z(\rho) = A + B\ln\rho$, where A and B are arbitrary constants. We do not expect E_z to have a singularity on the axis, hence $B = 0$, which means that the electric field does not depend on ρ, and the electric field on the axis is given by the same equation (12.19),

$$\tilde{E}_z|_{\rho=0} = \tilde{E}_z|_{\rho=a}. \tag{12.22}$$

We have also proven that in the ultrarelativistic case the longitudinal electric field inside the pipe is constant throughout the pipe cross section at any given z and t.

To find $E_z(z, t)$ we substitute Eq. (12.19) into the Fourier integral,

$$E_z(z, t) = \int_{-\infty}^{\infty} d\omega \tilde{E}_z e^{-i\omega(t-z/c)}, \tag{12.23}$$

which gives

$$E_z(z, t) = \sqrt{\frac{Z_0}{2c\sigma}}\frac{q_1}{4\pi^2 a}\int_{-\infty}^{\infty} d\omega[i\,\mathrm{sgn}(\omega) - 1]\sqrt{|\omega|}e^{-i\omega(t-z/c)}. \tag{12.24}$$

Strictly speaking, the integral in this equation diverges as $|\omega| \to \infty$. This happens because, as we will show below, our approach is not valid in the limit of very high frequencies. To regularize the integral, we will introduce a convergence factor $e^{-\epsilon|\omega|}$ with $\epsilon > 0$ into the integrand and then take the limit $\epsilon \to 0$ in the result. Using the mathematical identity[2]:

$$\lim_{\epsilon\to 0}\int_{-\infty}^{\infty} d\omega[i\,\mathrm{sgn}(\omega) - 1]\sqrt{|\omega|}e^{-i\omega\xi-\epsilon|\omega|} = \frac{\sqrt{2\pi}}{\xi^{3/2}}h(\xi), \tag{12.25}$$

where h is the step function, we obtain,

$$E_z(z, t) = \sqrt{\frac{Z_0}{2c\sigma}}\frac{q_1}{4\pi^2 a}\frac{\sqrt{2\pi}}{(t-z/c)^{3/2}}h(t-z/c) = \frac{q_1 c}{4\pi^{3/2}a}\sqrt{\frac{Z_0}{\sigma s^3}}h(s). \tag{12.26}$$

Here $s = ct - z$ is understood as the distance between the charges q_1 and q_2, and the step function indicates that the field does not propagate ahead of charge q_1. The positive sign of E_z means that a trailing charge (if it has the same sign as q_1) will be

---

[2]This identity can be established with the help of the following formula: $\int_0^\infty dx\sqrt{x}e^{\pm ix-\epsilon x} = \frac{1}{2}\sqrt{\pi}(\epsilon \mp i)^{-3/2}$.

accelerated in the wake. Recall from Sect. (11.1) that the longitudinal fields in vacuum scale as $1/\gamma^2$. Thus, for highly relativistic beams the impact of the resistive wall can be greater than the space charge fields that have been ignored in this derivation.

In our derivation we took the magnetic field on the wall to be the same as in the case of perfect conductivity which is the same as the magnetic field in free space, Eq. (12.15), where it is generated by the current associated with the moving charge q_1. We are now in a position to estimate when this approximation is valid. Accounting for the finite conductivity, the magnetic field changes because in addition to the moving charge there is an additional source — the displacement current $\partial \boldsymbol{D}/\partial t$ (see the Maxwell equations (A.1d)). The z-component of the displacement current density is

$$j_z^{\mathrm{disp}} = \epsilon_0 \frac{\partial E_z}{\partial t} \,. \tag{12.27}$$

To be able to neglect the contribution of j_z^{disp} to B_θ, we require the total displacement current $\pi a^2 j_z^{\mathrm{disp}}$ to be much smaller than the beam current. In the Fourier representation, the time derivative $\partial/\partial t$ is replaced by the multiplication factor $-i\omega$, and the requirement reduces to

$$\epsilon_0 \pi a^2 \omega |\tilde{E}_z| \ll |\tilde{I}| \,, \tag{12.28}$$

where $\tilde{I}$ is the Fourier component of the beam current. The latter is calculated by making the Fourier transformation of the equation $I = q_1 c \delta(z - ct)$ for the current of a point charge, yielding $|\tilde{I}| = q_1/2\pi$. Substituting this into Eq. (12.28) and using Eq. (12.19) for $\tilde{E}_z$ we obtain

$$\frac{\omega}{c} \ll \left(\frac{Z_0 \sigma}{a^2} \right)^{1/3} \,. \tag{12.29}$$

As was pointed out above, we find that indeed our solution in the frequency domain is limited to the range of frequencies defined by this inequality. In the space-time domain, a high spectral frequency ω corresponds to a small distance s behind the charge q_1, $s \sim c/\omega$. The inequality (12.29) means that our solution (12.26) is valid for distances $s \gg s_0$ where s_0 is of the order of the inverse right-hand side of Eq. (12.29):

$$s_0 \equiv \left(\frac{2a^2}{Z_0 \sigma} \right)^{1/3} \tag{12.30}$$

(the extra factor of 2 which has been added inside the cubic root is conventionally used in the literature). Values of s_0 are typically small for good conductors: for a pipe of radius $a = 5$ cm made of copper ($\sigma = 5.8 \times 10^7$ 1/Ohm·m) we have $s_0 = 60\,\mu\mathrm{m}$.

The solution E_z for small values of s, $s \lesssim s_0$, can be found in the literature, see [1]. In Fig. 12.3 we show a plot of E_z for s comparable to s_0. Note that the singularity $E_z \sim s^{-3/2}$ in Eq. (12.26) actually saturates for $s \lesssim s_0$, and the electric

Fig. 12.3 Longitudinal electric field as a function of distance s from the particle. The field is normalized by $q_1/4\pi\epsilon_0 a^2$, and the distance is normalized by s_0. The value of the normalized field at the origin is equal to -4

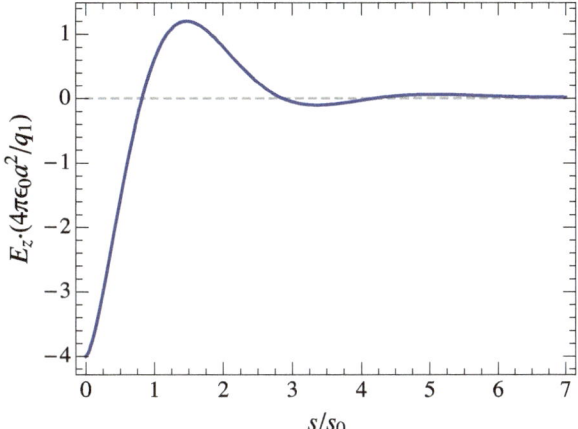

field normalized by q_1 changes sign here and is negative at $s = 0$. The negative sign of E_z/q_1 when $s \to 0$ is expected: it corresponds to deceleration of the charge q_1 and is explained by the fact that this charge loses energy which goes into the heating of the metal walls.

Worked Examples

Problem 12.1 *Calculate the skin depth in copper ($\sigma = 5.8 \times 10^7$ 1/Ohm·m) and stainless steel ($\sigma = 1.4 \times 10^6$ 1/Ohm·m) at the frequency of 5 GHz.*

Solution: The skin depth is given by $\delta = \sqrt{2/\mu_0\sigma\omega}$, with $\mu_0 = 1.3 \times 10^{-6}$ H/m. At $\omega = 2\pi f = 10\pi$ GHz, copper has $\delta = 0.9$ μm. For stainless steel, $\delta = 6$ μm.

Problem 12.2 *Given the tangential component $B_0 e^{-i\omega t}$ of the magnetic field on the surface, use the Poynting vector to find the averaged over time energy absorbed in the metal per unit time per unit area.*

Solution: To find the energy of the EM wave going into the metal, we will use the Poynting vector, $\mathbf{S} = \mathbf{E} \times \mathbf{H}$, which defines the energy flow per unit time per unit area. We have a tangential component of the magnetic field, $\mathbf{H} = B_0 e^{-i\omega t}/\mu$, and $\mathbf{E} = \zeta \mathbf{H} \times \mathbf{n}$, with $\zeta = (1 - i)/\sigma\delta(\omega)$ (here we assume $\omega > 0$). Before substituting the fields in the Poynting vector, we need to take their real parts,

$$\mathbf{S} = \mathrm{Re}\left[(\zeta \mathbf{H} \times \mathbf{n})\right] \times \mathrm{Re}\left[\mathbf{H}\right]$$

$$= \frac{1}{\sigma\delta(\omega)}\mathrm{Re}\left[(1 - i)\mathbf{H}\right] \cdot \mathrm{Re}\left[\mathbf{H}\right]\hat{z}$$

$$= \frac{B_0^2}{\mu^2\sigma\delta(\omega)}\left[(\cos\omega t - \sin\omega t)\cos\omega t\right]\hat{z} .$$

This can be thought of as the instantaneous rate of energy flow in and out of the surface. Averaging $\cos^2 \omega t$ over time gives $1/2$, while $(\sin \omega t \cos \omega t)$ averages to 0. Taking $\delta(\omega) = \sqrt{2/\mu\sigma\omega}$, we find

$$
\begin{aligned}
\langle S_z \rangle &= \frac{B_0^2}{2\mu^2 \sigma \delta(\omega)} \\
&= \frac{\omega \delta(\omega) B_0^2}{4\mu} \propto \sqrt{\omega} .
\end{aligned}
$$

For comparison, the electromagnetic energy density just outside the surface is $B_0^2/2\mu_0$.

Problem 12.3 *An ultra-relativistic bunch with charge Q propagates along the axis of a round metal pipe of radius a. Using Eq. (12.26) calculate the energy loss of the bunch per unit length. Assume a Gaussian distribution with the rms bunch length σ_z.*

Solution: The electric field E_z at longitudinal coordinate s_1, with $s_1 = ct - z$ indicating the relative position inside the bunch, can be computed as a superposition of fields generated by all preceding elementary charges. For this, we will use Eq. (12.26) in which q_1 is replaced by $Q\lambda(s_2)ds_2$, the distance is set to $s = s_2 - s_1$, and the formula is integrated over s_2. We obtain

$$
E_z(s_1) = \frac{Qc}{4\pi^{3/2}a} \sqrt{\frac{Z_0}{\sigma}} \int_{s_1}^{\infty} ds_2\, \lambda(s_2)(s_2 - s_1)^{-3/2} . \tag{12.31}
$$

This integral actually diverges as $s_2 \to s_1$. The reason for this divergence is that, as we discussed in Sect. 12.3, Eq. (12.26) becomes invalid in the limit $s \to 0$. This divergence can be eliminated if we express the field's dependence on s through a partial derivative using

$$
(s_2 - s_1)^{-3/2} = -2\frac{\partial}{\partial s_2}(s_2 - s_1)^{-1/2} ,
$$

which is valid everywhere except for $s_2 = s_1$. Integrating equation (12.31) by parts and discarding the boundary term we obtain

$$
E_z(s_1) = \frac{Qc}{2\pi^{3/2}a} \sqrt{\frac{Z_0}{\sigma}} \int_{s_1}^{\infty} ds_2\, \lambda'(s_2)(s_2 - s_1)^{-1/2} ,
$$

where the prime denotes derivative with the respect to the argument.[3]

---

[3]Getting rid of the singularity through the integration by parts is a subtle point. A careful analysis shows that it requires the point charge field $E_z(s)$ to have the property $\int_0^{\infty} ds\, E_z(s) = 0$. This is indeed satisfied by the field plotted in Fig. 12.3.

The energy loss per unit length is obtained by integrating $eE_z(s_1)$ with the distribution function over s_1,

$$\frac{d\mathcal{E}}{dz} = e \int_{-\infty}^{\infty} ds_1 E_z(s_1)\lambda(s_1)$$

$$= \frac{eQc}{2\pi^{3/2}a} \sqrt{\frac{Z_0}{\sigma}} \int_{-\infty}^{\infty} ds_1 \, \lambda(s_1) \int_{s_1}^{\infty} ds_2 \, \lambda'(s_2)(s_2 - s_1)^{-1/2} \, .$$

The two-dimensional integral can be simplified by replacing the second integration variable and changing the order of integration,

$$\int_{-\infty}^{\infty} ds_1 \, \lambda(s_1) \int_{s_1}^{\infty} ds_2 \, \lambda'(s_2)(s_2 - s_1)^{-1/2}$$

$$= \int_{-\infty}^{\infty} ds_1 \, \lambda(s_1) \int_{0}^{\infty} d\xi \, \lambda'(s_1 + \xi)\xi^{-1/2}$$

$$= \int_{0}^{\infty} d\xi \, \xi^{-1/2} \int_{-\infty}^{\infty} ds_1 \, \lambda(s_1)\lambda'(s_1 + \xi) \, .$$

For the Gaussian distribution function $\lambda(s) = (2\pi)^{-1/2}\sigma_z^{-1}e^{-s^2/2\sigma_z^2}$ the inner integral can be calculated analytically,

$$\int_{-\infty}^{\infty} ds_1 \, \lambda(s_1)\lambda'(s_1 + \xi) = -\frac{1}{4\sqrt{\pi}\sigma_z^3}\xi e^{-\xi^2/4\sigma_z^2} \, .$$

Using the mathematical identity

$$\int_{0}^{\infty} dt \, t^{1/2}e^{-t^2/4} = \sqrt{2}\,\Gamma\left(\frac{3}{4}\right) \, ,$$

where Γ is the gamma function, we arrive at the following result:

$$\frac{d\mathcal{E}}{dz} = -\frac{eQc}{2^{5/2}\pi^2 a\sigma_z^{3/2}} \sqrt{\frac{Z_0}{\sigma}}\,\Gamma\left(\frac{3}{4}\right) \, .$$

The negative sign in this formula indicates that the bunch is losing energy.

This result is valid if the bunch length is much longer than the parameter s_0 defined by Eq. (12.30), $\sigma_z \gg s_0$.

Reference

1. K.L.F. Bane, M. Sands, The short-range resistive wall wakefields. AIP Conf. Proc. **367**, 131 (1996)

Chapter 13
Plane Electromagnetic Waves and Gaussian Beams

Plane electromagnetic waves are solutions of the Maxwell equations that are unbounded in the plane perpendicular to the direction of propagation. They approximately describe local properties of the real field far from the source of radiation. They can also be used as building blocks from which a general solution of Maxwell's equations in free space can be constructed. An important practical example of electromagnetic radiation that finds many applications in accelerator physics and elsewhere is a focused laser beam. The distribution of the electromagnetic field in such light is characterized by Gaussian modes which can be considered as a superposition of plane waves propagating at small angles to the direction of the beam. Gaussian beams correctly describe the field structure near the focus and the diffraction of the beam as it propagates away from the focal region. In this chapter, we will briefly summarize the main properties of plane electromagnetic waves, and then derive the field in a Gaussian beam.

13.1   Plane Electromagnetic Waves

A plane electromagnetic wave is a solution of Maxwell's equations that describes propagation of electromagnetic fields in free space. In this solution, all components of the field depend only on one variable $\xi = z - ct$,

$$\boldsymbol{E}(\boldsymbol{r}, t) = \boldsymbol{F}(\xi), \qquad \boldsymbol{B}(\boldsymbol{r}, t) = \boldsymbol{G}(\xi), \tag{13.1}$$

where $\boldsymbol{F}$ and $\boldsymbol{G}$ are vector functions of ξ. From the equation $\nabla \cdot \boldsymbol{E} = 0$ it follows that $\partial F_z / \partial \xi = 0$, from which we conclude that $F_z = 0$ (a more general solution $F_z = \text{const}$ describes a constant longitudinal electric field and is not related to the wave). Similarly, the equation $\nabla \cdot \boldsymbol{B} = 0$ yields $G_z = 0$. We conclude that a plane wave is *transverse*: the electromagnetic field is perpendicular to the direction of propagation z.

G. Stupakov and G. Penn, *Classical Mechanics and Electromagnetism in Accelerator Physics*, Graduate Texts in Physics, https://doi.org/10.1007/978-3-319-90188-6_13

Applying the Maxwell equation $\partial \boldsymbol{B}/\partial t = -\nabla \times \boldsymbol{E}$ to the fields (13.1) we find

$$F'_x = c\,G'_y\,, \qquad F'_y = -c\,G'_x\,, \tag{13.2}$$

where the prime denotes differentiation with respect to the argument ξ. From these relations we conclude that $F_x = c\,G_y$ and $F_y = -c\,G_x$ (again, we neglect possible constants of integration that describe constant fields). In vector notation, these relations can be written as $\boldsymbol{F} = -c\boldsymbol{n} \times \boldsymbol{G}$, or

$$\boldsymbol{E} = -c\,\boldsymbol{n} \times \boldsymbol{B}\,, \tag{13.3}$$

where $\boldsymbol{n}$ is a unit vector in the direction of propagation (in our case $\boldsymbol{n}$ is directed along the z axis). Multiplying vectorially Eq. (13.3) by $\boldsymbol{n}$, we also obtain

$$\boldsymbol{B} = \frac{1}{c}\,\boldsymbol{n} \times \boldsymbol{E}\,. \tag{13.4}$$

The potentials ϕ and $\boldsymbol{A}$ for a plane wave also depend on ξ only: $\phi = \phi(\xi)$, $\boldsymbol{A} = \boldsymbol{A}(\xi)$. A useful formula that we will use later expresses the magnetic field in a plane wave through the time derivative of the vector potential $\boldsymbol{A}$:

$$\begin{aligned}
\boldsymbol{B} &= \nabla \times \boldsymbol{A} \\
&= -\hat{\boldsymbol{x}}A'_y + \hat{\boldsymbol{y}}A'_x \\
&= \boldsymbol{n} \times \boldsymbol{A}' \\
&= -\frac{1}{c}\,\boldsymbol{n} \times \frac{\partial \boldsymbol{A}}{\partial t}\,,
\end{aligned} \tag{13.5}$$

where on the last step we replaced the derivative $d/d\xi$ by the partial derivative $-c^{-1}\partial/\partial t$. After the magnetic field is found, the electric field can be obtained from Eq. (13.3).

In applications, a plane wave often has a sinusoidal time dependence with some frequency ω. For such waves, it is convenient to use the complex notation:

$$\boldsymbol{E} = \mathrm{Re}(\boldsymbol{E}_0 e^{-i\omega t + i\boldsymbol{k}\cdot\boldsymbol{r} + i\phi_0})\,, \qquad \boldsymbol{B} = \mathrm{Re}(\boldsymbol{B}_0 e^{-i\omega t + i\boldsymbol{k}\cdot\boldsymbol{r} + i\phi_0})\,, \tag{13.6}$$

where $\boldsymbol{E}_0$ and $\boldsymbol{B}_0$ are the amplitudes of the wave with $E_0 = cB_0$, $\boldsymbol{k} = \boldsymbol{n}\omega/c$ is the wavenumber that defines the direction of propagation, and ϕ_0 is the phase. In general, $\boldsymbol{E}_0$ and $\boldsymbol{B}_0$ can be complex vectors orthogonal to $\boldsymbol{k}$, i.e., $\boldsymbol{E}_0 = \boldsymbol{E}_0^{(r)} + i\boldsymbol{E}_0^{(i)}$ with $\boldsymbol{E}_0^{(r)}$ and $\boldsymbol{E}_0^{(i)}$ real. When $\boldsymbol{E}_0^{(r)}$ and $\boldsymbol{E}_0^{(i)}$ are collinear, the wave has a *linear polarization*; when they are orthogonal and equal in length, the wave is *circularly polarized*; in the most general case, the polarization is *elliptical*.

The energy flow in the plane wave is given by the Poynting vector,

$$\boldsymbol{S} = \boldsymbol{E} \times \boldsymbol{H} = \boldsymbol{n}\sqrt{\frac{\epsilon_0}{\mu_0}}\,E_0^2\cos^2(-\omega t + \boldsymbol{k}\cdot\boldsymbol{r} + \phi_0)\,, \tag{13.7}$$

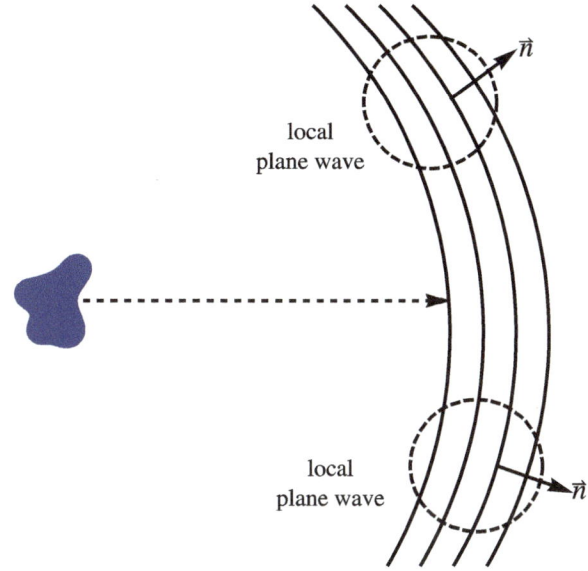

Fig. 13.1 Approximation of local plane waves far from a radiative system of charges shown by the blue blob. Solid lines indicate the phase fronts of the waves. Note that in different regions the directions n of the wave propagation are different

(in this formula E_0 is assumed real). As follows from this formula, the energy flows along n, in the direction of propagation of the wave. Denoting by $\bar{S}$ the energy flow averaged over time, we find

$$\bar{S} = \frac{n}{2}\sqrt{\frac{\epsilon_0}{\mu_0}}E_0^2 = n\frac{E_0^2}{2Z_0} = n\frac{c^2 B_0^2}{2Z_0}. \tag{13.8}$$

In reality one never deals with plane waves that occupy the whole space as is indicated by Eq. (13.6). Usually, a plane wave provides a useful approximation to a propagating electromagnetic field in a limited region of space. We will use this approximation in Chaps. 16–18 to study the radiation field of moving charges at large distances. Figure 13.1 illustrates how a plane-wave approximation is applied in different regions of the radiation field.

Another important aspect of sinusoidal plane waves is that an arbitrary solution of Maxwell's equations in free space (without charges) can be represented as a superposition of plane waves with various amplitudes and directions of propagation. This is illustrated by Problem 13.1.

13.2 Gaussian Beams

The plane-wave approximation from the previous section is not sufficient for describing a laser beam propagating towards its focus and then diffracting away from the focal point. A better representation of laser beams is provided by Gaussian modes in free space. In this section, we will derive the field in an axisymmetric fundamental Gaussian mode using the *paraxial* approximation, which assumes that the field is represented by a plane wave with an amplitude that slowly varies in space.

We start from the wave equation (A.3) for the x component of the electric field assuming a linear polarization of the laser light,

$$\frac{\partial^2 E_x}{\partial x^2} + \frac{\partial^2 E_x}{\partial y^2} + \frac{\partial^2 E_x}{\partial z^2} - \frac{1}{c^2}\frac{\partial^2 E_x}{\partial t^2} = 0 . \tag{13.9}$$

Let us seek a solution to this equation in the following form,

$$E_x(x, y, z, t) = u(x, y, z)e^{-i\omega t + ikz} , \tag{13.10}$$

where $\omega = ck$ and u is a slow varying function of its arguments mathematically expressed in the requirement

$$\left|\frac{1}{u}\frac{\partial u}{\partial z}\right| \ll k . \tag{13.11}$$

As we will see from the result, u also varies slowly in the transverse directions x and y. Note that a constant u corresponds to a sinusoidal plane electromagnetic wave propagating along the z axis; allowing u to vary in space takes us beyond the limits of the plane-wave approximation.

Putting Eq. (13.10) into (13.9) and neglecting the second derivative $\partial^2 u/\partial z^2$ in comparison with $k\partial u/\partial z$ yields

$$\frac{\partial^2 u}{\partial x^2} + \frac{\partial^2 u}{\partial y^2} + 2ik\frac{\partial u}{\partial z} = 0 . \tag{13.12}$$

We are looking for an axisymmetric solution to this equation that depends on $\rho = \sqrt{x^2 + y^2}$, so that $u = u(\rho, z)$. Eq. (13.12) then becomes

$$\frac{1}{\rho}\frac{\partial}{\partial \rho}\rho\frac{\partial u}{\partial \rho} + 2ik\frac{\partial u}{\partial z} = 0 . \tag{13.13}$$

The fundamental Gaussian mode has the following particular dependence on ρ and z:

$$u = A(z)e^{Q(z)\rho^2} , \tag{13.14}$$

where $A(z)$ and $Q(z)$ are as yet unknown functions. We will see that Q has a negative real part so that u exponentially decays in the radial direction.

Substituting Eq. (13.14) into (13.13) yields

$$4Q^2\rho^2 u + 4Qu + 2ik\left(\frac{A'}{A} + Q'\rho^2\right) = 0 , \tag{13.15}$$

where the prime denotes differentiation with respect to z. In order for this equation to be valid for arbitrary ρ, both the coefficient in front of ρ^2 and the terms that do not contain ρ should vanish,

$$2Q^2 + ikQ' = 0,$$

$$2Q + ik\frac{A'}{A} = 0.$$

(13.16)

The solution of the first equation can be written as

$$Q(z) = -\frac{1/w_0^2}{1 + 2iz/kw_0^2},$$

(13.17)

where w_0 is a constant of integration which has dimension of length and is called the *beam waist*. Substituting Q into the second of Eq. (13.16) and integrating it yields

$$A(z) = \frac{E_0}{1 + 2iz/kw_0^2},$$

(13.18)

where E_0 is another constant of integration that has a meaning of the field amplitude.

We now introduce two important geometrical parameters: the *Rayleigh length Z_R* and the angle θ:

$$Z_R = \frac{kw_0^2}{2}, \qquad \theta = \frac{w_0}{Z_R} = \frac{2}{kw_0}.$$

(13.19)

The Rayleigh length and the waist can be expressed through the angle θ and the reduced wavelength $\lambda = k^{-1} = c/\omega$:

$$Z_R = 2\frac{\lambda}{\theta^2}, \qquad w_0 = 2\frac{\lambda}{\theta}.$$

(13.20)

Rewriting Q and A as $Q = -w_0^{-2}/(1 + iz/Z_R)$ and $A = E_0/(1 + iz/Z_R)$ and substituting them into Eq. (13.14), we arrive at the following expression for the electric field:

$$u = \frac{E_0}{1 + iz/Z_R} e^{-\rho^2/w_0^2(1+iz/Z_R)}.$$

(13.21)

At a given coordinate z, the radial dependence of $|u|$ is $\propto e^{-\rho^2/w_0^2(1+z^2/Z_R^2)}$, which means that the quantity w_0 gives the minimal transverse size of the mode at the focal spot $z = 0$. On the axis of the beam, $\rho = 0$, we have $|u| = E_0/\sqrt{1 + z^2/Z_R^2}$, and hence Z_R is the characteristic length of the focal region along the z axis. The angle θ characterizes the divergence of the beam far from the focus. These geometrical characteristics of the mode are illustrated in Fig. 13.2.

Fig. 13.2 Geometric parameters of a Gaussian beam

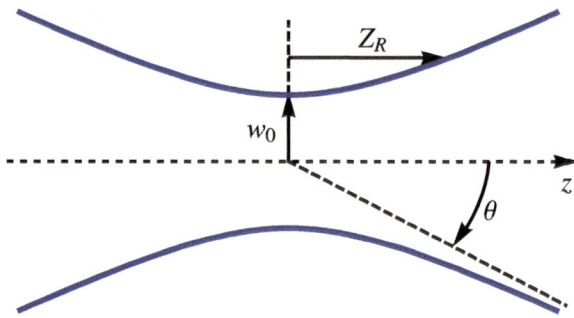

We can now derive conditions for the validity of the paraxial approximation. Evaluating $\partial^2 u/\partial z^2 \sim u/Z_R^2$ and $k\partial u/\partial z \sim ku/Z_R$ we see that in order to be able to neglect the second derivative in Eq. (13.12) we need to require $Z_R \gg \lambda$ or, equivalently, $\theta \ll 1$. The latter condition means that the waves that constitute a Gaussian beam propagate at small angles to the axis z, which is the origin of the term *paraxial*. A small θ also means that $\lambda \ll w_0 \ll Z_R$ which implies that the size of the focal spot w_0 is much larger than the reduced wavelength λ. Locally, on the scale of several λ, a Gaussian beam can be considered as a plane wave, but on a larger scale $\sim w_0$ in the transverse direction and $\sim Z_R$ in the longitudinal direction, the field substantially deviates from the plane-wave approximation.

The magnetic field in a Gaussian beam can be found in the lowest order using Eq. (13.4). In our case $\boldsymbol{n}$ is directed along z, hence the magnetic field is directed along y,

$$B_y = \frac{1}{c}E_x . \tag{13.22}$$

Let us now look at the field at a large distance from the focus. At $|z| \gg Z_R$ we can approximate Q and A as:

$$A(z) \approx -\frac{iE_0 k w_0^2}{2z} , \qquad Q(z) \approx \frac{ik}{2z} - \frac{k^2 w_0^2}{4z^2} . \tag{13.23}$$

From these equations it follows that the amplitude of the field on the axis decays as $1/z$, and the radial profile at coordinate z as given by the absolute value of the exponential factor is $e^{-k^2 w_0^2 \rho^2/4z^2}$. The region occupied by the field in the radial direction can be estimated as $\rho \sim 2z/kw_0 = \theta z$, it expands as the field propagates away from the focus. There is also a correction $\operatorname{Im} Q \rho^2 = k\rho^2/2z$ to the plane-wave phase kz, so that the total phase is

$$\phi = kz + \frac{k\rho^2}{2z} . \tag{13.24}$$

One can give this phase the following physical interpretation. Let us introduce the distance R from the focal point at the origin of the coordinate system to the observation point, $R = \sqrt{z^2 + \rho^2}$. In the limit of large z, assuming $z > 0$, we have $R \approx z + \rho^2/2z$, and we see that the phase (13.24) is approximately equal to kR. Such a phase dependence is characteristic for a *spherical wave*, and we conclude that at a large distance from the focus a Gaussian beam represents a spherical wave diverging away from the focus. The electric and magnetic fields in this wave are not constant on the wave front — they are localized within the angle θ defined by Eq. (13.19).

Worked Examples

Problem 13.1 *At time $t = 0$ the electromagnetic field in free space is given by functions $E_0(r)$ and $B_0(r)$ such that $\nabla \cdot E_0 = \nabla \cdot B_0 = 0$. Find the field at time t by representing $E(r, t)$ and $B(r, t)$ as integrals over plane waves.*

Solution: We represent the time-dependent electromagnetic field as a combination of waves propagating in both directions, along k and $-k$,

$$E(t, r) = \int d^3k (e^{-ir\cdot k + i\omega t} E_k^{(+)} + e^{-ir\cdot k - i\omega t} E_k^{(-)}) ,$$

$$B(t, r) = \int d^3k (e^{-ir\cdot k + i\omega t} B_k^{(+)} + e^{-ir\cdot k - i\omega t} B_k^{(-)}) , \tag{13.25}$$

with $\omega = c|k|$. From Eq. (13.4) applied to each wave it follows that

$$B_k^{(+)} = \frac{1}{\omega} k \times E_k^{(+)} ,$$

$$B_k^{(-)} = -\frac{1}{\omega} k \times E_k^{(-)} . \tag{13.26}$$

At time $t = 0$ we have

$$E(0, r) = \int d^3k e^{-ir\cdot k} (E_k^{(+)} + E_k^{(-)}) = E_0(r) ,$$

$$B(0, r) = \int d^3k e^{-ir\cdot k} (B_k^{(+)} + B_k^{(-)}) = B_0(r) ,$$

which means that the sums $E_k^{(+)} + E_k^{(-)}$ and $B_k^{(+)} + B_k^{(-)}$ are equal to the corresponding Fourier transforms of the initial fields:

$$E_k^{(+)} + E_k^{(-)} = \tilde{E}_{0,k} \equiv \frac{1}{(2\pi)^3} \int d^3r e^{ir\cdot k} E_0(r),$$

$$B_k^{(+)} + B_k^{(-)} = \tilde{B}_{0,k} \equiv \frac{1}{(2\pi)^3} \int d^3r e^{ir\cdot k} B_0(r) . \tag{13.27}$$

Note that $\tilde{E}_{0,k} \cdot k = \tilde{B}_{0,k} \cdot k = 0$. Solving this equation in combination with Eq. (13.26) we find

$$E_k^{(-)} = \frac{1}{2} \left(\tilde{E}_{0,k} + \frac{c^2}{\omega} k \times \tilde{B}_{0,k} \right) ,$$

$$E_k^{(+)} = \frac{1}{2} \left(\tilde{E}_{0,k} - \frac{c^2}{\omega} k \times \tilde{B}_{0,k} \right) .$$

Together with Eq. (13.26) this solves the problem.

Problem 13.2 *A plane electromagnetic wave propagates at some angle in a frame moving with velocity βc along the z axis. The magnitude of the Poynting vector at some location in the wave is equal to S'. Show that in the lab frame the magnitude of the Poynting vector at this location is given by the following equation:*

$$S = \frac{S'}{\gamma^2 (1 - \beta \cos \theta)^2} , \tag{13.28}$$

where θ is the angle between the direction of propagation in the lab frame and the z axis.

Solution: First, consider the plane wave in the lab frame where it is moving at an angle θ to the z axis. Assume the E field is in the x-z plane, so that $|B| = B_y = B_0$. The amplitude of the electric field is $E_0 = cB_0$. We have $E_x = E_0 \cos \theta$ and $E_z = -E_0 \sin \theta$. Transforming into the moving frame,

$$E_z' = E_z = -E_0 \sin \theta ,$$
$$E_x' = \gamma [E_x + (v \times B)_x] = \gamma E_0 (\cos \theta - \beta) ,$$

where $(v \times B)_x = -\beta E_0$ was used in the final line. According to Eq. (13.8) the Poynting vector in the moving frame is $S' = E'^2 / 2Z_0$. We have

$$
\begin{aligned}
S' &= \frac{E'^2}{2Z_0} = \frac{E_x'^2 + E_z'^2}{2Z_0} \\
&= \frac{E_0^2}{2Z_0} \left(\sin^2 \theta + \gamma^2 (\cos^2 \theta - 2\beta \cos \theta + \beta^2) \right) \\
&= \frac{E_0^2}{2Z_0} \gamma^2 (1 - \beta \cos \theta)^2 \\
&= S \gamma^2 (1 - \beta \cos \theta)^2 ,
\end{aligned}
$$

where $S = E_0^2 / 2Z_0$ is the Poynting vector in the lab frame and we have used the identity $1 + \gamma^2 \beta^2 = \gamma^2$. Note that it is necessary to start in the lab frame since we are given the angle, θ, in the lab frame. Alternatively, if we start in the moving frame, we find

$$S = S'\gamma^2 \left(1 + \beta \cos \theta'\right)^2 .$$

Then using $\cos \theta' = (\cos \theta - \beta)/(1 - \beta \cos \theta)$ we find

$$S = S'\gamma^2 \left(1 + \frac{\beta \cos \theta - \beta^2}{1 - \beta \cos \theta}\right)^2$$

$$= S' \frac{1}{\gamma^2(1 - \beta \cos \theta)^2} .$$

For a plane wave in which the electric field is polarized along y, one obtains the same result considering the Lorentz transformation for the magnetic field. Finally, for an arbitrary polarization, the Poynting vector is a sum of two orthogonal polarizations, and hence is also transformed according to Eq. (13.28).

Problem 13.3 *Calculate the longitudinal electric field E_z in the laser beam using the equation $\nabla \cdot E = 0$.*

Solution: From $\nabla \cdot E = 0$, for a Gaussian beam polarized along x we have $\partial E_z / \partial z = -\partial E_x / \partial x$. With $E_x = A(z)e^{Q(z)(x^2+y^2)}e^{-i\omega t + ikz}$, we have

$$\frac{\partial E_x}{\partial x} = 2x Q(z) E_x .$$

If we assume $E_z = v(x, y, z)e^{-i\omega t + ikz}$ with slowly varying function v, $\partial v/\partial z \ll ikv$, then $\partial E_z/\partial z = (v^{-1}\partial v/\partial z + ik)E_z \approx ikE_z$. We can then put together the two derivatives to find

$$E_z \approx -\frac{1}{ik}\frac{\partial E_x}{\partial x}$$

$$= \frac{-2x Q(z) E_x}{ik}$$

$$= \frac{2x}{ikw_0^2 - 2z} E_x .$$

For z small and $x \sim w_0$, this reduces to $E_z \sim 2E_x/ikw_0 \sim \theta E_x \ll E_x$, so we conclude $E_z \ll E_x$.

Problem 13.4 *Show that the energy flux in the laser beam (the Poynting vector integrated over the cross section of the beam) is equal to*

$$\frac{\pi}{4Z_0} E_0^2 w_0^2 .$$

Solution: We need to integrate over the beam cross section the time averaged Poynting vector $S = E_x^2/2Z_0$. At $z = 0$, we have $u = E_0 e^{-\rho^2/w_0^2}$, so

$$S = \frac{E_0^2}{2Z_0} \int 2\pi\rho \, d\rho \, e^{-2\rho^2/w_0^2} = \frac{\pi}{4Z_0} E_0^2 w_0^2 \, .$$

Problem 13.5 *Expand the Gaussian laser field over plane waves.*

Solution: The exact solution of Maxwell's equations corresponding to a Gaussian mode with the frequency ω can be represented as a superposition of plane electromagnetic waves with different wavenumbers k_x and k_y:

$$\boldsymbol{E} = e^{-i\omega t} \int \tilde{E}(k_x, k_y) e^{ik_x x + ik_y y + ik_z z} \left(\hat{\boldsymbol{x}} - \frac{k_x}{k_z}\hat{\boldsymbol{z}} \right) dk_x \, dk_y \, ,$$

where $k_z(k_x, k_y) \equiv (k^2 - k_x^2 - k_y^2)^{1/2}$ with $k = \omega/c$ and $\tilde{E}(k_x, k_y)$ is the wave amplitude with the wavenumbers k_x and k_y. Note that due to the particular form of the expression in the brackets in the integrand this field satisfies the equation $\nabla \cdot \boldsymbol{E} = 0$. To find the wave amplitudes $\tilde{E}(k_x, k_y)$ we take the x components of the field, set $z = 0$,

$$E_x|_{z=0} = e^{-i\omega t} \int dk_x \, dk_y \, \tilde{E}(k_x, k_y) e^{ik_x x + ik_y y} \, ,$$

and equate it to the Gaussian mode at $z = 0$,

$$E_0 e^{-i\omega t} e^{-(x^2+y^2)/w_0^2} \, .$$

From this equality, $\tilde{E}(k_x, k_y)$ can be found through the inverse Fourier transform:

$$\tilde{E}(k_x, k_y) = \frac{E_0}{(2\pi)^2} \int dx \, dy \, e^{-ik_x x - ik_y y} e^{-(x^2+y^2)/w_0^2} \, .$$

The integral on the right-hand side factors out into a product of one-dimensional integrals which are easily integrated,

$$\int_{-\infty}^{\infty} dx \, e^{-ik_x x - x^2/w_0^2} = \sqrt{\pi} w_0 e^{-k_x^2 w_0^2} \, ,$$

and we find

$$\tilde{E}(k_x, k_y) = \frac{E_0 w_0^2}{\pi} e^{-(k_x^2 + k_y^2)w_0^2} \, .$$

Note that the typical value for the angles relative to the z axis is $k_x/k \sim k_y/k \sim 1/w_0 k \sim \theta$.

Chapter 14
Waveguides and RF Cavities

In the previous chapter we studied propagation of electromagnetic waves in free space. Free space is an idealization which can be used when the waves propagate far from material boundaries. If this is not the case, one has to take into account the interaction of the electromagnetic field with the medium. In this chapter we will study one example that is especially important for accelerator applications, when the medium can be modeled as a perfectly conducting metal whose boundaries reflect the electromagnetic field without losses. The impact of resistive losses can often be treated perturbatively, as we have seen in Chap. 12. We will consider cylindrical waveguides and resonant cavities. We will also discuss how cavity eigenmodes can be excited by relativistic beams of charged particles.

14.1 TM Modes in Cylindrical Waveguides

We consider a cylindrical waveguide of radius a with perfectly conducting walls and choose a cylindrical coordinate system (r, ϕ, z) with the z-axis directed along the cylinder axis. Such a waveguide can serve as a conduit for the transportation of electromagnetic energy along its axis in the form of eigenmodes that have particular transverse and longitudinal field distributions. We will first focus our attention on the so-called *transverse magnetic modes*, or TM modes, that have zero longitudinal component of the magnetic field, $B_z = 0$, but nonzero E_z. To find E_z in TM modes, we will assume that

$$E_z(r, \phi, z, t) = \mathcal{E}(r)e^{-i\omega t - im\phi + i\varkappa z} , \tag{14.1}$$

where ω is the mode frequency, m is an integer number that defines the azimuthal variation of the field, and $\varkappa$ is the longitudinal wavenumber. The function E_z satisfies the wave equation (A.3) which in the cylindrical coordinate system r, ϕ, z reads

© Springer International Publishing AG, part of Springer Nature 2018
G. Stupakov and G. Penn, *Classical Mechanics and Electromagnetism in Accelerator Physics*, Graduate Texts in Physics,
https://doi.org/10.1007/978-3-319-90188-6_14

$$\frac{1}{r}\frac{d}{dr}r\frac{d\mathcal{E}}{dr} - \frac{m^2}{r^2}\mathcal{E} + \left(\frac{\omega^2}{c^2} - \varkappa^2\right)\mathcal{E} = 0. \tag{14.2}$$

The solution to this equation is

$$\mathcal{E} = E_0 J_m\left(k_\perp r\right), \tag{14.3}$$

where J_m is the Bessel function of mth order and $k_\perp = c^{-1}\sqrt{\omega^2 - c^2\varkappa^2}$. The boundary condition $E_z = 0$ at $r = a$ requires that $J_m\left(k_\perp a\right)$ be equal to zero. For each function J_m, there is an infinite sequence of zeros $j_{m,n}$ with $n = 1, 2, \ldots$, such that $J_m\left(j_{m,n}\right) = 0$. Hence $k_\perp = j_{m,n}/a$ for a given choice of m, n, and we can express the longitudinal wavenumber $\varkappa$ of that mode with the indices m, n in terms of the frequency as

$$\varkappa_{m,n} = \pm\left(\frac{\omega^2}{c^2} - \frac{j_{m,n}^2}{a^2}\right)^{1/2}. \tag{14.4}$$

We see from this equation that in order for a mode to have a real value of $\varkappa_{m,n}$, its frequency should be larger than the *cut-off frequency* $cj_{m,n}/a$. The term 'cut-off frequency' is generally used to refer to frequencies where the wave number goes to zero. The positive values of $\varkappa_{m,n}$ describe modes propagating in the direction of the z axis (assuming $\omega > 0$), and the negative ones correspond to modes propagating in the opposite direction. For $\omega < cj_{m,n}/a$ we deal with *evanescent* modes that, depending on the sign of Im $\varkappa_{m,n}$ exponentially decay or grow along the z-axis. Such modes play an important role in the formation of localized fields around obstacles inside a waveguide.

Given $E_z(r, \phi, z, t)$ defined by Eqs. (14.1) and (14.3) we can find all other components of the electric and magnetic fields using Maxwell's equations. They will all have the same dependence $e^{-i\omega t - im\phi + i\varkappa z}$ versus time, angle and z. The radial distribution of the four unknown components E_ϕ, E_r, B_ϕ and B_r (we recall that $B_z = 0$) are found from the four algebraic equations, which are the r and ϕ components of the two vector equations $\nabla \times \mathbf{E} = i\omega\mathbf{B}$ and $c^2\nabla \times \mathbf{B} = -i\omega\mathbf{E}$. A straightforward calculation yields:

$$E_r = E_0\frac{i\varkappa_{m,n}a}{j_{m,n}}J_m'\left(j_{m,n}\frac{r}{a}\right)e^{-i\omega t - im\phi + i\varkappa_{m,n}z}, \tag{14.5a}$$

$$E_\phi = -E_0\frac{m\varkappa_{m,n}a^2}{rj_{m,n}^2}J_m\left(j_{m,n}\frac{r}{a}\right)e^{-i\omega t - im\phi + i\varkappa_{m,n}z}, \tag{14.5b}$$

$$B_r = E_0\frac{m\omega a^2}{c^2rj_{m,n}^2}J_m\left(j_{m,n}\frac{r}{a}\right)e^{-i\omega t - im\phi + i\varkappa_{m,n}z}, \tag{14.5c}$$

$$B_\phi = E_0\frac{i\omega a}{c^2j_{m,n}}J_m'\left(j_{m,n}\frac{r}{a}\right)e^{-i\omega t - im\phi + i\varkappa_{m,n}z}, \tag{14.5d}$$

where J'_m is the derivative of the Bessel function of order m with respect to its argument. These modes are designated as TM_{mn}. Note that in addition to vanishing E_z on the wall, which we have satisfied by choosing $k_\perp = j_{m,n}/a$, we should also make sure that the other tangential component, E_ϕ, is equal to zero at $r = a$. This is indeed satisfied because the radial dependence of E_ϕ in (14.5b) is the same as E_z in (14.3).

Of course, only the real parts of Eqs. (14.1) and (14.5) have physical meaning. Since the longitudinal wavenumbers (14.4) do not depend on the sign of m, the modes with positive and negative values of m are degenerate—they have the same values of $\varkappa_{m,n}$. Often the sum and difference of m and $-m$ modes, which convert $e^{im\phi}$ and $e^{-im\phi}$ into $\cos m\phi$ and $\sin m\phi$, are used as an alternative choice for the set of fundamental eigenmodes in cylindrical waveguides.

14.2 TE Modes in Cylindrical Waveguides

Transverse electric modes, or TE modes, have $E_z = 0$ and nonzero longitudinal magnetic field B_z. They can be derived following closely the derivation of TM modes in the previous section. However, a simple observation of a special symmetry of Maxwell's equations allows one to obtain the fields in TE modes without an additional derivation.

Assuming the time dependence $\propto e^{-i\omega t}$ for all fields, Maxwell's equations in free space take the following form:

$$\nabla \times \boldsymbol{E} = i\omega\boldsymbol{B}, \quad c^2\nabla \times \boldsymbol{B} = -i\omega\boldsymbol{E}, \quad \nabla \cdot \boldsymbol{E} = 0, \quad \nabla \cdot \boldsymbol{B} = 0. \quad (14.6)$$

Note that the transformation

$$(\boldsymbol{E}, \boldsymbol{B}) \to (c\boldsymbol{B}, -\boldsymbol{E}/c) \quad (14.7)$$

does not change Eq. (14.6), or in other words Maxwell's equations are *invariant* with respect to this transformation. This means that having found any solution of Maxwell's equation one can obtain a second solution by means of the simple substitution (14.7), if the new solution satisfies proper boundary conditions. In the derivation of TM modes we satisfied the boundary condition of zero tangential electric field on the wall by choosing $k_\perp = j_{m,n}/a$. The boundary conditions for TE modes is different, so we replace $j_{m,n}$ in Eq. (14.5) by yet unknown numbers $j'_{m,n}$ and carry out the transformation (14.7) (also replacing E_0 by cB_0),

$$B_z = B_0 J_m \left(j'_{m,n} \frac{r}{a} \right) e^{-i\omega t - im\phi + i\varkappa_{m,n} z} ,$$

$$B_r = B_0 \frac{i \varkappa_{m,n} a}{j_{m,n}} J'_m \left(j'_{m,n} \frac{r}{a} \right) e^{-i\omega t - im\phi + i\varkappa_{m,n} z} ,$$

$$B_\phi = -B_0 \frac{m \varkappa_{m,n} a^2}{r j^2_{m,n}} J_m \left(j'_{m,n} \frac{r}{a} \right) e^{-i\omega t - im\phi + i\varkappa_{m,n} z} ,$$

$$E_r = -B_0 \frac{m \omega a^2}{r j^2_{m,n}} J_m \left(j'_{m,n} \frac{r}{a} \right) e^{-i\omega t - im\phi + i\varkappa_{m,n} z} ,$$

$$E_\phi = -B_0 \frac{i \omega a}{j_{m,n}} J'_m \left(j'_{m,n} \frac{r}{a} \right) e^{-i\omega t - im\phi + i\varkappa_{m,n} z} . \tag{14.8}$$

These modes are designated by TE$_{mn}$. The only tangential component of the electric field on the wall in TE modes is E_ϕ and in order for it to be equal to zero at $r = a$ we require

$$J'_m \left(j'_{m,n} \right) = 0 , \tag{14.9}$$

which means that $j'_{m,n}$ are the roots of the derivative J'_m of the Bessel function.

14.3   RF Modes in Cylindrical Resonators

A cylindrical resonator is a cylindrical waveguide with the ends closed by metallic walls perpendicular to its axis. The waveguide modes described in the previous section are now reflected by the end walls and become trapped in the resonator. The resonator modes can be obtained as linear combinations of waveguide modes propagating in the opposite directions.

In comparison with waveguides, a resonator has two additional boundary conditions—the vanishing tangential electric field on the end walls. Let us assume that the left wall is located at $z = 0$, and the right wall is at $z = L$, where L is the resonator length. We start with TM modes. To satisfy the boundary condition $E_r = E_\phi = 0$ at $z = 0$ we choose two waveguide modes with the same frequency and the same m and n indices but opposite values of $\varkappa_{m,n}$ (that is two identical waves propagating in the opposite directions), add them and divide the result by 2. With the help of the relations

$$\frac{1}{2}(e^{i\varkappa_{m,n} z} + e^{-i\varkappa_{m,n} z}) = \cos(\varkappa_{m,n} z), \tag{14.10}$$

$$\frac{1}{2}(\varkappa_{m,n} e^{i\varkappa_{m,n} z} - \varkappa_{m,n} e^{-i\varkappa_{m,n} z}) = i \varkappa_{m,n} \sin(\varkappa_{m,n} z) ,$$

it is easy to see that both E_r and E_ϕ acquire the factor $\sin(\varkappa_{m,n} z)$ and hence satisfy the boundary condition at $z = 0$. To satisfy the boundary condition at the opposite wall, $z = L$, we require $\varkappa_{m,n} L = l\pi$, where $l = 0, 1, 2, \ldots$, is an integer number. As a result we obtain the following expressions for the fields:

$$E_z = E_0 J_m \left(j_{m,n} \frac{r}{a} \right) \cos \left(\frac{l\pi z}{L} \right) e^{-i\omega t - im\phi} , \tag{14.11}$$

$$E_r = -E_0 \frac{l\pi a}{L j_{m,n}} J_m' \left(j_{m,n} \frac{r}{a} \right) \sin \left(\frac{l\pi z}{L} \right) e^{-i\omega t - im\phi} ,$$

$$E_\phi = -E_0 \frac{iml\pi a^2}{Lr j_{m,n}^2} J_m \left(j_{m,n} \frac{r}{a} \right) \sin \left(\frac{l\pi z}{L} \right) e^{-i\omega t - im\phi} ,$$

$$B_r = E_0 \frac{m\omega a^2}{c^2 r j_{m,n}^2} J_m \left(j_{m,n} \frac{r}{a} \right) \cos \left(\frac{l\pi z}{L} \right) e^{-i\omega t - im\phi} ,$$

$$B_\phi = E_0 \frac{i\omega a}{c^2 j_{m,n}} J_m' \left(j_{m,n} \frac{r}{a} \right) \cos \left(\frac{l\pi z}{L} \right) e^{-i\omega t - im\phi} .$$

Because $\varkappa_{m,n}$ is now determined by the boundary conditions on the end walls, Eq. (14.4) should now be solved for ω: replacing $\varkappa_{m,n}$ by $l\pi/L$, we find for the frequency ω of the mode:

$$\frac{\omega}{c} = \pm \left[\left(\frac{l\pi}{L} \right)^2 + \frac{j_{m,n}^2}{a^2} \right]^{1/2} . \tag{14.12}$$

The frequency depends on the indices m, n, and l as well as on the waveguide dimensions. The modes defined by Eqs. (14.11) and (14.12) are called the TM$_{mnl}$ modes.

A similar derivation can be carried out with the TE waveguide modes, but instead of adding, we need to subtract the mode with a negative $\varkappa_{m,n}$ from the mode with the positive $\varkappa_{m,n}$ and divide the result by $2i$. This gives the following expressions for the fields,

$$B_z = B_0 J_m \left(j_{m,n}' \frac{r}{a} \right) \sin \left(\frac{l\pi z}{L} \right) e^{-i\omega t - im\phi} , \tag{14.13}$$

$$B_r = B_0 \frac{l\pi a}{L j_{m,n}} J_m' \left(j_{m,n}' \frac{r}{a} \right) \cos \left(\frac{l\pi z}{L} \right) e^{-i\omega t - im\phi} ,$$

$$B_\phi = B_0 \frac{iml\pi a^2}{Lr j_{m,n}^2} J_m \left(j_{m,n}' \frac{r}{a} \right) \cos \left(\frac{l\pi z}{L} \right) e^{-i\omega t - im\phi} ,$$

$$E_r = -B_0 \frac{m\omega a^2}{r j_{m,n}^2} J_m \left(j_{m,n}' \frac{r}{a} \right) \sin \left(\frac{l\pi z}{L} \right) e^{-i\omega t - im\phi} ,$$

$$E_\phi = -B_0 \frac{i\omega a}{j_{m,n}} J_m' \left(j_{m,n}' \frac{r}{a} \right) \sin \left(\frac{l\pi z}{L} \right) e^{-i\omega t - im\phi} .$$

The frequency of the mode is

$$\frac{\omega}{c} = \pm \left[\left(\frac{l\pi}{L} \right)^2 + \left(\frac{j'_{m,n}}{a} \right)^2 \right]^{1/2} . \tag{14.14}$$

The modes given by these equations are called the TE$_{mnl}$ modes. Note that in TE modes l can not be equal to zero.

An important mode characteristic is the energy W of the electromagnetic field in the mode of a given amplitude. This energy is obtained by integrating $(\epsilon_0/2)(E^2 + c^2 B^2)$ over the volume of the cavity, where one has to take the real parts of the fields before squaring them.

For illustration, let us calculate the energy of the TM$_{010}$ mode. In this mode $m = l = 0$, and as follows from Eq. (14.11) only E_z and B_ϕ are not equal to zero. The calculation can be simplified if one notices that although E_z and B_ϕ depend on time, the energy W does not. Because there is a phase shift $\pi/2$ between these fields, at a moment when $B_\phi = 0$ the electric field reaches its maximum value equal to the absolute value of the complex expression $|E_z|$, and

$$\begin{aligned} W &= \frac{\epsilon_0}{2} \int dV |E_z|^2 = \frac{\epsilon_0}{2} \int_0^a 2\pi r\, dr \int_0^L dz |E_z|^2 \\ &= \frac{\epsilon_0}{2} \pi E_0^2 a^2 L J_1^2(j_{0,1}) , \end{aligned} \tag{14.15}$$

where we have used the integral $\int_0^1 x dx\, J_0^2(bx) = \frac{1}{2}[J_0^2(b) + J_1^2(b)]$ and we note that $J_0(j_{0,1}) = 0$.

In the analysis above we have assumed the perfect conductivity of the walls. In this approximation, there are no energy losses, and the modes have real frequencies. If one takes into account the finite wall conductivity, one finds that an initially excited mode decays with time because its energy is absorbed in the walls. This damping is reflected in the imaginary part γ in the mode frequency, $\omega = \omega' - i\gamma$, where ω' and γ are real and positive. A consistent derivation of the imaginary part of the frequency can be carried out by imposing the Leontovich boundary conditions on the surface of the metal instead of zero tangential electric field, as has been done in our analysis. Related to the damping is the *quality factor* Q of the cavity mode,

$$Q = \frac{\omega'}{2\gamma} . \tag{14.16}$$

Without derivation, we will give here the quality factor for the fundamental mode TM$_{010}$ of the cylindrical cavity,

$$Q = \frac{aL}{\delta(a + L)} , \tag{14.17}$$

where δ is the skin depth calculated at the frequency of the mode. For a crude estimate of the quality factor one can use $Q \sim l/\delta$, where l is a characteristic size of the cavity (assuming that all dimensions of the cavity are of the same order). Typical copper cavities used in accelerators have $Q \sim 10^4$; superconducting cavities may have quality factors as high as $Q \sim 10^{10}$.

14.4 Electromagnetic Field Pressure

An electromagnetic field confined by material surfaces exerts a force on these surfaces. In the most general formulation, this force can be derived from the so-called *Maxwell stress tensor*, see [1], Sect. 6.7. In this section, we will give a simplified derivation of this force for the case of a metal boundary.

If electric field lines are terminated on a metal plate as shown in Fig. 14.1, there are image charges on the surface of the metal with a surface charge density equal to $\epsilon_0 E_n$, where the subscript n indicates that the field is normal to the metal wall. To calculate the force, we need to consider in more detail the distribution of the electric field inside the metal. Let us assume that $z = 0$ corresponds to the surface of the metal and the metal occupies the region $z > 0$. Denote by $\rho(z)$ the charge density and $E_z(z)$ the electric field inside the metal. From the Maxwell equation $\nabla \cdot \boldsymbol{E} = \rho/\epsilon_0$ we obtain

$$\frac{dE_z}{dz} = \frac{\rho(z)}{\epsilon_0} . \tag{14.18}$$

The force per unit area of the metal is given by the integral

$$f_z^{(E)} = \int_0^\infty dz \, \rho E_z . \tag{14.19}$$

To take this integral, we express ρ from Eq. (14.18) to obtain

Fig. 14.1 Electric field lines are normal to the metal surface where they are terminated

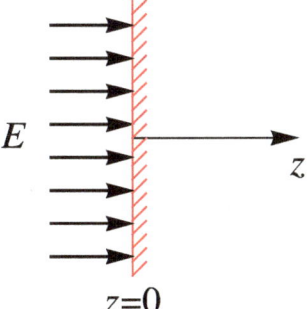

Fig. 14.2 Magnetic field
lines are parallel to the
lateral metal surface

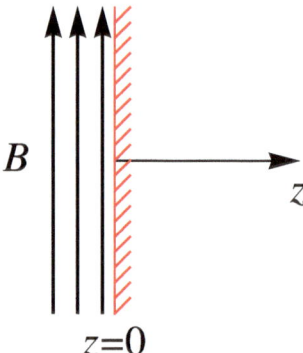

$$f_z^{(E)} = \epsilon_0 \int_0^\infty dz\, E_z \frac{dE_z}{dz} = -\frac{\epsilon_0}{2} E_n^2 , \qquad (14.20)$$

where we integrated by parts noting that $E_z\, dE_z/dz = (d/dz)E_z^2/2$ and took into
account that deep inside the metal the field vanishes, $E_z(\infty) = 0$, while on the
surface $E_z(0) = E_n$. The minus sign in this equation means that the electric field has
a "negative pressure"—it pulls the surface in the direction of free space. Remarkably,
the details of the charge distribution in Eq. (14.18) dropped out from the final result.

Similarly to the electric field, a magnetic field near a metal wall also exerts a
force on the metal, but the field is tangential to the surface as shown in Fig. 14.2. To
compute the force, we assume that the magnetic field $B_y(z)$ is directed along y, and
varies along z due to the current $j_x(z)$ flowing in the x direction inside the metal.
From the Maxwell equation

$$\frac{dH_y}{dz} = -j_x , \qquad (14.21)$$

with the expression for the force per unit area

$$f_z^{(M)} = \int_0^\infty dz\, j_x B_y , \qquad (14.22)$$

we find

$$f_z^{(M)} = -\int_0^\infty dz\, B_y \frac{dH_y}{dz} = \frac{1}{2\mu_0} B_t^2 , \qquad (14.23)$$

where we took into account that $B_y(\infty) = 0$, and on the surface $B_y(0) = B_t$ (we
neglect the magnetic properties of the metal and assume $\mu = \mu_0$). We see that $f_z^{(M)}$
is positive—it acts as a real pressure applied to the surface.

The effect of the electromagnetic pressure is usually small, however it causes a so-called *Lorentz detuning* in modern superconducting cavities as the resulting deformation increases with increasing amplitude of the mode, which should be compensated by a special control system (see [2], p. 580).

14.5   Slater's Formula

In practice, resonant cavities have complicated shapes and their resonant frequencies are calculated numerically using computer codes. In many cases it is important to be able to estimate how small deformations of the cavity shape, either due to manufacturing errors or from temporary distortions created, for example, by temperature changes, affect the frequency of the cavity mode. The resonant frequency can also be tuned by small amounts by applying an external force to shift the cavity walls. The electromagnetic forces derived in the previous section allow us to calculate the change in the frequency of a mode if the cavity shape is slightly distorted as shown in Fig. 14.3.

Let us assume that a mode with frequency ω is excited in the cavity, and the cavity walls are being slowly moved with the electromagnetic field oscillating inside. Due to the pressure of the electromagnetic field, to change the cavity shape requires some work, dA, (which can be positive or negative) to act against the pressure forces. From the energy balance, this work changes the electromagnetic energy of the mode, $\delta W = dA$. Since the distortions of the shape are assumed small, to calculate the force we can take the unperturbed distribution of the electric and magnetic fields on the walls, compute the sum of the electric and magnetic pressures $f_z^{(E)} + f_z^{(M)}$ and average them over the period of oscillations. This averaging introduces a factor $\frac{1}{2}$ in Eqs. (14.20) and (14.23). We then multiply the force by the displacement h of the surface, and integrate it over the area of the dent,

$$\delta W = \frac{1}{2} \int_S h \, dS \left(\frac{1}{2\mu_0} |B_t|^2 - \frac{\epsilon_0}{2} |E_n|^2 \right), \tag{14.24}$$

Fig. 14.3 Initial (black line) and distorted (blue line with shaded interior) cavity shapes

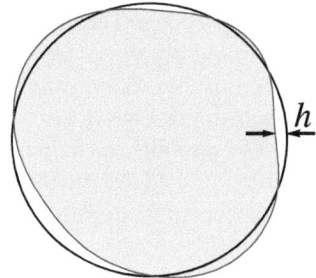

where h is assumed positive in the case when the volume of the cavity decreases, and negative otherwise. The quantities B_t and E_n are understood as the amplitude values of the fields on the surface. A positive (negative) value of δW means that the electromagnetic energy in the mode increases (decreases).

The easiest way to relate the change of the frequency of the mode to the energy change δW, is to think about the modes as a collection of photons of frequency ω in the cavity. The number of quanta of the electromagnetic field in the cavity does not change if the cavity reshaping occurs adiabatically slow. Because the photon energy is $\hbar\omega$, this number is proportional to the ratio of the electromagnetic energy to the frequency, W/ω, and by requiring $W/\omega = \text{const}$ we conclude that

$$
\frac{\delta\omega}{\omega} = \frac{\delta W}{W} \,. \tag{14.25}
$$

Using Eq. (14.24) we obtain

$$
\frac{\delta\omega}{\omega} = \frac{\epsilon_0}{4W} \int_{\Delta V} dV \left(c^2 |B_t|^2 - |E_n|^2 \right) , \tag{14.26}
$$

where the integration in the numerator goes over the volume of the dent. Note that the right-hand side of this formula does not depend on the amplitude of the mode: the integral is proportional to the amplitude of the field squared, but this dependence is canceled by the energy in the denominator that also scales as the square of the field. Equation (14.26) is often called *Slater's formula*.

As was emphasized in the derivation, for Eq. (14.26) to be valid, the change in the shape of the cavity should be small. In addition to this requirement, we also add that the perturbation should be smooth—otherwise the field variation near sharp edges of the dent would be large, and one could not use the unperturbed fields of the mode in the calculation of the energy (14.24).

14.6   Excitation of a Cavity Mode by a Beam

In accelerators, cavity resonators are predominantly used to accelerate beams of charged particles. The accelerating mode is excited by an external source that feeds radio-frequency power into the cavity through a coupling in the cavity wall. In addition to this controlled external excitation, a beam of accelerating particles passing through the resonator also contributes to the excitation of the mode. This effect, besides presenting a challenge in controlling the fields in accelerating cavities, can also be exploited to form the basis of cavity beam position monitors, which are highly accurate beam diagnostic tools [3].

   In this section we will calculate the mode excitation by a point charge using a method originally proposed by P. Wilson in Ref. [4]. We will assume that the charge moves with velocity close to the speed of light and use the approximation $v = c$. Our derivation is based on the principles of superposition and conservation of energy. Assume that a point charge q enters an empty cavity at time $t = 0$, and is moving along the z axis, $z = ct$. We represent the *real* longitudinal component of the electric field of the mode under consideration as

$$E_z(z) = E_0 e(z), \tag{14.27}$$

where E_0 is the amplitude and $e(z)$ gives the distribution of the field in the mode along the z axis at a given time. Note that the function $e(z)$, being a solution of the homogeneous wave equation, is defined with an arbitrary normalization factor. We will choose this normalization factor so that the total electromagnetic energy W of the mode is equal to E_0^2,

$$W = E_0^2. \tag{14.28}$$

When the particle passes through the cavity, the amplitude of the mode E_0 varies with time, $E_0(t)$, starting from an initial zero value, $E_0(0) = 0$. As the particle moves from $z = ct$ to $z + dz = ct + cdt$ due to the interaction with the field of the mode it changes its amplitude by an infinitesimal value dE_0. We can find dE_0 from energy conservation: the change in the mode energy dW is equal in magnitude, but opposite in sign, to the work of the electric field of the mode on the charge,

$$dW = -q E_z dz = -q E_0 e(ct) c\, dt. \tag{14.29}$$

On the other hand, from Eq. (14.28) we have $dW = 2E_0 dE_0$. Equating this expression to Eq. (14.29) we find the change in the mode amplitude,

$$dE_0 = -\frac{q}{2} e(ct) c\, dt. \tag{14.30}$$

To find the final electric field E_0 at the moment when the charge exits the cavity at $z = L$, $t_0 = L/c$, we need to add all the infinitesimal contributions (14.30) along the particle orbit. It is important to take into account that after an infinitesimal amplitude dE_0 is excited at time t it oscillates with the frequency of the mode ω and at time t_0 evolves to $dE_0 \exp[-i\omega(t_0 - t)]$ (in this formula we implicitly assume that the excited field starts the oscillations with a zero phase). Integrating equation (14.30) with proper phase factors we find

$$E_0(t_0) = -\frac{qc}{2} \int_0^{L/c} dt\, e^{-i\omega(t_0 - t)} e(ct) = -\frac{qV}{2} e^{-i\omega t_0}, \tag{14.31}$$

where

$$V = \int_0^L dz e^{i\omega z/c} e(z) \,. \tag{14.32}$$

An important characteristic of the mode excitation is the total energy deposited by the charge to the cavity. This energy, normalized per unit charge, is called the *loss factor*. Since the final amplitude of the field (14.31) is a complex number, for the mode energy we need to replace the square of the field in Eq. (14.28) with the absolute value of the field, $W = |E_0(t_0)|^2$, which gives for the loss factor k_{loss}

$$k_{loss} = \frac{1}{q^2}|E_0(t_0)|^2 = \frac{|V^2|}{4} \,. \tag{14.33}$$

Equation (14.33) was derived assuming a special normalization of the mode (14.28). We can drop this constraint if we redefine V by replacing $e(z)$ on the right-hand side of Eq. (14.32) by $E_z(z)$. This introduces an extra factor E_0 into V which we cancel by putting $E_0^2 = W$ into the denominator of the right-hand side of Eq. (14.33),

$$k_{loss} = \frac{1}{4W} \left| \int_0^L dz e^{i\omega z/c} E_z(z) \right|^2 \,. \tag{14.34}$$

The advantage of this formula is that now k_{loss} does not depend on the normalization of $E_z(z)$: multiplying $E_z(z)$ by a factor of λ adds a factor of λ^2 to both the numerator and denominator of the fraction and cancel each other.

Worked Examples

Problem 14.1 *Calculate the TM modes in a rectangular waveguide with cross section $a \times b$.*

Solution: Choosing the coordinate system so that the waveguide area occupies the region $0 < x < a$, $0 < y < b$, we require $E_z = 0$ on the surfaces $x = 0$, $x = a$, $y = 0$, and $y = b$. We can use simple plane waves, and choose the horizontal and vertical terms to be proportional to sin functions with appropriate k_x and k_y:

$$E_z(x, y, z, t) = E_0 \sin\left(\frac{m\pi x}{a}\right) \sin\left(\frac{n\pi y}{b}\right) e^{-i\omega t + i\varkappa z} \,.$$

The dispersion relation is

$$\varkappa_{m,n} = \pm \left(\frac{\omega^2}{c^2} - \frac{m^2\pi^2}{a^2} - \frac{n^2\pi^2}{b^2}\right)^{1/2} \,.$$

We calculate the horizontal electric field, first by taking the horizontal component of $c^2 \nabla \times \boldsymbol{B} = -i\omega \boldsymbol{E}$, and recalling that $B_z \equiv 0$:

$$-i\omega E_x = c^2 \left(\frac{\partial B_z}{\partial y} - \frac{\partial B_y}{\partial z} \right) = -c^2 \frac{\partial B_y}{\partial z} .$$

Then, from the vertical component of $\nabla \times \boldsymbol{E} = i\omega \boldsymbol{B}$,

$$i\omega B_y = \left(\frac{\partial E_x}{\partial z} - \frac{\partial E_z}{\partial x} \right) .$$

Combining these expressions yields

$$\left(\frac{\omega^2}{c^2} + \frac{\partial^2}{\partial z^2} \right) E_x = \frac{\partial^2 E_z}{\partial z \, \partial x} .$$

Using the initial expression for E_z and assuming that the horizontal component has the same z-dependence, $E_x \propto e^{i \varkappa_{m,n} z}$, we find that

$$\left(\frac{\omega^2}{c^2} - \varkappa_{m,n}^2 \right) E_x = \frac{\partial^2 E_z}{\partial x \partial z} = i \varkappa_{m,n} \frac{m\pi}{a} E_0 \cos \left(\frac{m\pi x}{a} \right) \sin \left(\frac{n\pi y}{b} \right) .$$

A similar calculation applies for E_y, or we can simply use $\partial E_y / \partial x - \partial E_x / \partial y = i\omega B_z = 0$. The result for all of the other components is

$$E_x = i E_0 \varkappa_{m,n} \frac{m\pi/a}{(m\pi/a)^2 + (n\pi/b)^2} \cos \left(\frac{m\pi x}{a} \right) \sin \left(\frac{n\pi y}{b} \right) e^{-i\omega t + i\varkappa z} ,$$

$$E_y = i E_0 \varkappa_{m,n} \frac{n\pi/b}{(m\pi/a)^2 + (n\pi/b)^2} \sin \left(\frac{m\pi x}{a} \right) \cos \left(\frac{n\pi y}{b} \right) e^{-i\omega t + i\varkappa z} ,$$

$$B_x = -i E_0 \frac{\omega}{c^2} \frac{n\pi/b}{(m\pi/a)^2 + (n\pi/b)^2} \sin \left(\frac{m\pi x}{a} \right) \cos \left(\frac{n\pi y}{b} \right) e^{-i\omega t + i\varkappa z} ,$$

$$B_y = i E_0 \frac{\omega}{c^2} \frac{m\pi/a}{(m\pi/a)^2 + (n\pi/b)^2} \cos \left(\frac{m\pi x}{a} \right) \sin \left(\frac{n\pi y}{b} \right) e^{-i\omega t + i\varkappa z} .$$

Problem 14.2 *Follow up on Problem 14.1 and derive TE modes in a rectangular waveguide by applying the transformation of Eq. (14.7) to TM modes and satisfying the boundary conditions on the wall.*

Solution: We will write B_z in the same form as E_z, but add as-yet unknown phase terms ϕ_x and ϕ_y. Together with the transformations (14.7) we find

$$B_z = B_0 \sin\left(\frac{m\pi x}{a} + \phi_x\right) \sin\left(\frac{n\pi y}{b} + \phi_y\right) e^{-i\omega t + i \varkappa z},$$

$$B_x = i B_0 \varkappa_{m,n} \frac{m\pi/a}{(m\pi/a)^2 + (n\pi/b)^2} \cos\left(\frac{m\pi x}{a} + \phi_x\right) \sin\left(\frac{n\pi y}{b} + \phi_y\right) e^{-i\omega t + i \varkappa z},$$

$$B_y = i B_0 \varkappa_{m,n} \frac{n\pi/b}{(m\pi/a)^2 + (n\pi/b)^2} \sin\left(\frac{m\pi x}{a} + \phi_x\right) \cos\left(\frac{n\pi y}{b} + \phi_y\right) e^{-i\omega t + i \varkappa z},$$

$$E_x = i B_0 \omega \frac{n\pi/b}{(m\pi/a)^2 + (n\pi/b)^2} \sin\left(\frac{m\pi x}{a} + \phi_x\right) \cos\left(\frac{n\pi y}{b} + \phi_y\right) e^{-i\omega t + i \varkappa z},$$

$$E_y = -i B_0 \omega \frac{m\pi/a}{(m\pi/a)^2 + (n\pi/b)^2} \cos\left(\frac{m\pi x}{a} + \phi_x\right) \sin\left(\frac{n\pi y}{b} + \phi_y\right) e^{-i\omega t + i \varkappa z}.$$

The boundary conditions for the electric field are: $E_x = 0$ when $y = 0$ or $y = b$, and $E_y = 0$ when $x = 0$ or $x = a$. This requires both ϕ_x and ϕ_y to be $\pi/2$ (or $-\pi/2$ but that just changes the overall sign of all the fields). The resulting field terms are:

$$B_z = B_0 \cos\left(\frac{m\pi x}{a}\right) \cos\left(\frac{n\pi y}{b}\right) e^{-i\omega t + i \varkappa z},$$

$$B_x = -i B_0 \varkappa_{m,n} \frac{m\pi/a}{(m\pi/a)^2 + (n\pi/b)^2} \sin\left(\frac{m\pi x}{a}\right) \cos\left(\frac{n\pi y}{b}\right) e^{-i\omega t + i \varkappa z},$$

$$B_y = -i B_0 \varkappa_{m,n} \frac{n\pi/b}{(m\pi/a)^2 + (n\pi/b)^2} \cos\left(\frac{m\pi x}{a}\right) \sin\left(\frac{n\pi y}{b}\right) e^{-i\omega t + i \varkappa z},$$

$$E_x = -i B_0 \omega \frac{n\pi/b}{(m\pi/a)^2 + (n\pi/b)^2} \cos\left(\frac{m\pi x}{a}\right) \sin\left(\frac{n\pi y}{b}\right) e^{-i\omega t + i \varkappa z},$$

$$E_y = i B_0 \omega \frac{m\pi/a}{(m\pi/a)^2 + (n\pi/b)^2} \sin\left(\frac{m\pi x}{a}\right) \cos\left(\frac{n\pi y}{b}\right) e^{-i\omega t + i \varkappa z}.$$

Problem 14.3 *Consider a point charge passing through a cylindrical cavity where the fundamental mode is excited with amplitude E_0. Calculate the maximum energy gain for the charge.*

Solution: The fundamental mode for a cylindrical cavity is the TM_{010} mode, which has a physical electric field given by

$$E_z = E_0 J_0\left(\frac{\omega r}{c}\right) \cos(\omega t - \phi),$$

where the field amplitude E_0 is assumed to be positive. This field is localized inside the cavity, $0 < z < L$. The charged particle moves on a trajectory $z = v(t - t_0)$, where t_0 is the time when the particle enters the cavity. We set $r = 0$ where the Bessel function is unity and assume that the velocity does not change significantly during the acceleration.

The instantaneous force on the particle is $q E_z$, so the total energy gained across the whole cavity is $W = \int dz \, q E_z$. Thus we obtain

$$W = q \int_0^L dz \, E_z = q E_0 \int_0^L dz \, \cos(\omega z/v + \omega t_0 - \phi)$$

$$= 2q E_0 \frac{v}{\omega} \sin\left(\frac{L\omega}{2v}\right) \cos\psi \,,$$

where we defined $\psi = \omega t_0 - \phi + \omega L/2v$.

As a function of ψ, the maximum energy gain occurs when $\psi = 0$ for positive q, or when $\psi = \pi$ for negative q. As a function of the cavity length L, the maximum possible energy gain in one cavity is $2|q|E_0 v/\omega$ and occurs if $L = \pi v/\omega$. The most effective rate of acceleration comes from using multiple cavities, with phases adjusted from cavity to cavity.

Problem 14.4 *The radius of a cylindrical cavity is changed by a small quantity δa, and the length is changed by δL. Consider this as a deformation of the cavity shape and find the frequency change of the fundamental mode TM_{010} in the cavity using Slater's formula. Verify that the result agrees with Eq. (14.12).*

Solution: As discussed in this chapter, during an adiabatic change to a cavity the number of photons should remain constant, so $\delta\omega/\omega = \delta W/W$. We calculate δW resulting from expanding the volume of a cylinder by δa (radius) and δL (length). The radial expansion has no contribution from the E field, because there is no E field normal to the radial walls. So for the E field we get only a contribution from the longitudinal term, given by the integral over the surface area, multiplied by the length of expansion, $h = -\delta L$. We may assume that only one side of cylinder, with surface S_1 at a fixed longitudinal position, is shifted by the full amount δL to make the cavity longer. By definition, h is negative for an expansion (because we are calculating work done *to* the field), so we find

$$\delta W_E = -\frac{\epsilon_0}{4}(-\delta L) \int_{S_1} dS \, |E_z(r)|^2$$

$$= \frac{\epsilon_0}{4} \delta L \int_0^a 2\pi r dr \, E_0^2 J_0^2(\omega r/c)$$

$$= \frac{\epsilon_0}{4} \delta L \pi E_0^2 a^2 J_1^2(\omega a/c) \,,$$

following the derivation of Eq. (14.15) and setting the frequency $\omega = j_{0,1}c/a$. Then for the B field, $B_\phi = E_0(i/c)J_1(\omega r/c)$, we have contributions from both δL and δa because the B field is tangential to the surface everywhere:

$$\delta W_B = \frac{\epsilon_0 c^2}{4}\left[(-\delta L)\int_{S_1} dS \, |B_\phi(r)|^2\right] + (-\delta a)\int_{S_2} dS \, |B_\phi(r)|^2$$

$$= -\frac{\epsilon_0 c^2}{4}\left[\delta L \int_0^a 2\pi r dr \, E_0^2 J_1^2(\omega r/c)/c^2 + 2\pi a \delta a L E_0^2 J_1^2(\omega a/c)/c^2\right]$$

$$= -\frac{\epsilon_0}{4} \delta L \pi E_0^2 a^2 J_1^2(\omega a/c) - \frac{\epsilon_0}{4} 2\pi a \delta a L E_0^2 J_1^2(\omega a/c) \,,$$

where S_2 is the radial boundary of the cavity, $r = a$, so the magnetic field magnitude is constant everywhere along S_2. We also used the identity $\int_0^1 x dx\, J_1^2(bx) = \frac{1}{2}[J_1^2(b) - J_0(b)J_2(b)]$.

We can see that the effect of the E field cancels the first term from the B field, so in total

$$\delta W = \delta W_E + \delta W_B = -\frac{\epsilon_0}{4} 2\pi a \delta a L E_0^2 J_1^2(j_1) \,.$$

Using for the mode energy Eq. (14.15) we find

$$\delta\omega = \frac{\delta W}{W}\omega = -\frac{c\,\delta a}{a^2} j_{0,1} \,.$$

This result agrees with the explicit formula for the frequency of the TM_{010} mode $\omega = cj_{0,1}/a$ from which it follows that $\delta\omega = -cj_{0,1}\delta a/a^2$.

Problem 14.5 *Find the loss factor for the fundamental mode TM_{010} of the cylindrical cavity. Assume $r = 0$.*

Solution: The longitudinal electric field E_z in the TM_{010} mode does not depend on z, $E_z = E_0$. The loss factor is given by $k = |V^2|/4W$, with

$$V = \int_0^L e^{i\omega z/c} E_z(z) dz$$

$$= E_0 \int_0^L e^{i\omega z/c} dz$$

$$= E_0 \frac{1}{ik}(e^{ikL} - 1) \,.$$

So $|V|^2 = 4E_0^2 \sin^2(kL/2)/k^2 = E_0^2 L^2 \text{sinc}^2(kL/2)$. The energy in the mode is given by Eq. (14.15), $W = (\epsilon_0/2)\pi E_0^2 a^2 L J_1^2(j_{0,1})$, so we find

$$k = \frac{L\,\text{sinc}^2(kL/2)}{2\pi\epsilon_0 a^2 J_1^2(j_{0,1})} \,.$$

References

1. J.D. Jackson, *Classical Electrodynamics*, 3rd edn. (Wiley, New York, 1999)
2. A.W. Chao, M. Tigner, *Handbook of Accelerator Physics and Engineering*, Third printing edition (World Scientific Publishing, Singapore, 2006)

3. Sean Walston, Stewart Boogert, Carl Chung, Pete Fitsos, Joe Frisch, Jeff Gronberg, Hitoshi Hayano, Yosuke Honda, Yury Kolomensky, Alexey Lyapin, Stephen Malton, Justin May, Douglas McCormick, Robert Meller, David Miller, Toyoko Orimoto, Marc Ross, Mark Slater, Steve Smith, Tonee Smith, Nobuhiro Terunuma, Mark Thomson, Junji Urakawa, Vladimir Vogel, David Ward, Glen White, Performance of a high resolution cavity beam position monitor system. Nucl. Instrum. Methods Phys. Res. A **578**, 1–22 (2007)
4. P. Wilson. High energy electron linacs: Applications to storage ring rf systems and linear colliders. Report SLAC-AP-2884 (Rev.), SLAC (1991)

Chapter 15
Radiation and Retarded Potentials

In this chapter, based on simple intuitive arguments we derive the Liénard–Wiechert potentials that solve the problem of the electromagnetic field of a point charge moving in free space.

15.1 Radiation Field

Let us consider a point charge that has been at rest for $t < 0$. At $t = 0$ it is rapidly accelerated to velocity v_0 and moves with this velocity from then on. The dependence of the charge velocity versus time is shown in Fig. 15.1a. Before the acceleration, the charge has a spherically symmetric Coulomb electric field with straight field lines emanating from the origin. We would like to understand how the field lines of the electric field look like after the acceleration. From the wave equation for the electromagnetic field we know that the field perturbations propagate with the speed of light. Hence, outside of the sphere of radius $r = ct$, the field "does not know" that the charge has been set in motion, and the field remains the same Coulomb field that it had been before the acceleration. Inside of the sphere, the electromagnetic field changes into the field of a moving charge described by Eq. (11.4). In a thin spherical sheath in the vicinity of the radius $r = ct$ the Coulomb field transitions from the field of a charge at rest to the field of a moving charge, as shown in Fig. 15.2a. This transitional layer expanding with the speed of light is the *radiation* electromagnetic field in this case.

If the charge is accelerated two times, at $t = 0$ and then at $t = t_1$, as shown in Fig. 15.1b, the field lines at time $t > t_1$ would look like shown in Fig. 15.2b. In addition to the original spherical shell $r = ct$, there is another one of radius $r = c(t - t_1)$ with the center at $z = v_0 t_1$, that is at the location of the charge at the moment of the second acceleration. We now have two spherical shells of the radiation field.

© Springer International Publishing AG, part of Springer Nature 2018
G. Stupakov and G. Penn, *Classical Mechanics and Electromagnetism in Accelerator Physics*, Graduate Texts in Physics,
https://doi.org/10.1007/978-3-319-90188-6_15

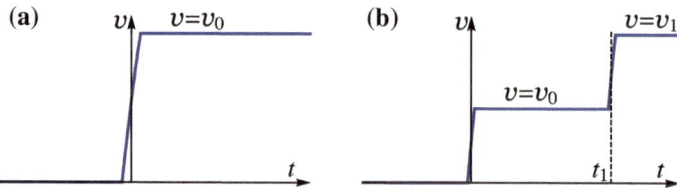

Fig. 15.1 The velocity versus time of an abruptly accelerated charge

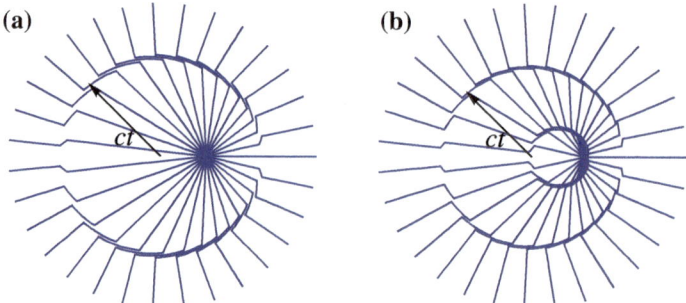

Fig. 15.2 Field lines corresponding to the acceleration shown in Fig. 15.1a, b, respectively. The spherical shell in figure **a** and the bigger sphere in figure **b** have their origin at $z = 0$. The smaller sphere in figure **b** is centered at $z = v_0 t_1$

One can now easily imagine that a continuously accelerating charge is constantly radiating spherical shells at every moment, and the shells are expanding with the speed of light eventually filling the whole space with radiation.

For a quantitative description of the radiation process, we first need to relate a point on a given expanding sphere to the time and location of the charge when this particular sphere was emitted. This time, t_{ret}, is called the *retarded* time and the position of the charge at this moment — see Fig. 15.3 — the *retarded* position. If the trajectory of the charge is given by a vector function $r_0(t)$, and we make an observation at time t at point r, then the retarded time t_{ret} is determined from the equation

$$c(t - t_{\mathrm{ret}}) = |r - r_0(t_{\mathrm{ret}})|, \qquad\qquad (15.1)$$

which equates the radius of the sphere $|r - r_0(t_{\mathrm{ret}})|$ to the time of expansion $t - t_{\mathrm{ret}}$ from the moment of the emission multiplied by the speed of light. The retarded position is $r_0(t_{\mathrm{ret}})$. Note that both t_{ret} and $r_0(t_{\mathrm{ret}})$, for a given path (determined by the function $r_0(t)$), are functions of variables t and r.

Fig. 15.3 Illustration characterizing the definition of the retarded time and retarded position. The magenta dot shows the current position of the charge, and the green dot is its retarded position. The red circle is the radiation sphere at time t corresponding to the same t_{ret}

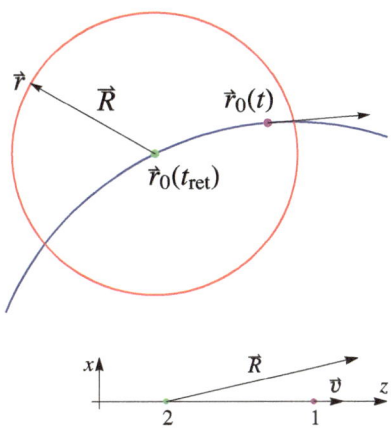

Fig. 15.4 Point charge moving with constant velocity along the z-axis. Point 1 (magenta) corresponds to the charge position at time t and point 2 (green) is its position at the retarded time. Vector R connects point 2 with the observation point

15.2 Retarded Time and Position of a Particle Moving with Constant Velocity

If a charged particle emits spherical shells of electromagnetic field in the process of acceleration, what happens if the acceleration is zero? The answer to this question is that even in this case we can formally consider the field of a particle moving with a constant velocity as a result of emission of spherical shells. We will now give a mathematical representation of this process using the expressions for the field of a moving charge from Sect. 11.1.

Consider a point charge moving with constant velocity v along the z axis. Its trajectory is given by

$$\mathbf{r}_0(t) = (0, 0, vt).\tag{15.2}$$

Let the observation point at time t have coordinates x, y, z. We denote by $\mathbf{R}$ the vector connecting the retarded position with the observation point, as shown in Fig. 15.4, $\mathbf{R} = \mathbf{r} - \mathbf{r}_0(t_{\text{ret}})$, and $\mathbf{n} = \mathbf{R}/R$. We first need to find the retarded time t_{ret}. Squaring Equation (15.1) yields

$$c^2(t - t_{\text{ret}})^2 = (z - vt_{\text{ret}})^2 + x^2 + y^2.\tag{15.3}$$

This is a quadratic equation for t_{ret} which can easily be solved. It has two solutions, one of which is an *advanced* solution with $t < t_{\text{ret}}$, the other one is our retarded solution with $t > t_{\text{ret}}$. The advanced solution should be discarded because it does not satisfy the original Equation (15.1) which requires $t > t_{ret}$; for the retarded one we have

$$c(t - t_{ret}) = \gamma^2 \left[\beta(z - vt) + \sqrt{(z - vt)^2 + \gamma^{-2}(x^2 + y^2)} \right], \qquad (15.4)$$

with $\beta = v/c$. Noticing that the square root in this expression is the same as in $\mathcal{R}$ in Eq. (11.5) and that $c(t - t_{ret}) = R$, we can try to express $\mathcal{R}$ through the vector $\boldsymbol{R}$. Let us show that

$$\mathcal{R} = R(1 - \boldsymbol{\beta} \cdot \boldsymbol{n}) = R - \boldsymbol{\beta} \cdot \boldsymbol{R}, \qquad (15.5)$$

where vector $\boldsymbol{\beta}$ is $\boldsymbol{\beta} = (0, 0, v/c)$. To prove it we take the square of Eq. (15.5),

$$\mathcal{R}^2 = (R - \boldsymbol{\beta} \cdot \boldsymbol{R})^2, \qquad (15.6)$$

or using coordinates:

$$(z - vt)^2 + \frac{x^2 + y^2}{\gamma^2} = [c(t - t_{ret}) - \beta(z - vt_{ret})]^2, \qquad (15.7)$$

where we have used $R = c(t - t_{ret})$ and $\boldsymbol{\beta} \cdot \boldsymbol{R} = \beta(z - vt_{ret})$. Substituting $x^2 + y^2 = c^2(t - t_{ret})^2 - (z - vt_{ret})^2$ from Eq. (15.3) we arrive at

$$(z - vt)^2 + \frac{c^2(t - t_{ret})^2 - (z - vt_{ret})^2}{\gamma^2} = [c(t - t_{ret}) - \beta(z - vt_{ret})]^2. \qquad (15.8)$$

Taking into account that $\gamma^{-2} = 1 - \beta^2$ it is now a straightforward calculation to check that the above equation is an identity, which proves Eq. (15.5).

The potentials (11.9) for a particle moving with a constant velocity can now be written as

$$\phi = \frac{1}{4\pi\epsilon_0} \frac{q}{R(1 - \boldsymbol{\beta} \cdot \boldsymbol{n})}, \qquad \boldsymbol{A} = \frac{Z_0}{4\pi} \boldsymbol{\beta} \frac{q}{R(1 - \boldsymbol{\beta} \cdot \boldsymbol{n})}. \qquad (15.9)$$

Remembering that R involves the retarded position of the particle, and noting that $\boldsymbol{\beta}$ does not depend on time, we can formally consider $\boldsymbol{\beta}$ as also taken at the retarded time. With this new interpretation of Eq. (15.9), not only do we maintain consistency with the previous solution for a charged particle moving with constant velocity, but it turns out that these expressions are valid for arbitrary motion of a point charge, even when the charge is being accelerated. They are known in classical electrodynamics under the name of *Liénard–Wiechert potentials*.

15.3 Liénard–Wiechert Potentials

The *Liénard–Wiechert potentials* define the electrostatic potential ϕ and the vector potential A at the observation point r at time t for arbitrary motion of the point charge:

$$\phi(r, t) = \frac{1}{4\pi\epsilon_0} \frac{q}{R(1 - \beta_{ret} \cdot n)} \, ,$$

$$A(r, t) = \frac{Z_0}{4\pi} \frac{q\beta_{ret}}{R(1 - \beta_{ret} \cdot n)} \, . \tag{15.10}$$

Here the particle velocity β should be taken at the retarded time, $\beta_{ret} = \beta(t_{ret})$, and we remind the reader that $R = r - r_0(t_{ret})$ is a vector drawn from the retarded position of the particle to the observation point, and n is a unit vector in the direction of R.

Differentiating the potentials (15.10) we can derive expressions for the electric and magnetic fields, $E = -\nabla\phi - \partial A/\partial t$, $B = \nabla \times A$. When taking the derivatives it is important to remember that $t_{ret} = t_{ret}(r, t)$ is a function of r and t and should be differentiated over these variables. As an example, we will calculate the partial derivative $\partial t_{ret}/\partial t$. We begin by squaring both sides of Eq. (15.1) and taking the time derivative of both sides of the equality:

$$\frac{\partial}{\partial t} c^2 (t - t_{ret})^2 = \frac{\partial}{\partial t} [r - r_0(t_{ret})]^2 \, , \tag{15.11}$$

which reduces to

$$-2c^2(t - t_{ret}) \left(\frac{\partial t_{ret}}{\partial t} - 1 \right) = -2(r - r_0(t_{ret})) \cdot \frac{\partial r_0}{\partial t_{ret}} \frac{\partial t_{ret}}{\partial t} \, . \tag{15.12}$$

Noting that $\partial r_0/\partial t_{ret} = \beta_{ret}$ we find

$$\frac{\partial t_{ret}}{\partial t} = \frac{1}{1 - \beta_{ret} \cdot n} \, . \tag{15.13}$$

In a similar fashion, one can calculate the spatial derivatives of t_{ret}, $c\nabla t_{ret} = -n/(1 - \beta_{ret} \cdot n)$. It is then straightforward to calculate the electric and magnetic fields from the Liénard–Wiechert potentials:

$$E = \frac{q}{4\pi\epsilon_0} \frac{n - \beta_{ret}}{\gamma^2 R^2 (1 - \beta_{ret} \cdot n)^3} + \frac{q}{4\pi\epsilon_0 c} \frac{n \times \{(n - \beta_{ret}) \times \dot{\beta}_{ret}\}}{R(1 - \beta_{ret} \cdot n)^3} \, ,$$

$$B = \frac{1}{c} n \times E \, , \tag{15.14}$$

where $\dot{\beta}_{ret}$ is the acceleration (normalized by the speed of light) taken at the retarded time, and $\gamma = (1 - \beta_{ret}^2)^{-1/2}$.

The two terms on the right-hand side of Eq. (15.14) scale differently with distance R. The first one decays as R^{-2} and is usually associated with the Coulomb field modified by the relativistic motion of the particle. It is the same field as given by Eq. (11.4), but expressed in terms of the vector R and the retarded velocity. The second one scales as inverse distance, $\propto R^{-1}$; this is the radiation field. This term is only present when a particle is being accelerated, $\dot{\beta} \neq 0$.

15.4 Radiation in the Ultra-Relativistic Limit

In the limit $\gamma \gg 1$ the behavior of the radiation from Eq. (15.14) is dominated by the $(1 - \beta_{\mathrm{ret}} \cdot n)$ term in the denominator. Denoting the angle between n and β_{ret} as θ, we can write for small angles

$$1 - \beta_{\mathrm{ret}} \cdot n \simeq 1 - \beta \left(1 - \frac{\theta^2}{2} \right) = 1 - \beta + \frac{1}{2}\beta\theta^2 \simeq \frac{1}{2}\left(\frac{1}{\gamma^2} + \theta^2 \right). \qquad (15.15)$$

The factor $(1 - \beta_{\mathrm{ret}} \cdot n)^{-3}$ makes the radiation far away from the charges localized within an angle $1/\gamma$ around the particle velocity, as shown in Fig. 15.5.

When the acceleration is perpendicular to the motion, $\dot{\beta}_{\mathrm{ret}} \perp \beta_{\mathrm{ret}}$, and the observation point is in the forward direction, $\theta = 0$, the radiation field is directed opposite to the acceleration and is equal to

$$E_{\mathrm{rad}} = -\frac{q}{4\pi\epsilon_0 c}4\dot{\beta}_{\mathrm{ret}}\frac{\gamma^4}{R}. \qquad (15.16)$$

The radiation fields at large angles, or traveling backwards relative to the particle velocity, are suppressed by a factor of $1/\gamma^4$ in comparison with Eq. (15.16).

Fig. 15.5 Expanding radiation sphere of a relativistic particle. The red area around the direction of velocity v shows high intensity of the radiation field. The magenta dot indicate the current position of the particle (1), and the green dot its retarded position (2)

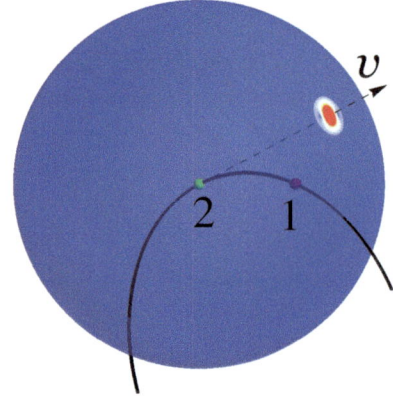

15.5 Retarded Potentials for an Ensemble of Particles

The Liénard–Wiechert potentials given by Eq. (15.10) are convenient for calculation of fields of a moving point charge. What if we are given continuous, time-dependent current and charge distributions $\rho(r, t)$ and $j(r, t)$? Can we integrate the Liénard–Wiechert potentials over all space to obtain the result for such a distribution?

Naively, one might think that to obtain the potential for a continuous distribution one has to replace the charge q by the infinitesimal charge $\rho(r', t)d^3r'$ in the elementary volume d^3r' and integrate over the space with vector R directed from r' to the observation point r, $R = r - r'$. The expression $q[R(1 - \beta_{\text{ret}} \cdot n)]^{-1}$ on the right-hand side of Eq. (15.10) would be replaced by

$$\int \frac{\rho(r', t_{\text{ret}})d^3r'}{|r - r'|(1 - \beta_{\text{ret}} \cdot n)} , \tag{15.17}$$

where $n = (r - r')/|r - r'|$ and the retarded time is now defined by the equation

$$c(t - t_{\text{ret}}) = |r - r'| . \tag{15.18}$$

This logic, however, is flawed. Indeed, if we want, using Eq. (15.17), to recover the original Liénard–Wiechert potentials for a point charge we need to do the integral with $\rho(r, t) = q\delta(r - r_0(t))$:

$$\int \frac{\delta(r' - r_0(t_{\text{ret}}(r, r', t)))}{|r - r'|(1 - \beta_{\text{ret}} \cdot n)}d^3r' = \frac{1}{(1 - v_{\text{ret}} \cdot \nabla_{r'}t_{\text{ret}})} \frac{1}{|r - r'|(1 - \beta_{\text{ret}} \cdot n)} , \tag{15.19}$$

where we have explicitly indicated the arguments of the retarded time and used standard rules for integration of the delta functions[1]. The gradient of the retarded time with respect to the vector r' can be found from Eq. (15.18), and is equal to $\nabla_{r'}t_{\text{ret}} = n/c$. Hence, we conclude that

$$\int \frac{\delta(r' - r_0(t_{\text{ret}}))d^3r'}{|r - r'|(1 - \beta_{\text{ret}} \cdot n)} = \frac{1}{|r - r_0(t_{\text{ret}})|(1 - \beta_{\text{ret}} \cdot n)^2} , \tag{15.20}$$

and we get an extra factor $(1 - \beta_{\text{ret}} \cdot n)$ in the denominator in comparison with Eq. (15.10). The origin of this discrepancy can be traced to the fact that because the second argument in ρ is t_{ret}, the integral of the charge density over the volume is not equal to the total charge q, $\int \rho(r, t_{\text{ret}})d^3r \neq q$. Hence, the replacement of charge q in the Liénard–Wiechert potentials by $\rho(r', t_{\text{ret}})d^3r'$ with the subsequent integration over the space cannot be justified.

---

[1] An integral of a three-dimensional delta function $\int \delta(f(x, y, z), g(x, y, z), h(x, y, z))dxdydz$ is equal to $|J|^{-1}$ where J is a Jacobian of the transformation from x, y, z to f, g, h taken at the point where $f = g = h = 0$.

It is now clear, however, how we can correct for this extra factor in the transition from the point charge to a continuous distribution. We just need to use the rule

$$\frac{q}{1 - \boldsymbol{\beta}_{\text{ret}} \cdot \boldsymbol{n}} \rightarrow \rho(\boldsymbol{r}', t_{\text{ret}}) d^3 r',$$

which gives

$$\phi(\boldsymbol{r}, t) = \frac{1}{4\pi\epsilon_0} \int \frac{\rho(\boldsymbol{r}', t_{\text{ret}})}{|\boldsymbol{r} - \boldsymbol{r}'|} d^3 r',$$

$$\boldsymbol{A}(\boldsymbol{r}, t) = \frac{Z_0}{4\pi c} \int \frac{\boldsymbol{j}(\boldsymbol{r}', t_{\text{ret}})}{|\boldsymbol{r} - \boldsymbol{r}'|} d^3 r'. \tag{15.21}$$

These integrals are called the *retarded* potentials. They descrive the radiation field in free space of a system of charges represented by continuous distributions of the charge density ρ and the current density j.

Worked Examples

Problem 15.1 *Show that*

$$\nabla t_{\text{ret}} = -\frac{\boldsymbol{n}}{c(1 - \boldsymbol{n} \cdot \boldsymbol{\beta}_{\text{ret}})}.$$

Find $\partial R/\partial t$ and ∇R. The operator ∇ here is understood as $\hat{\boldsymbol{x}}\partial/\partial x + \hat{\boldsymbol{y}}\partial/\partial y + \hat{\boldsymbol{z}}\partial/\partial z$.

Solution: To simplify notation, for ∇t_{ret}, we will only calculate the x component for now, using $\hat{\boldsymbol{x}} \cdot \nabla = \partial/\partial x$. We square both sides of Eq. (15.1) and differentiate it with respect to x,

$$-2Rc\frac{\partial t_{\text{ret}}}{\partial x} = 2\left[\boldsymbol{r} - \boldsymbol{r}_0(t_{\text{ret}})\right] \cdot \left(\hat{\boldsymbol{x}} - \frac{d\boldsymbol{r}_0}{dt_{\text{ret}}}\frac{\partial t_{\text{ret}}}{\partial x}\right)$$

$$= 2\left(\boldsymbol{R} \cdot \hat{\boldsymbol{x}} - \boldsymbol{R} \cdot c\boldsymbol{\beta}_{\text{ret}}\frac{\partial t_{\text{ret}}}{\partial x}\right),$$

where we used the notation $\boldsymbol{R} = \boldsymbol{r} - \boldsymbol{r}_0(t_{\text{ret}})$ and $R = |\boldsymbol{R}| = c(t - t_{\text{ret}})$. Solving this equation for $\partial t_{\text{ret}}/\partial x$ we find

$$\frac{\partial t_{\text{ret}}}{\partial x} = \frac{-\hat{\boldsymbol{x}} \cdot \boldsymbol{n}}{c(1 - \boldsymbol{\beta} \cdot \boldsymbol{n})}.$$

The vertical and longitudinal gradients are calculated analogously but with $\hat{x} \cdot \boldsymbol{n}$ replaced with $\hat{y} \cdot \boldsymbol{n}$ and $\hat{z} \cdot \boldsymbol{n}$ respectively. Therefore

$$\nabla t_{\mathrm{ret}} = -\frac{\boldsymbol{n}}{c(1 - \boldsymbol{\beta}_{\mathrm{ret}} \cdot \boldsymbol{n})} \, .$$

The derivatives $\partial R / \partial t$ and ∇R are now easily found:

$$\begin{aligned}
\frac{\partial R}{\partial t} &= c\frac{\partial}{\partial t}(t - t_{\mathrm{ret}}) = c\left(1 - \frac{\partial t_{\mathrm{ret}}}{\partial t}\right) \\
&= c\left(1 - \frac{1}{1 - \boldsymbol{\beta}_{\mathrm{ret}} \cdot \boldsymbol{n}}\right) = c\frac{\boldsymbol{\beta}_{\mathrm{ret}} \cdot \boldsymbol{n}}{1 - \boldsymbol{\beta}_{\mathrm{ret}} \cdot \boldsymbol{n}} \, ,
\end{aligned}$$

and

$$\nabla R = -c\nabla t_{\mathrm{ret}} = \frac{\boldsymbol{n}}{1 - \boldsymbol{\beta}_{\mathrm{ret}} \cdot \boldsymbol{n}} \, .$$

Problem 15.2 *Find the angular dependence of the radiation electric field $\boldsymbol{E}_{\mathrm{rad}}$ for a relativistic particle assuming that $\dot{\boldsymbol{\beta}}_{\mathrm{ret}}$ is collinear to $\boldsymbol{\beta}_{\mathrm{ret}}$. Consider small angles between $\boldsymbol{n}$ and $\boldsymbol{\beta}_{\mathrm{ret}}$.*

Solution: The radiation field is given by the second term in the expression for the electric field in Eq. (15.14),

$$\boldsymbol{E}_{\mathrm{rad}} = \frac{q}{4\pi\epsilon_0 c}\frac{\boldsymbol{n} \times \{(\boldsymbol{n} - \boldsymbol{\beta}) \times \dot{\boldsymbol{\beta}}\}}{R(1 - \boldsymbol{\beta} \cdot \boldsymbol{n})^3} \, , \tag{15.22}$$

where to simplify the notation we dropped the subscript $_{\mathrm{ret}}$. Choosing the coordinate system with axis z directed along the velocity we have $\boldsymbol{\beta} = \hat{z}\beta$, $\dot{\boldsymbol{\beta}} = \hat{z}\dot{\beta}$ and $\boldsymbol{n} = \hat{x}n_x + \hat{y}n_y + \hat{z}n_z$. In the spherical coordinate system with the angles θ and ϕ we can also write $n_x = \sin(\theta)\cos(\phi)$, $n_y = \sin(\theta)\sin(\phi)$ and $n_z = \cos(\theta)$. For a relativistic particles we can use $\beta \approx 1 - 1/2\gamma^2$, and for a small angle θ we replace $\sin(\theta) \approx \theta$ and $\cos(\theta) \approx 1 - \theta^2/2$ in the expressions for $\boldsymbol{n}$. Using these approximations, the numerator and the denominator in Eq. (15.22) become

$$\boldsymbol{n} \times \{(\boldsymbol{n} - \boldsymbol{\beta}) \times \dot{\boldsymbol{\beta}}\} \approx \dot{\beta}(\hat{x}n_x + \hat{y}n_y)$$

$$(1 - \boldsymbol{\beta} \cdot \boldsymbol{n})^3 \approx \frac{1}{8}(\theta^2 + \gamma^{-2})^3 \, ,$$

and we obtain for the electric field

$$\boldsymbol{E}_{\mathrm{rad}} = \frac{2q\dot{\beta}}{\pi\epsilon_0 c R}\frac{\theta}{(\theta^2 + \gamma^{-2})^3}[\hat{x}\cos(\phi) + \hat{y}\sin(\phi)] \, .$$

This field has a radial polarization and vanishes in the forward direction, $\theta = 0$.

Chapter 16
Dipole Radiation and Scattering of Electromagnetic Waves

In this chapter we consider scattering of an electromagnetic wave on a free charged particle — a process often referred to as Thomson scattering. This interaction induces a damping force on the charged particle called the radiation reaction force, which maintains the energy balance. Additionally, the radiation exerts another force in the direction of propagation called the light pressure, which is responsible for conserving the combined momentum of the particles and fields. We also briefly discuss inverse Compton scattering off a point charge moving with a relativistic velocity.

16.1 Dipole Radiation of a Linear Oscillator

The general formula (15.14) for the electromagnetic field of a point charge can be considerably simplified if the charge is moving with a non-relativistic velocity, $\beta \ll 1$. Neglecting in (15.14) β_{ret} in comparison with vectors of unit length, for the radiation field we obtain

$$
E = \frac{q}{4\pi\epsilon_0 c R} n \times \left(n \times \dot{\beta}_{\mathrm{ret}} \right) ,
$$
$$
B = \frac{1}{c} n \times E = -\frac{Z_0 q}{4\pi c R} n \times \dot{\beta}_{\mathrm{ret}} . \tag{16.1}
$$

The distance R in this equation is defined as the distance from the retarded position of the particle to the observation point r, $R = |r - r_0(t_{\mathrm{ret}})|$, and n is the unit vector in this direction.

An important example of non-relativistic radiation is the harmonic oscillator for which $v(t) = \mathrm{Re}(v_0 e^{-i\omega t})$, where ω is the oscillation frequency and $v_0 \ll c$ is the velocity amplitude. Due to the linearity of Eq. (16.1) involved in the calculation of the fields, it is convenient to work with complex numbers omitting the symbol for the real value, with the understanding that all physically meaningful quantities are

© Springer International Publishing AG, part of Springer Nature 2018   201
G. Stupakov and G. Penn, *Classical Mechanics and Electromagnetism in Accelerator Physics*, Graduate Texts in Physics, https://doi.org/10.1007/978-3-319-90188-6_16

obtained by taking the real parts of these complex quantities. The equation for the velocity then reads

$$v(t) = v_0 e^{-i\omega t} .$$ (16.2)

Integrating the velocity over time gives the position of the oscillator $r_0(t)$,

$$r_0(t) = \frac{i}{\omega} v_0 e^{-i\omega t} ,$$ (16.3)

where we have chosen the constant of integration so that the averaged over time position of the particle is at the origin of the coordinate system. From the condition $v_0 \ll c$ it follows that $r_0 \omega / c \ll 1$, or

$$r_0 \ll \lambda ,$$ (16.4)

where the reduced wavelength $\lambda = c/\omega = \lambda/2\pi$, with λ being the wavelength of the electromagnetic radiation with frequency ω.

Having calculated the particle motion, we can now find its radiation. The radiation field becomes dominant at distances larger than a wavelength and the size of the source of the radiation, so we have

$$r_0 \ll \lambda \ll r .$$ (16.5)

This leads us to the *dipole approximation*, where we can approximate $R \approx r$ in Eq. (16.1) by neglecting $r_0(t)$ in comparison with r. Correspondingly, for the retarded time we have

$$t_{\text{ret}} = t - \frac{r}{c} .$$ (16.6)

By evaluating the acceleration at this value for t_{ret}, we obtain for the magnetic field of the radiation

$$B = \frac{Z_0 q i \omega}{4\pi c r} n \times \beta_0 e^{-i\omega t + ikr} ,$$ (16.7)

where $k = \omega/c$. This is a *spherical electromagnetic wave*, whose amplitude decays with distance as $1/r$.

The intensity of the radiation is the energy flow given by the Poynting vector (A.7). This vector is directed along n and the time average of this component, which we denote by $\bar{S}$, is given by the same equations as the Poynting vector in the plane electromagnetic wave (13.8):

Fig. 16.1 The three-dimensional angular distribution of the dipole radiation. The diagram is axially symmetric with respect to the vector β_0. The length of the blue vectors in the cutout is proportional to the radiation intensity in the direction of the vectors

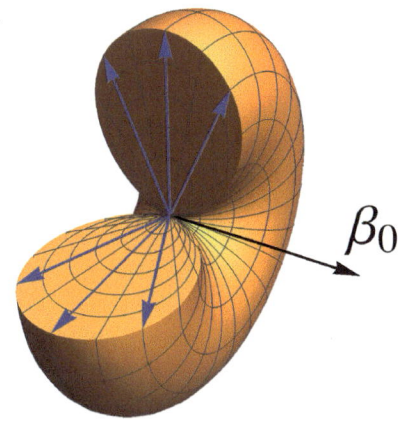

$$\bar{S} = \frac{1}{2Z_0}|E|^2 = \frac{c^2}{2Z_0}|B|^2$$
$$= \frac{Z_0}{32\pi^2}\frac{q^2\omega^2}{r^2}|\boldsymbol{n}\times\boldsymbol{\beta_0}|^2 \,. \tag{16.8}$$

The Poynting vector gives the energy flow through a unit area. Another convenient quantity is $\bar{S}r^2$ which describes the power radiated per unit solid angle. We denote this quantity by $d\mathcal{P}/d\Omega$. We have

$$\frac{d\mathcal{P}}{d\Omega} = \bar{S}r^2 = \frac{Z_0}{32\pi^2}q^2\omega^2\beta_0^2\sin^2\psi \,, \tag{16.9}$$

where ψ is the angle between the direction of the observation $\boldsymbol{n}$ and β_0. The angular distribution of the dipole radiation is shown in Fig. 16.1. Integrating $d\mathcal{P}/d\Omega$ over the solid angle gives the total power of radiation:

$$\mathcal{P} = \int \frac{d\mathcal{P}}{d\Omega}d\Omega$$
$$= \frac{Z_0}{32\pi^2}q^2\omega^2\beta_0^2\int_0^\pi \sin^2\psi \cdot 2\pi\sin\psi \, d\psi$$
$$= \frac{Z_0}{32\pi^2}\left(\frac{8\pi}{3}\right)q^2\omega^2\beta_0^2 \,. \tag{16.10}$$

16.2 Radiation Reaction Force

Because the charge is losing energy to radiation, it should feel a force that, on average, works against the velocity to keep the energy balance in the system. Indeed, such a force exists, and is called the *radiation reaction* force. If the charge executes free oscillations with the frequency ω, this force would cause the amplitude of the oscillations to decay with time and would eventually terminate the oscillations with all of the kinetic energy converted into radiation. If the oscillations are driven by an external force at a constant amplitude, then this force is responsible for taking the energy from the external source and supplying it to the radiated waves.

Let us calculate this force, f_r, equating the radiated power $\mathcal{P}$ to the work of the force with the negative sign,

$$\mathcal{P} = -\langle f_r \cdot v \rangle, \tag{16.11}$$

where the angular brackets indicate averaging over time. We take the expression for the velocity Eq. (16.2) in real form

$$v = v_0 \cos \omega t, \tag{16.12}$$

and assume that f_r acts in the direction opposite to the velocity,

$$f_r = -A v. \tag{16.13}$$

We then have

$$\langle f_r \cdot v \rangle = -\frac{1}{2} A v_0^2. \tag{16.14}$$

Equating this expression to $\mathcal{P}$ given by Eq. (16.10), we find A and the force,

$$f_r = -\frac{Z_0}{6\pi} \frac{q^2 \omega^2}{c^2} v. \tag{16.15}$$

Because for an oscillator $\omega^2 v = -\ddot{v}$ we can also write

$$f_r = \frac{1}{6\pi\epsilon_0} \frac{q^2}{c^3} \ddot{v} = \frac{2}{3} \frac{r_c m}{c} \ddot{v}, \tag{16.16}$$

where r_c is the *classical radius*,

$$r_c = \frac{1}{4\pi\epsilon_0} \frac{q^2}{mc^2}; \tag{16.17}$$

for electrons, $r_c = 2.8 \times 10^{-15}$ m. The expression (16.16), as it turns out, is more general than our derivation assumes — it is valid for arbitrary nonrelativistic motion of a point charge.

As we emphasized above, the radiation reaction force is responsible for the energy balance in the radiation process. There is one subtle issue with the application of this force [1, 2] which is clearly seen if we equate this force to the particle acceleration,

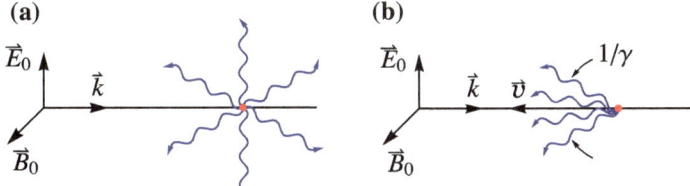

Fig. 16.2 **a** In Thomson scattering a plane electromagnetic wave illuminates a charge at rest. Due to oscillations of the charge in the incident wave, secondary waves will be radiated shown by blue wiggly arrows. **b** When the charge is moving towards the wave with relativistic velocity, the radiation is localized in a narrow angle $\sim 1/\gamma$

$\dot{v} = f_r/m$, in the absence of external fields. It is easy to solve this differential equation and find an exponentially growing solution $v \propto \exp(2ct/3r_c)$, which is clearly unphysical. The origin of this solution is in the infinite Coulomb energy of a point charge. To avoid such kinds of singularities one has to use Eq. (16.16) only in situations where a particle is accelerated by external electromagnetic fields and substitute for $\ddot{v}$ on the right-hand side of this equation the time derivative of the acceleration caused by these fields. In other words, it cannot be applied in a fully self-consistent manner, but has to be evaluated on the basis of all forces excluding the radiation reaction force itself[1].

In Chap. 22 we will consider the implications of this force for the case of synchrotron radiation of a relativistic particle.

16.3 Thomson Scattering

As an application of the results of Sect. 16.1, we now consider Thomson scattering — radiation of a point charge caused by motion in a plane electromagnetic wave. We assume that an electron that was initially at rest is illuminated by a plane electromagnetic wave with frequency ω. This electron starts to oscillate in the wave and to radiate secondary electromagnetic waves (Fig. 16.2).

We first need to find the electron motion in the incident wave. We will assume that this wave is not too strong so that the electron velocity in the wave is nonrelativistic, $v \ll c$. The field in the wave is given by Eq. (13.6) and the equation of motion for the electron in complex number notation is

$$m\frac{dv}{dt} = qE_0 e^{-i\omega t + ik\cdot r} , \tag{16.18}$$

---

[1]This holds true even though in some cases the radiation reaction force for a relativistic particle can be larger than the Lorentz force from an external electromagnetic field [3].

(here we have assumed the phase $\phi_0 = 0$). The magnetic force in this equation is omitted because in the limit $v \ll c$ it is much smaller than the electric one. Assuming that the electron is located near the origin of the coordinate system, $r \approx 0$, we drop the term $i\mathbf{k} \cdot \mathbf{r}$ on the right-hand side of Eq. (16.18),

$$m\frac{d\mathbf{v}}{dt} = q\mathbf{E}_0 e^{-i\omega t} . \tag{16.19}$$

The smallness of the parameter $\mathbf{k} \cdot \mathbf{r}$ is justified by the inequality (16.4) valid for oscillations with a non-relativistic velocity $\mathbf{v}$. Integration over time gives

$$\mathbf{v} = \frac{iq}{m\omega}\mathbf{E}_0 e^{-i\omega t} . \tag{16.20}$$

Comparing this expression with Eq. (16.2) we relate $\mathbf{v}_0$ to the electric field in the plane wave,

$$\mathbf{v}_0 = \frac{iq}{m\omega}\mathbf{E}_0 . \tag{16.21}$$

Using this equation, all the results of Sect. 16.1 can now be expressed in terms of the amplitude of the electric field, E_0. For examples, the total radiation power (16.10) becomes

$$\mathcal{P} = \frac{Z_0}{32\pi^2}\left(\frac{8\pi}{3}\right)\frac{q^4 E_0^2}{m^2 c^2} . \tag{16.22}$$

From the condition $v \ll c$ it follows that

$$a \equiv \frac{qE_0}{m\omega c} \ll 1 , \tag{16.23}$$

where we have introduced the dimensionless parameter a, which characterizes the strength of the electromagnetic field. This is a very important condition, which we will meet again, in a different context, in Chap. 19.

If we divide the radiated power by the average energy flow in the incident wave, we obtain a quantity that has dimension of length squared. This quantity can be interpreted as a scattering cross section, and is called the *Thomson cross section*:

$$\sigma_T = \frac{\mathcal{P}}{E_0^2/2Z_0} = \left(\frac{1}{4\pi\epsilon_0}\right)^2\frac{8\pi q^4}{3m^2 c^4} = \frac{8\pi}{3}r_c^2 . \tag{16.24}$$

In Chap. 19 we will need the intensity of the radiation written in terms of spherical coordinates, in which the wave propagates in the z direction and the electric field is directed along x. We introduce the polar angle θ measured relative to the z axis and

Fig. 16.3 The spherical coordinate system. Vector $\mathbf{n}$ shown in red is defined by its two angles, θ and ϕ. The wave propagates along the z axis, and the electric field in the wave is directed along the x axis

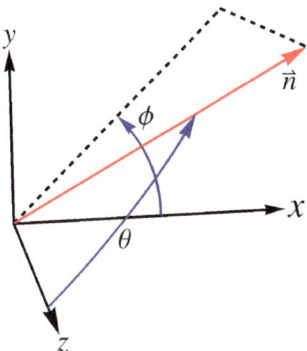

the azimuthal angle ϕ measured in the $x - y$ plane, see Fig. 16.3. In terms of these variables the (x, y, z) components of the propagation direction $\mathbf{n}$ are given by

$$\mathbf{n} = (\sin\theta \cos\phi, \sin\theta \sin\phi, \cos\theta), \qquad (16.25)$$

and with $\mathbf{E}_0 = (E_0, 0, 0)$ we find

$$|\mathbf{n} \times \mathbf{E}_0|^2 = E_0^2(1 - \sin^2\theta \cos^2\phi). \qquad (16.26)$$

Returning to the definition $d\mathcal{P}/d\Omega = \bar{S}r^2$ and using Eq. (16.8) with Eqs. (16.21) and (16.26) we obtain

$$\frac{d\mathcal{P}}{d\Omega} = \frac{Z_0}{32\pi^2} \frac{q^4 E_0^2}{m^2 c^2}(1 - \sin^2\theta \cos^2\phi). \qquad (16.27)$$

While Thomson scattering is a classical phenomenon and as such is perfectly well described by classical electrodynamics[2], it is instructive to convert the radiation power (16.22) into quantum language. In quantum theory the electromagnetic field is represented by photons, so we want to calculate the number of photons emitted by the charge per unit time. This number, $\dot{N}_p$, is obtained by dividing the radiation power $\mathcal{P}$ by the photon energy $\hbar\omega$,

$$\dot{N}_p = \frac{\mathcal{P}}{\hbar\omega} = \frac{1}{12\pi\epsilon_0} \frac{q^2 E_0^2}{m^2 c^2 \omega^2} \omega \frac{q^2}{\hbar c}$$
$$= \frac{1}{3}a^2 \omega\alpha, \qquad (16.28)$$

where a is defined by Eq. (16.23) and α is the fine structure constant,

---

[2]We will estimate the conditions when quantum effects start to play a role in electromagnetic scattering in the next section.

$$\alpha = \frac{1}{4\pi\epsilon_0}\frac{q^2}{\hbar c} \approx \frac{1}{137}.$$ (16.29)

If we multiply Eq. (16.28) by $2\pi/\omega$, we find that the number of photons scattered by an electron in one period of oscillations is approximately $2\alpha a^2$.

16.4 Light Pressure

The quantum picture of Thomson scattering helps to understand the important physical effect of light pressure. The photons of the incident wave within the Thomson cross section are scattered off in different directions. The incident photons carry momentum in the direction of k, that is the direction of the wave propagation. Because the scattered photons are equally distributed between the forward and backward directions, the averaged momentum of the scattered radiation is zero. Hence the initial momentum of the incident photons is transferred to the scatterer, which means that there is a force, f_p, exerted on the charge in the direction of the incident wave propagation.

It takes only one more step to calculate this force from the quantum viewpoint. We need to multiply the number of photons scattered per unit time (16.28) by the photon momentum $\hbar k$,

$$f_p = \hbar k \dot{N}_p = \frac{\mathcal{P}}{c} = \sigma_{\mathrm{T}}\frac{E_0^2}{2cZ_0}.$$ (16.30)

It is clear that this force is directed along k. Note that although we used quantum theory in the derivation, our final result does not involve the Planck constant and hence is classical, and can also be derived in classical electromagnetism.

In the classical derivation of f_p we first need to modify the equation of motion (16.19) by adding on the right-hand side the radiation reaction force (16.16),

$$m\frac{d\boldsymbol{v}}{dt} = q\boldsymbol{E}_0 e^{-i\omega t} + \frac{2}{3}\frac{r_c m}{c}\frac{d^2\boldsymbol{v}}{dt^2}.$$ (16.31)

With this modification, the solution Eq. (16.20) acquires an additional term $\boldsymbol{v}_1$,

$$\boldsymbol{v} = \frac{iq}{m\omega}\boldsymbol{E}_0 e^{-i\omega t} + \boldsymbol{v}_1.$$ (16.32)

We expect that the correction $\boldsymbol{v}_1$ is small and substituting this expression into Eq. (16.31) we neglect $d^2\boldsymbol{v}_1/dt^2$ on the right-hand side, which then gives

$$m\frac{d\boldsymbol{v}_1}{dt} = -\frac{2}{3}\frac{iqr_c\omega}{c}\boldsymbol{E}_0 e^{-i\omega t}.$$ (16.33)

The solution of this equation is

$$v_1 = \frac{2}{3} \frac{q r_c}{mc} E_0 e^{-i\omega t}.$$ (16.34)

The final step in the derivation is to calculate the average over time magnetic force $q v_1 \times B$ which arises from the cross product of the velocity v_1 and the magnetic field in the incident wave, $B = (k/ck) \times E_0 e^{-i\omega t}$. To average the force, we have to take the real parts of v_1 and B, with the result

$$q \langle \mathrm{Re}\, v_1 \times \mathrm{Re}\, B \rangle = \frac{1}{3} \frac{q^2 r_c}{mc^2 k} E_0 \times (k \times E_0) = \frac{1}{3} \frac{q^2 r_c E_0^2}{mc^2} \frac{k}{k}.$$ (16.35)

We see that this force acts in the direction of the wave propagation. The net force is associated with a small change in the phase offset between the velocity and the electromagnetic fields, from $\pi/2$ to $\pi/2 - 2 r_c \omega / 3c$. It is easy to check that the magnitude of the force is exactly equal to the expression (16.30) obtained in the quantum derivation.

We conclude this section with an estimate of when the classical theory cannot be applied to the scattering. If the photon energy in the incident electromagnetic wave $\hbar \omega$ is so large that it is comparable to the rest energy mc^2, $\hbar \omega \sim mc^2$, the classical description breaks down. In this case, a part of the incident photon energy goes into the recoil momentum of the charge, and the frequency of the scattered photons ω' becomes smaller than the incident frequency, $\omega' < \omega$. The process in which scattering leads to a change of frequency is usually referred to as *Compton scattering*.

16.5 Inverse Compton Scattering

Thomson scattering, analyzed in Sect. 16.3, can be generalized to the case of a moving scatterer. In practice, the most interesting situation occurs when the scatterer is relativistic and is moving towards the wave as shown in Fig. 16.2b. This is usually refered to as *inverse Compton scattering*. The process is most easily calculated through a Lorentz transformation from the lab frame to the particle frame of reference where the average velocity of the scatterer is equal to zero and we can use the results of Sect. 16.3. The radiation in that reference frame should then be transformed back to the lab frame using the Lorentz transformations for the fields. In this section, we will limit our analysis to the frequency of the scattered radiation in the lab frame when v is close to the speed of light, $\gamma \gg 1$.

Let us choose the coordinate system with the z axis directed along the velocity v and denote the incident wave frequency in the lab frame by ω_0. The wave propagates in the negative-z direction. In the particle frame of reference, the frequency ω' of the incident wave is different from ω_0 — it can be obtained from the Lorentz transformation (B.15) setting $\theta = \pi$,

$$\omega' = 2\gamma\omega_0 . \tag{16.36}$$

This is also the frequency of the scattered radiation in the particle frame. To transform it back to the lab frame we will assume small angles θ (θ is the angle between a scattered photon and the z axis) and use Eq. (B.16),

$$\omega = \frac{4\gamma^2\omega_0}{1 + \gamma^2\theta^2} . \tag{16.37}$$

We see that the frequency of the scattered radiation now depends on the direction of observation, and the frequncy *spectrum* has a maximum frequency equal to $4\gamma^2\omega_0$, which for $\gamma \gg 1$ is much larger than the incident wave frequency ω_0. It follows from Eq. (16.37) that the high frequencies, $\omega \sim \gamma^2\omega_0$, are localized within a narrow range of angles $\theta \sim 1/\gamma$ around the direction of the velocity v.

   We know that in the beam frame the photons are radiated more or less in all directions. The angles in the lab frame θ are related to the angles in the beam frame through Eq. (B.12). In the limit $\gamma \gg 1$, this equation maps most of the interval $0 < \theta' < \pi$ into a narrow range of angles $\theta \sim 1/\gamma$ around the velocity v (to be more precise, for θ to be $\sim 1/\gamma$, the angle θ' should not be very close to π, $\pi - \theta' \gg 1/\gamma$). This means that almost all photons in the lab frame propagate within the angle $1/\gamma$ in the direction of the beam motion. The number of photons will be the same as in the beam frame, but their energy in the lab frame is $\sim \gamma^2$ times larger than the energy of the incident photons. The fact that the dominant part of the scattering is localized within the angle $\sim 1/\gamma$ is a manifestation of the general properties of the radiation of relativistic particles discussed in Sect. 15.4.

Worked Examples

Problem 16.1 *Prove that the ratio $q E_0/m\omega c$ is a Lorentz invariant — it does not change under the Lorentz transformation (in other words, it is the same in any coordinate system moving relative to the laboratory reference frame). Assume that the frames move in the direction of propagation of the wave.*

   Solution: It is sufficient to establish the invariance of the ratio E_0/ω. From Appendix B we know that when performing a Lorentz boost in the same direction as the radiation propagation the frequency becomes $\omega' = \gamma(1 - \beta)\omega$.

   For a radiation field, where $\boldsymbol{B} = (1/\omega)\boldsymbol{k} \times \boldsymbol{E}$, the electric field transforms as

$$\boldsymbol{E} = \gamma(\boldsymbol{E}' - \boldsymbol{v} \times \boldsymbol{B}') = \gamma(1 + \beta)\boldsymbol{E}' .$$

This gives $\boldsymbol{E}' = \boldsymbol{E}/\gamma(1 + \beta) = \gamma(1 - \beta)\boldsymbol{E}$. So we find $E_0'/\omega' = E/\omega$, showing that the ratio is Lorentz invariant.

Problem 16.2 *Consider scattering of an electromagnetic wave on a charge q that is attached to an immobile point through a spring, and can oscillate with the frequency ω_0. Find the scattering cross section as a function of frequency of the incident wave ω assuming that the electric field in the wave is directed along the line of oscillations of the charge.*

Solution: The charge motion is determined by

$$\ddot{x} + \omega_0^2 x = \frac{qE_0}{m} e^{-i\omega t + i\mathbf{k}\cdot\mathbf{r}} \approx \frac{qE_0}{m} e^{-i\omega t}.$$

We will look for solutions of the form $x = Ae^{-i\omega t}$. Substituting into the above equation we find

$$A = \frac{qE_0}{m(\omega_0^2 - \omega^2)}.$$

The charge velocity is

$$v_x = -i\omega x = \frac{iqE_0}{m\omega} \frac{\omega^2}{(\omega^2 - \omega_0^2)} e^{-i\omega t}.$$

The power radiated, $\mathcal{P}$, is proportional to v_x^2. Comparing our expression for v_x to that of a free particle, Eq. (16.20), we see that the spring modifies the velocity by the term $\omega^2/(\omega^2 - \omega_0^2)$, so our power is modified by $\omega^4/(\omega^2 - \omega_0^2)^2$, and the new cross section is

$$\sigma = \frac{\mathcal{P}}{E_0^2/2Z_0} = \sigma_T \left(\frac{\omega^2}{\omega^2 - \omega_0^2}\right)^2.$$

When the frequency ω approaches ω_0, the cross section becomes large, $\sigma \gg \sigma_T$.

Problem 16.3 *In the derivation of the light pressure we neglected the term $q\mathbf{v} \times \mathbf{B}$ where $\mathbf{v}$ is given by the real part of Eq. (16.20). Show that $\langle \mathbf{v} \times \mathbf{B} \rangle = 0$.*

   Solution: The velocity $\mathbf{v} \propto \mathrm{Re}\,(i\mathbf{E}_0 e^{-i\omega t})$ and $\mathbf{B} \propto \mathrm{Re}[\mathbf{n} \times \mathbf{E}_0 e^{-i\omega t}]$. When we average the product over time, we first take the real part of $\mathbf{v}$ and $\mathbf{B}$ to find

$$\langle \mathbf{v} \times \mathbf{B} \rangle \propto \langle \sin(\omega t)\cos(\omega t) \rangle = 0.$$

Problem 16.4 *It has been proposed to use Thomson scattering in a compact electron ring as a source of intense X-ray radiation (e.g., Phys. Rev. Lett., **80**, 976, (1998)). Consider the following parameters of such a system: the electron energy in the ring is 8 MeV, the number of electrons in the bunch is $N_e = 1.1 \times 10^{10}$, the laser energy is 20 mJ, the laser pulse length is 1 mm, and the laser is focused to the spot size 25 micron. Estimate the number of photons from a single collision of the laser pulse with the electron beam.*

Solution: The photon flux of the laser beam at 800 nm ($h\nu = 2.5 \times 10^{-19}$ J), with laser energy $E_L = 20$ mJ and waist $w = 25$ µm is

$$\frac{E_L}{\pi w_L^2 h\nu} = \frac{2 \times 10^{-2}}{\pi \times 25^2 \times 10^{-12} \times 2.5 \times 10^{-19}} \text{ photons/m}^2$$

$$= 4 \times 10^{25} \text{ photons/m}^2 \,.$$

The total scattering cross section of all electrons in the bunch is $N_e \sigma_T = 6.7 \times 10^{-19}$ m^2. So a single interaction scatters 3×10^7 photons.

Problem 16.5 *An electromagnetic wave with frequency ω and amplitude of the electric field E_0 occupies a volume with dimensions $L_x \times L_y \times L_z$. It propagates along the z axis with fields $E_x = cB_y$. Using the results of Problem 13.2, find the electromagnetic energy W of the wave in the lab frame and the energy W' in a frame K' moving with velocity v along the z axis relative to the lab frame. Show that $W/\omega = W'/\omega'$, where ω' is the frequency of the wave in K'.*

Solution: The averaged energy density w in a plane wave is equal to $w = \epsilon_0 E_0^2/2$. Comparing this expression with the Poynting vector (13.8) we see that $w = S/c$. We can now write the energy in the lab frame as $W = wL_xL_yL_z = SL_xL_yL_z/c$. To find the energy in the moving frame we need to transform S and L_z.

The electromagnetic field has the same number of periods in each frame, so L_z must transform as the wavelength λ, which transforms as the inverse of $\omega' = \gamma(1 - \beta)\omega$ as in (B.15). For the radiation field, we know from Eq. (B.17) that $E_x = \gamma(1 + \beta)E_x'$ and $B_y = \gamma(1 + \beta)B_y'$. We then find

$$\frac{W'}{\omega'} = \frac{S'L_x'L_y'L_z'}{c\omega'}$$

$$= \frac{SL_xL_yL_z}{c\gamma^4(1 - \beta)^2(1 + \beta)^2\omega}$$

$$= \frac{SL_xL_yL_z}{c\omega} \frac{(1 - \beta^2)^2}{(1 - \beta)^2(1 + \beta)^2}$$

$$= \frac{W}{\omega} \,.$$

Since the ratio W/ω is proportional to the number of photons, its invariance tells us that the number of photons in both frames is the same.

References

1. W. Panofsky, M. Phillips. *Classical Electricity and Magnetism*, 2nd edn. (Addison-Wesley 1962)
2. J.D. Jackson, *Classical Electrodynamics*, 3rd edn. (Wiley, New York, 1999)
3. L.D. Landau, E.M. Lifshitz. *The classical theory of fields, volume 2 of Course of Theoretical Physics*, 4th edn. (Elsevier Butterworth-Heinemann, Burlington MA, 1980) (translated from Russian)

Chapter 17
Transition and Diffraction Radiation

Transition radiation occurs when a moving charged particle crosses a boundary between two media with different electrodynamic properties. In its simplest form, which is commonly used in experiments, transition radiation is generated by sending a beam through a metallic foil. In this chapter, we will derive the spectrum and angular distribution of the transition radiation when a particle crosses a foil at normal incidence. We will also discuss radiation generated by the beam when it passes through a hole in a metal foil — the so-called diffraction radiation.

17.1 Transition Radiation

We will calculate the transition radiation generated when a plane metal surface is hit by a point charge moving with a constant velocity v in the direction perpendicular to the surface, as shown in Fig. 17.1a. We choose the coordinate system with the origin located at the entrance to the metal in such a way that the particle is moving along the z axis in the positive direction. The metal occupies the region $z > 0$ with $z = 0$ being the metal boundary.

To find the electromagnetic field at $z < 0$ one has to solve Maxwell's equation with the charge and current density corresponding to the moving point charge q, $\rho = q\delta(z - vt)$, $j = v\rho$, at $t < 0$. At time $t = 0$ the charge q enters the metal and gets shielded by the charges inside the metal; there are no charges at $t > 0$ in the upper half space. Assuming the perfect conductivity of the metal, we need to solve Maxwell's equations with the boundary condition of zero tangential electric field at $z = 0$. The solution is greatly simplified if one uses the method of image charges. In this method, one replaces the metal with an image charge of the opposite sign, moving with velocity v in the opposite direction, as shown in Fig. 17.1b. In what follows, we will refer to the original charge q by index 1, and to the image charge $-q$ by index 2. Using Eq. (11.4) from Chap. 11, it is easy to verify that in the plane $z = 0$, the components of the electric field tangential to the metal surface (that is, perpendicular

© Springer International Publishing AG, part of Springer Nature 2018
G. Stupakov and G. Penn, *Classical Mechanics and Electromagnetism in Accelerator Physics*, Graduate Texts in Physics, https://doi.org/10.1007/978-3-319-90188-6_17

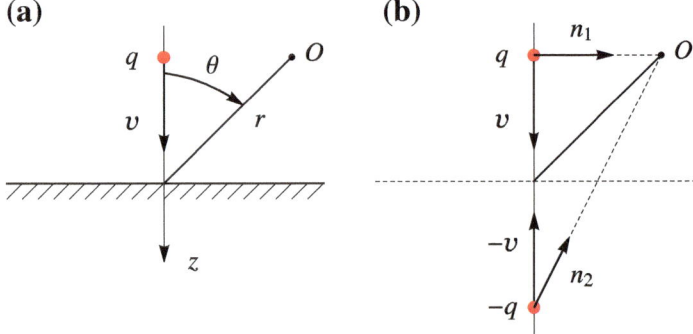

Fig. 17.1 A charge moving perpendicular to the metal surface (**a**), and the two charges in the method of image charges (**b**). The observation point O is a distance r from where the charge will strike the surface, and the line connecting these two points forms an angle θ with the direction normal to the surface. Panel (**b**) also shows the unit vectors $\boldsymbol{n}_1$ and $\boldsymbol{n}_2$ from Eq. (17.1)

to the z direction) of the two charges cancel each other and the boundary conditions $E_x = E_y = 0$ are automatically satisfied. At time $t = 0$ the two charges meet at point $z = 0$ where they annihilate so that at time $t > 0$ there are no charges in the system.

While we assume that the metal occupies the half space $z > 0$, we would obtain the same electromagnetic field in the region $z < 0$ in the case of a metal foil occupying the region $0 < z < h$, where h is the thickness of the foil. This follows from the fact that the boundary condition of zero tangential field at $z = 0$ does not change for the foil (again, assuming the perfect conductivity of the metal). Hence our results for the transition radiation will be also valid for the metal foil.[1] In either case, the fields from the real particle are shielded by the metal for $t > 0$.

To calculate the radiation field, we need to find the vector potential $\boldsymbol{A}(\boldsymbol{r}, t)$ at the observation point $\boldsymbol{r}$ at time t. The trajectories of the two charges, 1 and 2, for $t < 0$ are given by the equations $\boldsymbol{r}_1(t) = (0, 0, vt)$ and $\boldsymbol{r}_2(t) = (0, 0, -vt)$ respectively. Because the charges are moving with a constant velocity, we can use Eq. (15.9) for the potentials. The retarded times for both particles, $t_{\text{ret}}^{(1)}(\boldsymbol{r}, t)$ and $t_{\text{ret}}^{(2)}(\boldsymbol{r}, t)$, satisfy the equations $c(t - t_{\text{ret}}^{(1)}) = |\boldsymbol{r} - \boldsymbol{r}_1(t_{\text{ret}}^{(1)})|$ and $c(t - t_{\text{ret}}^{(2)}) = |\boldsymbol{r} - \boldsymbol{r}_2(t_{\text{ret}}^{(2)})|$, respectively (see Eq. (15.1)), again for $t_{\text{ret}}^{(1)} < 0$ and $t_{\text{ret}}^{(2)} < 0$. Note that the moment $t_{\text{ret}}^{(1)} = t_{\text{ret}}^{(2)} = 0$ corresponds to $t = r/c$ — this is a spherical shell in the upper half space expanding with the speed of light, $r = ct$. Because at $t_{\text{ret}}^{(1)} = t_{\text{ret}}^{(2)} = 0$ the charges collide and disappear, the potentials are zero for $t_{\text{ret}}^{(1)}, t_{\text{ret}}^{(2)} > 0$, or equivalently for $t > r/c$. Hence the vector potential $\boldsymbol{A}$ can be obtained as a combination of expressions (15.9) with the step function $h(r/c - t)$ which guarantees the zero value of $\boldsymbol{A}$ for $t > r/c$:

---

[1] If the charge passes through the foil and exits to the region $z > h$, there is transition radiation in this region of space as well. It can be solved by the same method of image charges.

$$A = \frac{Z_0}{4\pi} \left(\beta \frac{q}{R_1(t_{\text{ret}}^{(1)})(1 - \beta \cdot n_1)} + (-\beta) \frac{(-q)}{R_2(t_{\text{ret}}^{(2)})(1 + \beta \cdot n_2)} \right) h \left(\frac{r}{c} - t \right),$$

$$(17.1)$$

where

$$R_1(t) = \sqrt{(z - vt)^2 + x^2 + y^2},$$
$$R_2(t) = \sqrt{(z + vt)^2 + x^2 + y^2},$$

$$(17.2)$$

and h is equal to one for a positive argument and zero otherwise.

Let us now assume that the radiation is observed at a large distance from the origin: we can then neglect the difference between n_1 and n_2 and assume that they both are equal to the unit vector n directed from the origin of the coordinate system to the observation point. The magnetic field of the radiation can be calculated using Eq. (13.5):

$$B = -\frac{1}{c} n \times \frac{\partial A}{\partial t}.$$

$$(17.3)$$

When we differentiate (17.1) with respect to time, we only need to differentiate the function h — differentiating R_1 and R_2 would give a field that decays faster than $1/r$, which is simply the static field of each moving charge. The result of the differentiation is

$$B = \frac{Z_0}{4\pi} \frac{q}{c} \delta \left(\frac{r}{c} - t \right) \left(\frac{1}{R_1(0)(1 + \beta \cos\theta)} + \frac{1}{R_2(0)(1 - \beta \cos\theta)} \right) n \times \beta,$$

$$(17.4)$$

where θ is the angle between n and the normal to the metal, see Fig. 17.1. Because of the delta function factor, the values of R_1 and R_2 in this equation are taken at the retarded times $t_{\text{ret}}^{(1)} = t_{\text{ret}}^{(2)} = 0$:

$$R_1(0) = R_2(0) = \sqrt{z^2 + x^2 + y^2} = r,$$

$$(17.5)$$

which simplifies the formula for the magnetic field to yield

$$B = \frac{Z_0}{4\pi} \frac{2q}{rc} \delta \left(\frac{r}{c} - t \right) \frac{n \times \beta}{1 - \beta^2 \cos^2\theta}.$$

$$(17.6)$$

We see that the radiation field consists of an infinitely thin spherical shell expanding away from the point where the charge enters the metal.

17.2   Fourier Transformation of the Radiation Field and the Radiated Power

When a radiation field is a pulse of finite duration, as derived in the preceding section, it makes sense to try to calculate the total radiation energy and its *spectral distribution*, or *spectrum*. Let us denote by $B(t)$ the radiation magnetic field as a function of time at a distance r from the radiation source. This field also depends on the direction of the vector $\boldsymbol{n}$ and decays with distance as $1/r$. The energy radiated per unit solid angle, $dW/d\Omega$, is given by the time integral of the product of the Poynting vector $S = |\boldsymbol{E} \times \boldsymbol{H}|$ and the squared distance r^2:

$$\frac{dW}{d\Omega} = r^2 \int_{-\infty}^{\infty} dt\, S(t) = \frac{c^2 r^2}{Z_0} \int_{-\infty}^{\infty} dt\, B(t)^2 \,. \tag{17.7}$$

In the last equation we have used the fact that in the radiation zone the field can be approximated locally by a plane wave in which $E = cB$ and the electric and magnetic fields are perpendicular to each other.

The spectrum is obtained by taking the Fourier transform of the field and representing the radiated energy as an integral over the frequencies ω. Using Parseval's theorem from Fourier analysis we express the time integral of the square of the magnetic field through the frequency integral of the square of the absolute value of its Fourier transform,

$$\int_{-\infty}^{\infty} dt\, B(t)^2 = \frac{1}{2\pi} \int_{-\infty}^{\infty} d\omega\, |\tilde{B}(\omega)|^2$$
$$= \frac{1}{\pi} \int_{0}^{\infty} d\omega\, |\tilde{B}(\omega)|^2 \,, \tag{17.8}$$

where

$$\tilde{B}(\omega) = \int_{-\infty}^{\infty} dt\, B(t) e^{i\omega t} \,. \tag{17.9}$$

At this point, we introduce the energy radiated per unit frequency interval per unit solid angle, $d^2W/d\omega d\Omega$, as

$$\frac{d^2W}{d\omega d\Omega} = \frac{c^2 r^2}{\pi Z_0} |\tilde{B}(\omega)|^2 \,, \tag{17.10}$$

so that the total energy radiated per unit solid angle can be represented as

$$\frac{dW}{d\Omega} = \int_{0}^{\infty} d\omega \frac{d^2W}{d\omega d\Omega} \,. \tag{17.11}$$

Fig. 17.2 Angular
distribution of the transition
radiation for a relativistic
particle

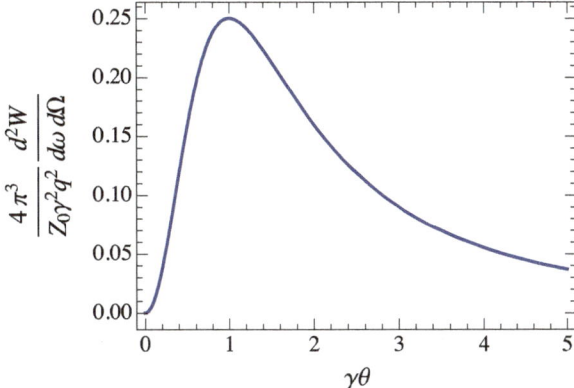

To find the spectrum of the transition radiation with the magnetic field given by
Eq. (17.6) we first make the Fourier transform,

$$\tilde{B}(\omega) = \int_{-\infty}^{\infty} dt \, B(t) e^{i\omega t} = \frac{Z_0}{4\pi} \frac{2q e^{i\omega r/c}}{rc} \frac{n \times \beta}{1 - \beta^2 \cos^2 \theta}. \tag{17.12}$$

Since the Fourier transform of the delta function is equal to one, the spectrum does not
depend on the frequency. Again, this is one of the consequences of our assumption
of the perfect conductivity of the metal — in reality at very high frequencies the
approximation of the perfect conductivity fails and the intensity of the radiation
decreases. For the angular distribution of the spectral power we have

$$\frac{d^2W}{d\omega d\Omega} = \frac{c^2 r^2}{\pi Z_0} |\tilde{B}(\omega)|^2 = \frac{Z_0 q^2}{4\pi^3} \frac{\beta^2 \sin^2 \theta}{(1 - \beta^2 \cos^2 \theta)^2}. \tag{17.13}$$

It follows from this equation that for a relativistic particle the dominant part of the
radiation goes in the backward direction, $\theta \approx 0$. Using $\beta^2 = 1 - \gamma^{-2}$ and approxi-
mating $\sin \theta \approx \theta$ and $\cos^2 \theta \approx 1 - \theta^2$ we find

$$\frac{d^2W}{d\omega d\Omega} \approx \frac{Z_0 q^2}{4\pi^3} \frac{\theta^2}{(\gamma^{-2} + \theta^2)^2}. \tag{17.14}$$

The plot of this function is shown in Fig. 17.2 — the maximum intensity of the
radiation is emitted at angle $\theta = 1/\gamma$, while the intensity is zero at $\theta = 0$.
    One can integrate Eq. (17.13) over all angles to find the spectrum of the transition
radiation,

$$\frac{dW}{d\omega} = 2\pi \int_0^{\pi/2} \sin \theta d\theta \frac{d^2W}{d\omega d\Omega} = \frac{Z_0 q^2}{4\pi^2} \left[\left(\frac{1}{\beta} + \beta \right) \text{arctanh}(\beta) - 1 \right]. \tag{17.15}$$

Because the spectrum of the radiation does not depend on frequency, the integral over ω from zero to infinity diverges which formally means that the total radiated energy is infinite. This is in agreement with the fact that for the magnetic field proportional to the delta function of time, the integral (17.7) diverges as well. In reality, as was already mentioned above, the energy is finite because at the very high frequencies metals lose their capability of being almost perfect conductors, and the spectrum eventually tends to zero. In the time domain, this means that the particle does not instantaneously become shielded by the metal as it crosses $z = 0$, but instead takes some time for the shielding to be complete.

The transition radiation is often used in accelerators for measuring the transverse size and position of the beam by observing an image of the radiation produced when it strikes a metal foil in its path.

17.3   Diffraction Radiation

An interception of a beam with a metal foil to generate the transition radiation either destroys the beam or deteriorates its properties, making this technique unsuitable for many applications. In some cases one would like to generate radiation without strongly perturbing the beam. This can be achieved if the beam passes through a small hole in a metal foil — it then generates the so-called *diffraction radiation*, which can serve for diagnostic purposes while avoiding significant disruption of the electron beam.

A complete electromagnetic solution of the diffraction radiation in the general case is extremely complicated and is beyond the scope of this book. A simplified treatment [1] that uses methods from the theory of diffraction can be carried out for a round hole in the limit when the radiation wavelength is much smaller than the hole radius a. In the limit $\gamma \gg 1$, and for small observation angles, $\theta \ll 1$, this analysis results in the following formula for the angular spectral distribution:

$$\frac{d^2 \mathcal{W}}{d\omega d\Omega} \approx \frac{Z_0 q^2}{4\pi^3} \frac{\theta^2}{(\gamma^{-2} + \theta^2)^2} F\left(\frac{\omega a \theta}{c}, \frac{\omega a}{c\gamma}\right), \tag{17.16}$$

where

$$F(x, y) = \left[y J_2(x) K_1(y) - \frac{y^2}{x} J_1(x) K_2(y) \right]^2, \tag{17.17}$$

with $J_{1,2}$ the Bessel functions and $K_{1,2}$ the modified Bessel functions. In the limit $x \to 0$ and $y \to 0$, corresponding to the zero size of the hole a, the function $F(x, y)$ tends to 1 and we recover the result for transition radiation (17.14). This also means that, at a given frequency ω, the hole has a small effect on the transition radiation if $a \ll c\gamma/\omega$. In Fig. 17.3 we plot the spectral intensity of the radiation as a function of the angle θ for several values of the parameter $a\omega/c\gamma$.

Fig. 17.3 Angular distribution of the diffraction radiation for various values of the parameter $a\omega/c\gamma$. The dashed line shows the limit $a \to 0$, corresponding to the case of transition radiation

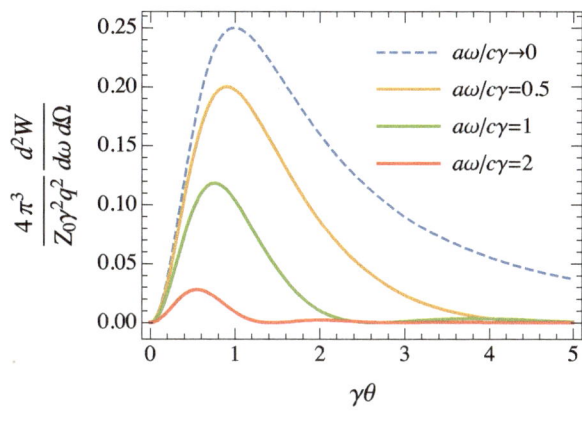

Fig. 17.4 Transition radiation with foil tilted at $45°$

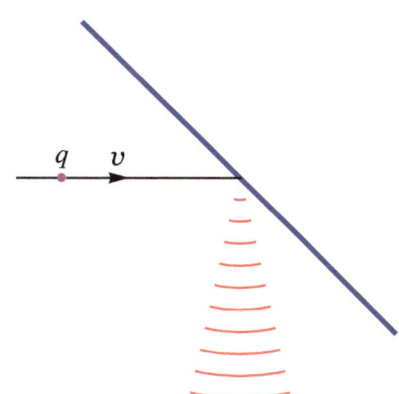

In contrast to transition radiation, the total energy in diffraction radiation is finite even in the perfect conducting limit. It can be obtained by integrating Eq. (17.16) over the frequency ω and the solid angle θ.

Worked Examples

Problem 17.1 *The usual experimental setup for an optical transition radiation (OTR) diagnostic is shown in Fig. 17.4: the beam passes through a metal foil tilted at the angle 45 degrees relative to the beam orbit. Show that in this case the radiation propagates predominantly in the direction perpendicular to the orbit. How can this problem be solved using the method of image charges?*

Solution: The image charge for $t < 0$ (before the charge hits the foil) has charge $-q$ and moves downwards vertically. At the point of contact, it looks like the image charge annihilates with the real charge. The disappearance of the image charge produces the radiation propagating vertically in the direction that the image particle was moving in.

Instead of appealing to image charges, one can equivalently consider the transition radiation as being reflected off of the foil. Both descriptions are only accurate for frequencies sufficiently low that the foil can be treated as a good conductor.

If the particle passes completely through the foil then there will be another burst of radiation on the other side, this time moving horizontally.

Problem 17.2 *Calculate the total energy of the diffraction radiation by integrating Eq. (17.16) over the angle and the frequency.*

Solution: The diffraction radiation is localized at small angles, so we can approximate $\sin\theta \approx \theta$ and extend the integration range in the integral over the angle to infinity:

$$
\begin{aligned}
\mathcal{W} &= 2\pi \int_0^\infty \theta d\theta \int_0^\infty d\omega \frac{d^2\mathcal{W}}{d\omega d\Omega} \\
&= \frac{Z_0 q^2}{2\pi^2} \int_0^\infty \frac{\theta^3 d\theta}{(\gamma^{-2}+\theta^2)^2} \int_0^\infty d\omega F\left(\frac{\omega a \theta}{c}, \frac{\omega a}{c\gamma}\right) .
\end{aligned}
$$

Introducing the integration variables $u = \gamma\theta$ and $v = a\omega/c\gamma$ we obtain

$$
\mathcal{W} = A \frac{Z_0 q^2 c\gamma}{2\pi^2 a} ,
$$

where the numerical factor A is

$$
A = \int_0^\infty \frac{u^3 du}{(1+u^2)^2} \int_0^\infty dv F(uv, v) .
$$

The two-dimensional integral can be calculated numerically and gives

$$
A \approx 0.90 .
$$

Reference

1. Z. Huang, G. Stupakov, M. Zolotorev, Calculation and optimization of laser acceleration in vacuum. Phys. Rev. ST Accel. Beams **7**, 011302 (2004)

Chapter 18
Synchrotron Radiation

A relativistic charge moving along a circular orbit emits *synchrotron radiation*. In this chapter we will calculate the intensity of the radiation in the limit $\gamma \gg 1$. Using the Liénard–Wiechert potentials we first derive the fields at a large distance from the charge in the plane of the orbit and find the radiation spectrum. We then discuss angular and spectral distributions of the synchrotron radiation for arbitrary angles.

18.1 Synchrotron Radiation Pulse in the Plane of the Orbit

To simplify calculations, we begin our analysis with the particular case of the radiation emitted in the plane of the orbit. The layout for our calculation is shown in Fig. 18.1. A point charge q moves on a circular orbit of radius ρ with constant magnitude of the velocity v. When the orbit is the result of a transverse magnetic field, ρ is given by Eq. (5.2). An observer is located at point O in the plane of the orbit far from the origin of the radiation. The observer will see a periodic sequence of pulses of electromagnetic radiation repeating with the revolution period of the particle in the ring. Each pulse originates from the region $x \approx z \approx 0$ in the coordinate system shown in the figure (the origin of the coordinate system is located at the point where the observation line touches the circle).

To calculate the radiation electromagnetic field, we will use the Liénard–Wiechert potentials (15.10). Because the observer is located at a large distance from the particle, we can replace R in the denominator of Eq. (15.10) by r, the distance from the observation point to the origin of the coordinate system:

$$A(r, t) = \frac{Z_0 q}{4\pi r} \frac{\beta(t_{\text{ret}})}{1 - \beta(t_{\text{ret}}) \cdot n} . \tag{18.1}$$

© Springer International Publishing AG, part of Springer Nature 2018
G. Stupakov and G. Penn, *Classical Mechanics and Electromagnetism in Accelerator Physics*, Graduate Texts in Physics,
https://doi.org/10.1007/978-3-319-90188-6_18

Fig. 18.1 A schematic
showing the particle's orbit
and the observation point.
The *y* axis is directed out of
the page

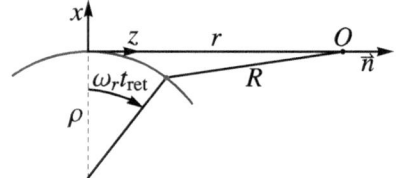

We will also use the fact that locally the radiation field can be represented by a plane
wave for which, according to Eq. (13.5), the magnetic field can be found from the
vector potential,

$$\boldsymbol{B} = -\frac{1}{c}\hat{z} \times \frac{\partial \boldsymbol{A}}{\partial t}, \tag{18.2}$$

where we have replaced $\boldsymbol{n}$ in Eq. (13.5) by the unit vector in the z direction.

We set the time $t = 0$ at the moment when the particle passes through the origin
of the coordinate system; the position of the particle on the circle at time t_{ret} is given
by the angle $\omega_r t_{\text{ret}}$, as shown in Fig. 18.1, with the angular revolution frequency $\omega_r = \beta c/\rho$. As we will see later, the main contribution to the radiation pulse comes from a
small part of the particle trajectory where the angle $\omega_r t_{\text{ret}}$ is small, $\omega_r t_{\text{ret}} \ll 1$. Due to
the smallness of this angle we can use an approximation $R(t_{\text{ret}}) \approx r - \rho \sin(\omega_r t_{\text{ret}})$
which upon substitution into the equation $R(t_{\text{ret}}) = c(t - t_{\text{ret}})$ gives

$$r - \rho \sin(\omega_r t_{\text{ret}}) = c(t - t_{\text{ret}}). \tag{18.3}$$

In what follows, we will replace t_{ret} by the dimensionless variable ξ, $\xi = t_{\text{ret}} c/\rho \approx \omega_r t_{\text{ret}}$. We see that ξ is approximately equal to the angle between the position of the
particle on the circle and the x-axis, which, by assumption, is much smaller than one,
$|\xi| \ll 1$. Equation (18.3) can now be written as

$$ct - r = \rho[\xi - \sin(\beta\xi)]. \tag{18.4}$$

For the particle velocity at the retarded time we have

$$\beta_x(t_{\text{ret}}) = -\beta \sin(\omega_r t_{\text{ret}}) = -\beta \sin(\beta\xi),$$
$$\beta_z(t_{\text{ret}}) = \beta \cos(\omega_r t_{\text{ret}}) = \beta \cos(\beta\xi). \tag{18.5}$$

Because the vector potential A has only x and z components, it follows from Eq. (18.2)
that $\boldsymbol{B}$ is directed along y,

$$cB_y = -\frac{\partial A_x}{\partial t} = -\frac{\partial A_x/\partial \xi}{\partial t/\partial \xi}. \tag{18.6}$$

The x-component of the vector potential (18.1) as a function of ξ is given by

Fig. 18.2 The radiation pulse of the electromagnetic field in dimensionless variables

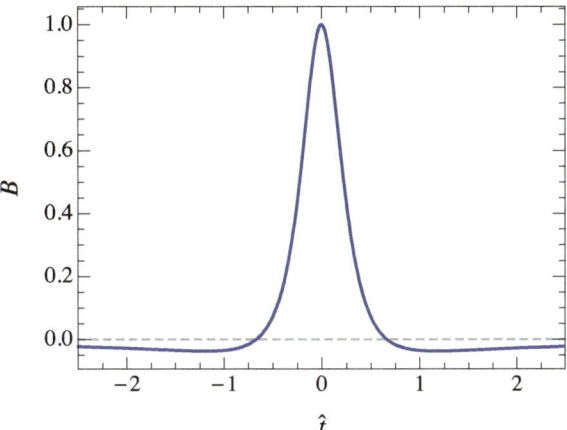

$$A_x = \frac{Z_0 q}{4\pi r} \frac{\beta_x(t_{\mathrm{ret}})}{1 - \beta_z(t_{\mathrm{ret}})} = -\frac{Z_0 q}{4\pi r} \frac{\beta \sin(\beta\xi)}{1 - \beta \cos(\beta\xi)}, \tag{18.7}$$

and the function $t(\xi)$, the derivative of which appears in the denominator of Eq. (18.6), is defined by Eq. (18.4). Using the smallness of ξ we expand the trigonometric functions,

$$\xi - \sin(\beta\xi) \approx \xi(1 - \beta) + \frac{1}{6}\xi^3 \approx \frac{1}{2\gamma^2}\xi + \frac{1}{6}\xi^3$$

$$1 - \beta\cos(\beta\xi) \approx 1 - \beta + \frac{1}{2}\xi^2 \approx \frac{1}{2\gamma^2} + \frac{1}{2}\xi^2. \tag{18.8}$$

Substituting these expressions into Eqs. (18.4) and (18.7) we find

$$A_x = -\frac{Z_0 q}{4\pi r} \frac{2\xi}{\gamma^{-2} + \xi^2}, \qquad t = \frac{r}{c} + \frac{\rho}{c}\left(\frac{1}{2\gamma^2}\xi + \frac{1}{6}\xi^3\right), \tag{18.9}$$

with the magnetic field

$$B_y = \frac{Z_0 q}{\pi r \rho} \frac{\gamma^{-2} - \xi^2}{(\xi^2 + \gamma^{-2})^3}. \tag{18.10}$$

The equation for the magnetic field is further simplified if we use the dimensionless time variable $\hat{t} = (c\gamma^3/\rho)(t - r/c)$ and the dimensionless magnetic field $\hat{B} = (\pi r \rho / Z_0 q \gamma^4) B_y$. We then have

$$\hat{B} = \frac{1 - \zeta^2}{(\zeta^2 + 1)^3}, \qquad \hat{t} = \frac{1}{2}\zeta + \frac{1}{6}\zeta^3, \tag{18.11}$$

where $\zeta = \gamma\xi$. These two equations implicitly define the function $\hat{B}(\hat{t})$ whose plot is shown in Fig. 18.2. We see from this plot that the characteristic width of the pulse $\Delta\hat{t} \sim 1$, which means that the duration of the pulse in physical units is

$$\Delta t \sim \frac{\rho}{c\gamma^3} .$$ (18.12)

If we make the Fourier transformation of the magnetic field, the width of the spectrum can be estimated as $\Delta\omega \sim 1/\Delta t \sim c\gamma^3/\rho$. In the next section we will study this spectrum in more detail.

18.2   Radiation Spectrum in the Plane of the Orbit

Using the results of the previous section we can now calculate the energy radiated in a unit solid angle $d\Omega$ along the x-z plane of the orbit and the spectrum of the radiation, as discussed in Sect. 17.2. This energy is given by the time integral (17.7) of the Poynting vector, and the angular distribution of the spectral intensity and the spectrum are defined by Eqs. (17.10) and (17.11), respectively. The magnetic field B in these equations now has only a B_y component.

To calculate $\tilde{B}_y(\omega)$ it is convenient to start from Eq. (18.6) which in the Fourier representation becomes

$$c\tilde{B}_y(\omega) = i\omega\tilde{A}_x(\omega) .$$ (18.13)

We then use Eq. (18.9) to find the Fourier transform of A_x,

$$\begin{aligned}
\tilde{A}_x(\omega) &= \int_{-\infty}^{\infty} A_x(t)e^{i\omega t}\,dt \\
&= \int_{-\infty}^{\infty} A_x(\xi)e^{i\omega t(\xi)}\frac{dt}{d\xi}\,d\xi \\
&= -\frac{Z_0 q}{4\pi r}\frac{\rho}{c}e^{i\omega r/c}\int_{-\infty}^{\infty} \xi e^{i(\omega\rho/2c)(\gamma^{-2}\xi+\xi^3/3)}\,d\xi .
\end{aligned}$$ (18.14)

Introducing the new integration variable $\zeta = \gamma\xi$ and the *critical frequency*

$$\omega_c = \frac{3c\gamma^3}{2\rho} ,$$ (18.15)

we find

$$\tilde{A}_x(\omega) = -i\frac{Z_0 q}{4\pi r}\frac{\rho}{c\gamma^2}e^{i\omega r/c}F\left(\frac{3\omega}{4\omega_c}\right) ,$$ (18.16)

where

$$F(x) = \text{Im} \int_{-\infty}^{\infty} \zeta e^{ix(\zeta + \zeta^3/3)} d\zeta = \frac{2}{\sqrt{3}} K_{2/3}\left(\frac{2x}{3}\right), \tag{18.17}$$

with $K_{2/3}$ the modified Bessel function of the second type of order $\frac{2}{3}$. Note that the real part of the integral in Eq. (18.17) is equal to zero because of the symmetry of the integrand. This gives for the spectrum the following expression:

$$\frac{d^2\mathcal{W}}{d\omega d\Omega} = \frac{Z_0 q^2}{12\pi^3}\left(\frac{\omega\rho}{c}\right)^2\left(\frac{1}{\gamma^2}\right)^2 K_{2/3}^2\left(\frac{\omega}{2\omega_c}\right). \tag{18.18}$$

18.3 Synchrotron Radiation Out of the Orbit Plane

Our analysis in the previous two sections was carried out for the radiation in the plane of the orbit. A more general consideration is needed to treat the radiation propagating out of plane, at an angle $\psi \neq 0$, as shown in Fig. 18.3. While conceptually similar to what has been done above, these calculations are more involved, and we will not try to reproduce them here giving only the final result for the radiation spectrum.

A formula that generalizes Eq. (18.18) for the case $\psi \neq 0$ is

$$\frac{d^2\mathcal{W}}{d\omega d\Omega} = \frac{Z_0 q^2}{12\pi^3}\left(\frac{\omega\rho}{c}\right)^2\left(\frac{1}{\gamma^2} + \psi^2\right)^2\left[K_{2/3}^2(\chi) + \frac{\psi^2}{1/\gamma^2 + \psi^2}K_{1/3}^2(\chi)\right], \tag{18.19}$$

where

$$\chi = \frac{\omega\rho}{3c}\left(\frac{1}{\gamma^2} + \psi^2\right)^{3/2} = \frac{\omega}{2\omega_c}\left(1 + \gamma^2\psi^2\right)^{3/2}. \tag{18.20}$$

It is easy to see that setting $\psi = 0$ reduces this expression to Eq. (18.18).

Fig. 18.3 The particle's orbit and the coordinate system for the off-axis radiation shown by the red wiggling line

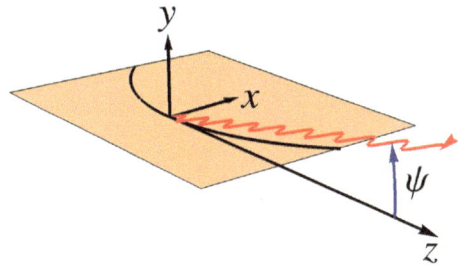

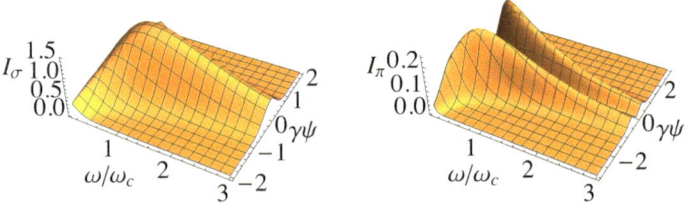

Fig. 18.4 Intensity of σ and π modes in arbitrary units

The modified Bessel functions of the second type $K_{2/3}(\chi)$ and $K_{1/3}(\chi)$ decay exponentially when their argument becomes much larger than unity. This means that in the range of frequencies $\omega \sim \omega_c$ the angular spread of the radiation can be estimated from $\chi \sim 1$ as $|\psi| \sim 1/\gamma$. This estimate is in agreement with our general conclusion in Sect. 15.4 that the bulk of the radiation of a relativistic particle is localized within an angle $1/\gamma$. Radiation at lower frequencies, $\omega \ll \omega_c$, has a larger angular spread $\psi \sim (\omega_c/\omega)^{1/3}/\gamma \sim (c/\omega\rho)^{1/3}$.

The two terms in the square brackets of Eq. (18.19) correspond to different polarizations of the radiation. The first one is the so-called σ mode which has nonzero E_x and B_y; this is the only polarization that survives in the limit $\psi = 0$. The second term has fields with nonzero E_y and B_x; this polarization is called the π mode. The angular distribution of the spectral intensity, $d^2W/d\omega d\Omega$, for these two modes is shown in Fig. 18.4. Note that the intensity of the π mode is zero in the plane of the orbit, $\psi = 0$.

18.4   Integral Characteristics of the Synchrotron Radiation

To find the total spectral energy $dW/d\omega$ radiated in one revolution into all angles, we need to integrate (18.19) over the solid angle $d\Omega$. This integration consists of two steps. First, we need to integrate over the angle ψ. Due to the fast convergence of the integral over ψ in the region $|\psi| \ll 1$, the limits of integration can be extended from minus to plus infinity. Second, we need to take into account that the radiation intensity does not depend on the angle in the plane of the orbit, which we will denote by θ. The integration over the angle θ extends from 0 to 2π covering all possible directions into which a particle radiates in one turn on the circle,

$$\frac{dW}{d\omega} = \int d\Omega \frac{d^2W}{d\omega d\Omega} = \int_0^{2\pi} d\theta \int_{-\infty}^{\infty} d\psi \frac{d^2W}{d\omega d\Omega} = 2\pi \int_{-\infty}^{\infty} d\psi \frac{d^2W}{d\omega d\Omega}. \quad (18.21)$$

The result can be written in the following form:

Fig. 18.5 S function

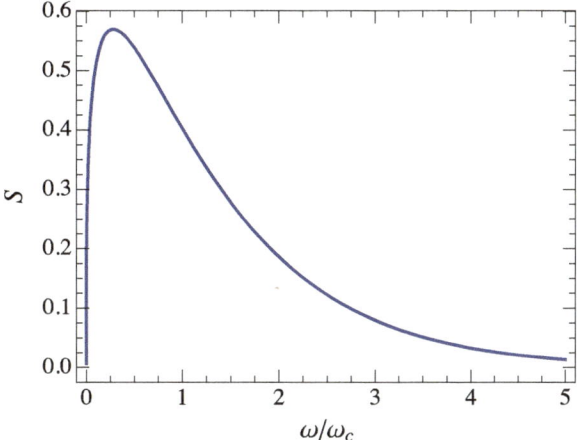

$$\frac{dW}{d\omega} = \frac{2\pi\rho}{c}\frac{Z_0 q^2 c\gamma}{9\pi\rho} S\left(\frac{\omega}{\omega_c}\right), \qquad (18.22)$$

where, substituting $\tau = \gamma\psi$,

$$S(x) = \frac{27x^2}{16\pi^2}\int_{-\infty}^{\infty} d\tau\,(1+\tau^2)^2$$

$$\times \left[K_{2/3}^2\left(\frac{x}{2}(1+\tau^2)^{3/2}\right) + \frac{\tau^2}{1+\tau^2}K_{1/3}^2\left(\frac{x}{2}(1+\tau^2)^{3/2}\right) \right]. \qquad (18.23)$$

The function S is normalized to one, $\int_0^\infty dx\, S(x) = 1$; it is plotted in Fig. 18.5. One can show (with some difficulty) that the function S can also be written as

$$S(x) = \frac{9\sqrt{3}}{8\pi}x\int_x^\infty K_{5/3}(y)dy. \qquad (18.24)$$

This form of S is the most common definition used and is convenient for establishing its asymptotic expressions in the limits of small and large values of the argument x:

$$S(x) \simeq \frac{27}{8\pi}\frac{\sqrt{3}}{2^{1/3}}\Gamma\left(\frac{5}{3}\right)x^{1/3}, \qquad\qquad x \ll 1, \qquad (18.25a)$$

$$S(x) \simeq \frac{9}{8}\sqrt{\frac{3}{2\pi}}\sqrt{x}e^{-x}, \qquad\qquad x \gg 1. \qquad (18.25b)$$

Integrating $dW/d\omega$ over all frequencies, we find the total energy W radiated in one revolution:

$$W_r = \int_0^\infty d\omega \frac{dW}{d\omega} = \frac{2\pi\rho}{c} \frac{Z_0 q^2 c\gamma}{9\pi\rho} \omega_c . \tag{18.26}$$

The radiation power (the energy radiated per unit time) by a point charge is

$$\mathcal{P} = \frac{W_r}{2\pi\rho/c} = \frac{Z_0 q^2 c\gamma}{9\pi\rho} \omega_c = \frac{Z_0 q^2 c^2 \gamma^4}{6\pi\rho^2} = \frac{2 r_c m c^3 \gamma^4}{3\rho^2} , \tag{18.27}$$

where $r_c = q^2/(4\pi\epsilon_0 mc^2)$ is the classical radius of the particle.

In quantum electrodynamics the radiation is represented by photons travelling with the speed of light, where each photon of a given frequency carries energy $h\nu = \hbar\omega$. Estimating the typical frequency as ω_c, we can write the number of photons emitted per unit length for a relativistic particle as

$$\frac{dN}{dz} \approx \frac{P}{\hbar\omega_c c} = \frac{Z_0 q^2}{9\pi\hbar} \frac{\gamma}{\rho} = \frac{4}{9} \alpha \left(\frac{\rho}{\gamma}\right)^{-1} , \tag{18.28}$$

taking q as the electron charge and where $\alpha = e^2/4\pi\epsilon_0\hbar c$ is the fine structure constant, $\alpha \approx 1/137$. The length scale ρ/γ corresponds to a circular arc of $1/\gamma$ radians, which is the instantaneous angular spread of the radiated beam as it crosses a fixed location a large distance away. We can be a little more precise by noting that the typical photon frequency weighted by *photon number* instead of power is about $0.31\,\omega_c$. This gives for dN/dz about $1.44\,\alpha/(\rho/\gamma)$.

Worked Examples

Problem 18.1 *Find the asymptotic dependence $B_y(t)$ for $|t - r/c| \gg \rho/c\gamma^3$.*

Solution: For $|t - r/c| \gg \rho/c\gamma^3$, we have $|\hat{t}| = |t - r/c|/(\rho/c\gamma^3) \gg 1$. The equation $\hat{t} = \zeta/2 + \zeta^3/6$ then implies $|\zeta| \gg 1$ and $\hat{t} \approx \zeta^3/6$. For the dimensionless magnetic field $\hat{B}$ we obtain

$$\hat{B} = \frac{1 - \zeta^2}{(\zeta^2 + 1)^3} \approx -\zeta^{-4} \approx -(6\hat{t})^{-4/3} .$$

We recall that $B_y \propto \hat{B}$.

Problem 18.2 *Prove that the area under the curve $B_y(t)$ is equal to zero (that is $\int_{-\infty}^\infty B_y(t)dt = 0$).*

Solution: The area under the curve $B_y(t)$ is

$$\int_{-\infty}^\infty B_y(t)dt \propto \int_{-\infty}^\infty \hat{B} d\hat{t} = \int_{-\infty}^\infty \hat{B} \frac{d\hat{t}}{d\zeta} d\zeta$$

$$= \int_{-\infty}^\infty \frac{(1 - \zeta^2)(1 + \zeta^2)}{2(1 + \zeta^2)^3} d\zeta$$

$$= \left. \frac{\zeta/2}{1+\zeta^2} \right|_{-\infty}^{\infty} = 0 \, ,$$

where we used $\hat{t} = \zeta/2 + \zeta^3/6$ to find $d\hat{t}/d\zeta$ and we note that the integration is the reverse of the derivative of A_x used to calculate B_y in Eq. (18.10). We could have seen this more simply by noting that $B_y(t) \propto \partial A_x/\partial t$, so $\int_{-\infty}^{\infty} B_y dt \propto A_x|_{-\infty}^{\infty} = 0$ for $A_x \propto \xi/(\gamma^{-2} + \xi^2)$.

Problem 18.3 *Simplify Eq. (18.19) in the limit $\psi \gg 1/\gamma$. Make a plot of the quantity $\omega^{-2/3} d^2 W/(d\omega d\Omega)$ versus the quantity $\omega\rho\psi^3/c$. Infer from these equations that the angular spread of the radiation at frequency $\omega \ll \omega_c$ is of order of $(c/\omega\rho)^{1/3}$.*

Solution: We are interested in the angular dependence of the radiation at large angles (outside the main peak, $\psi \gg 1/\gamma$). Equation (18.19) becomes

$$\frac{d^2 W}{d\omega d\Omega} \propto \left(\frac{\omega\rho}{c} \right)^2 \psi^4 \left[K_{2/3}^2(\chi) + K_{1/3}^2(\chi) \right] \, ,$$

and

$$\chi \simeq \frac{\omega\rho}{3c} \psi^3 = \frac{\omega}{2\omega_c} \gamma^3 \psi^3 \, .$$

Then we find

$$\frac{d^2 W}{d\omega d\Omega} \propto \left(\frac{\omega\rho}{c} \right)^{2/3} \left(\frac{\omega\rho}{c} \right)^{4/3} \psi^4 \left[K_{2/3}^2(\chi) + K_{1/3}^2(\chi) \right]$$

$$\propto \left(\frac{\omega\rho}{c} \right)^{2/3} \chi^{4/3} \left[K_{2/3}^2(\chi) + K_{1/3}^2(\chi) \right] \, .$$

Figure 18.6 shows the function $\chi^{4/3}[K_{2/3}^2(\chi) + K_{1/3}^2(\chi)]$. It reaches a peak of roughly 1.4 at $\chi \simeq 0.1$, but even for $\chi \ll 1$ it remains above unity in this approximation. At $\chi = 2$ this function is already below 0.1. So we conclude that the radiation is localized in the range of angles $\omega\rho\psi^3/c \lesssim 1$ and the angular spread is

$$\Delta\psi \sim \left(\frac{\lambda}{\rho} \right)^{1/3} \, .$$

This approximation only makes sense for frequencies well below ω_c, otherwise the range of angles is set by $1/\gamma$.

Problem 18.4 *Calculate the RF power needed to compensate the synchrotron radiation in the Advanced Light Source. The parameters are: energy is 1.9 GeV, $I = 0.56$ A, circumference is 200 m.*

Solution: The power from synchrotron radiation of a single electron is $P = 2r_e mc^2 \gamma^4 c/3\rho^2 = cC_\gamma E^4/2\pi\rho^2$, where r_e is the classical electron radius and $C_\gamma \equiv 4\pi r_e/3(mc^2)^3$. With $r_e = 2.82 \times 10^{-15}$ m and $m_e c^2 = 5.11 \times 10^{-4}$ GeV, we find $C_\gamma = 8.9 \times 10^{-5}$ m/GeV³. Thus the power radiated is 54 GeV/s.

Fig. 18.6 Plot of
$\chi^{4/3}\left[K_{2/3}^2(\chi) + K_{1/3}^2(\chi)\right]$.

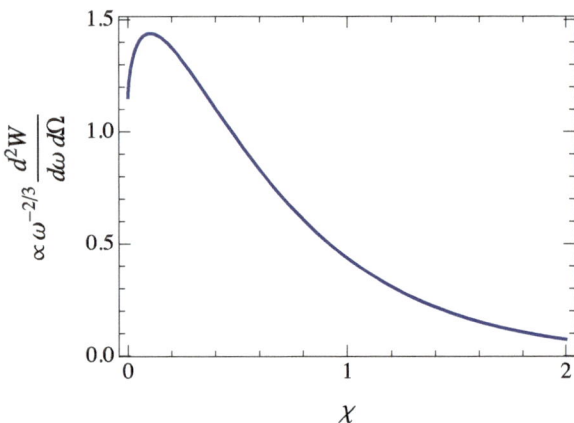

The energy lost in one revolution is

$$\Delta E = \int P dt = \frac{2\pi\rho P}{c} = \frac{C_\gamma E^4}{\rho}$$
$$= 36\,\text{keV}.$$

The total power from all of the electrons in the ring is

$$P_T = (\Delta E)I/e = (36\,\text{kV})(0.56\,\text{A}) = 20\,\text{kW}.$$

Chapter 19
Undulator Radiation

Undulators and wigglers are widely used in modern accelerator-based light sources. The significant amount of radiated power they can emit also make them important components for electron or positron damping rings. We derive the properties of the undulator radiation using the solution of the Thomson scattering problem from Chap. 16.

19.1 Undulators and Wigglers

A plane undulator is shown in Fig. 19.1. The magnetic field in the undulator is directed along the y axis and sinusoidally varies in the z direction,

$$B_y(z) = B_0 \cos k_u z, \tag{19.1}$$

where $k_u = 2\pi/\lambda_u$ with λ_u the undulator period and B_0 the amplitude of the magnetic field. We will denote by N_u the number of periods in the undulator.

We assume that a relativistic beam propagates along the z axis with velocity v. The equation of motion in the horizontal (x–z) plane is obtained by equating the Lorentz force in the x direction to the product of the relativistic mass $m\gamma$ and the acceleration $\ddot{x}$:

$$
\begin{aligned}
m\gamma\ddot{x} &= -qvB_0 \cos k_u z \\
&\approx -qcB_0 \cos k_u ct, \tag{19.2}
\end{aligned}
$$

where we have used the approximation $z \approx vt \approx ct$. The solution of this equation is

© Springer International Publishing AG, part of Springer Nature 2018
G. Stupakov and G. Penn, *Classical Mechanics and Electromagnetism in Accelerator Physics*, Graduate Texts in Physics,
https://doi.org/10.1007/978-3-319-90188-6_19

Fig. 19.1 Magnetic field
and an electron trajectory in
a plane undulator

$$x \approx \frac{q B_0}{m \gamma k_u^2 c} \cos k_u c t \approx \frac{q B_0}{m \gamma k_u^2 c} \cos k_u z \,, \tag{19.3}$$

where we have set two integration constants equal to zero (these constants can be eliminated by the proper choice of the offset and the direction of the z axis). The sinusoidal orbit (19.3) is characterized by the maximum deflection angle

$$\left. \frac{dx}{dz} \right|_{\max} = \frac{q B_0}{m \gamma k_u c} \,. \tag{19.4}$$

Comparing this angle with the inverse Lorentz factor γ^{-1} we introduce an important quantity, the *undulator parameter* K:

$$K = \gamma \left. \frac{dx}{dz} \right|_{\max} = \frac{q B_0}{k_u m c} \simeq 0.934 \, \lambda_u[\mathrm{cm}] \, B_0[\mathrm{Tesla}] \,, \tag{19.5}$$

where for m we have used the value of the electron mass. A device with $K \lesssim 1$ is usually referred to as an *undulator*, and the term *wiggler* is used in the case of large K. We will use the term undulator in the general case when K is not specified.

With the particle orbit in the undulator found above, we could calculate its radiation using the retarded potential formalism as we did for the synchrotron radiation in Chap. 18. Here, however, we will use a different approach, and calculate the spectrum of the radiation by applying the Lorentz transformation to the solution of the Thomson scattering problem studied in Chap. 16.

19.2   Undulator Radiation for $K \ll 1$

Let us consider a long undulator with $K \ll 1$ and a large number of periods, and neglect the effects associated with the entrance to and the exit from the undulator. To calculate the undulator radiation we will transform into the frame of reference in which the particle is on average at rest. We will use the primes to denote quantities in this reference frame.

In the particle frame, the undulator is moving in the negative direction of the z axis with velocity $\mathbf{v} = (0, 0, -v)$. Using the Lorentz transformations for the fields and assuming $\gamma \gg 1$ we find that in this frame the undulator field has both electric and magnetic components perpendicular to the z axis:

$$E'_x = \gamma v B_0 \cos k_u z \approx \gamma c B_0 \cos [k_u \gamma(z' - ct')],$$
$$B'_y = \gamma B_0 \cos k_u z \approx \gamma B_0 \cos [k_u \gamma(z' - ct')], \tag{19.6}$$

where on the last step we have used the Lorentz transformation to express the coordinate z through the coordinate z' in the particle frame and replaced β by 1,

$$z = \gamma(z' - \beta ct') \approx \gamma(z' - ct'). \tag{19.7}$$

In Eq. (19.6) we recognize a plane electromagnetic wave propagating in the negative z direction with the frequency $\omega' = \gamma k_u c$ and the field γ times larger than the lab field of the undulator. This field would cause oscillations of the particle and would result in the Thomson scattering that we studied in Chap. 16. To calculate the parameter a defined by Eq. (16.23), we substitute for the electric field $E_0 \to \gamma c B_0$ and for the frequency $\omega \to \omega' = \gamma k_u c$ and find that a is exactly equal to the undulator parameter (19.5). Because the results of Chap. 16 are valid in the limit $a \ll 1$, we require that $K \ll 1$. We shall also see in Sect. 19.4 that only in the limit $K \ll 1$ is the electron velocity v a good approximation for the velocity used in the Lorentz transformation.

The incident electromagnetic wave is scattered by the electron and the intensity of the scattered radiation is given by Eq. (16.27), which we rewrite here using the new notation,

$$\frac{d\mathcal{P}'}{d\Omega'} = \frac{Z_0}{32\pi^2} \frac{q^4 \gamma^2 B_0^2}{m^2} (1 - \sin^2 \theta' \cos^2 \phi). \tag{19.8}$$

Note that the angle ϕ' in the $x'-y'$ plane is the same as the angle ϕ. We now need to transform the quantities $d\mathcal{P}'$, $d\Omega'$, θ', as well as ω' into the lab frame. For the transformation of the angle θ we can use Eq. (B.13) from the Doppler effect,

$$\sin \theta' = \frac{\sin \theta}{\gamma(1 - \beta \cos \theta)} \approx \frac{2\gamma\theta}{1 + \gamma^2\theta^2}, \tag{19.9}$$

where we have assumed that $\theta \ll 1$, expanded $\cos \theta \approx 1 - \theta^2/2$, and used $1 - \beta \approx 1/2\gamma^2$. The relativistic frequency transformation Eq. (B.16) gives

$$\omega \approx \frac{2\gamma\omega'}{1 + \gamma^2\theta^2} = \frac{2\gamma^2 k_u c}{1 + \gamma^2\theta^2}. \tag{19.10}$$

Note that the frequency now varies with the angle θ and hence depends on the direction of the radiation. We denote by ω_0 the maximum frequency of the radiation which goes in the forward direction, $\theta = 0$. This frequency is equal to

$$\omega_0 = 2\gamma\omega' = 2\gamma^2 k_u c . \tag{19.11}$$

For the differential of the elementary solid angle $d\Omega'$ we write it as

$$\begin{aligned} d\Omega' &= \sin(\theta')d\theta'd\phi \\ &= |d\cos(\theta')|d\phi , \end{aligned} \tag{19.12}$$

and then use Eq. (B.13) to obtain

$$d\Omega' = \frac{1-\beta^2}{(1-\beta\cos\theta)^2}|d\cos(\theta)|d\phi \approx \frac{4\gamma^2}{(1+\gamma^2\theta^2)^2}d\Omega , \tag{19.13}$$

where $d\Omega$ is the solid angle in the lab frame corresponding to $d\Omega'$ in the beam frame. Finally, we need to transform into the lab frame the differential $d\mathcal{P}'$ which has the meaning of the radiated energy of the electromagnetic field per unit time, $d\mathcal{P}' = dE'/dt'$. We know the transformation of time, $dt' = dt/\gamma$, which is the time dilation effect. The easiest way to figure out how to transform the energy is to consider radiation as a collection of photons. In quantum language, the photon energy is equal to $\hbar\omega$, and the number of photons N_{ph} is the same in any reference frame, $N_{ph}' = N_{ph}$ (cf. Problem 16.5). Hence the energy of an ensemble of photons with the same frequency is transformed as the frequency, $dE' = N_{ph}\hbar\omega' = dE(\omega'/\omega)$. We now have

$$\begin{aligned} \frac{d\mathcal{P}}{d\Omega} &= \frac{dE}{d\Omega dt} \\ &= \frac{dE'}{d\Omega'dt'}\frac{\omega}{\omega'}\frac{1}{\gamma}\frac{4\gamma^2}{(1+\gamma^2\theta^2)^2} \\ &= \frac{d\mathcal{P}'}{d\Omega'}\frac{8\gamma^2}{(1+\gamma^2\theta^2)^3} \\ &= \frac{Z_0}{32\pi^2}\frac{q^4\gamma^2 B_0^2}{m^2}\left[1 - \frac{4\gamma^2\theta^2}{(1+\gamma^2\theta^2)^2}\cos^2\phi\right]\frac{8\gamma^2}{(1+\gamma^2\theta^2)^3} \\ &= \frac{Z_0}{4\pi^2}\frac{q^4\gamma^4 B_0^2}{m^2}\frac{(1+\gamma^2\theta^2)^2 - 4\gamma^2\theta^2\cos^2\phi}{(1+\gamma^2\theta^2)^5} . \end{aligned} \tag{19.14}$$

From this equation it follows that the dominant part of the radiation energy is emitted within the angle $\theta \sim 1/\gamma$, in agreement with the general principles of the radiation of relativistic particles discussed in Sect. 15.4.

In Sect. 17.2 we introduced the angular distribution of the spectral intensity of the radiation, Eq. (17.11), so that the integral over the frequency gives the energy radiated per unit solid angle. We can apply the same notion to the undulator radiation and introduce the angular distribution of the spectral power of the radiation, $d\mathcal{P}/d\Omega d\omega$, such that $\int_0^\infty (d\mathcal{P}/d\Omega d\omega)d\omega = d\mathcal{P}/d\Omega$. In the model of an infinitely long undulator that we assume here the frequency is uniquely related to the angle through Eq. (19.10).

Mathematically this one-to-one correspondence between the frequency and the angle can be expressed through a delta function that indicates an infinitely narrow spectrum at each given angle:

$$\frac{dP}{d\Omega d\omega} = \frac{dP}{d\Omega} \delta \left(\omega - \frac{2\gamma^2 k_u c}{1 + \gamma^2 \theta^2} \right). \tag{19.15}$$

It is clear that the integration of $dP/d\Omega d\omega$ over the frequency gives the angular distribution of the power $dP/d\Omega$. The infinitely narrow frequency spectrum here is due to the assumption of an infinitely long undulator. Taking into account a finite undulator length, and hence a finite time of flight through the undulator, introduces a nonzero width of the spectrum, as we will see in the next section.

To find the energy radiated per unit time in all directions we integrate equation (19.14) over Ω using the approximation $\sin \theta \approx \theta$ for small angles,

$$\begin{aligned}
\mathcal{P}_0 &= \int \frac{dP}{d\Omega} d\Omega \approx \int_0^\infty \theta d\theta \int_0^{2\pi} d\phi \frac{dP}{d\Omega} \\
&= \frac{Z_0}{12\pi} \frac{q^4 \gamma^2 B_0^2}{m^2},
\end{aligned} \tag{19.16}$$

where we have used the mathematical identity $\int_0^\infty (1 + x^2)(1 + x)^{-5} dx = 1/3$. If we rewrite in this expression the square of the peak amplitude of the magnetic field B_0^2 in terms of the square of the field averaged over the undulator period, $B_0^2 = 2\langle B^2 \rangle$, and compare it with the intensity of the synchrotron radiation (18.27) (remembering that $\rho = \gamma m c / q B$), we find that they are equal. Hence the radiated power from the undulator, per unit time, is equal to the radiated power from a bending magnet with the same averaged square of the magnetic field. In contrast to the synchrotron radiation, where the radiation spectrum is broad, in an undulator with a small K the radiation at a given angle θ is localized within a narrow range around the frequency (19.10).

19.3   Effects of Finite Length of the Undulator

Taking now into account the finite length of the undulator, we will assume that the number of periods in the undulator N_u is large, $N_u \gg 1$. As was pointed out in the previous section, the finite number of periods in the undulator results in a nonzero width of the spectrum of the radiation. The shape of the spectrum can be rather easily established if one looks at the time dependence of the electric field in the radiation pulse. In the particle reference frame an undulator with N_u periods is represented by the same plane wave (19.6) which now has N_u periods. The particle executes N_u oscillations in the wave and emits a scattered radiation pulse with the same N_u number of periods. An example of such pulse is shown in Fig. 19.2 for $N_u = 10$. One can see that the pulse in this case is a piece of a sinusoidal function with the number of oscillations equal to the number of periods in the undulator; the oscillation frequency is given by the same Eq. (19.10), which we now denote by $\omega_1(\theta)$,

$$\omega_1(\theta) = \frac{2\gamma^2 k_u c}{1 + \gamma^2\theta^2} . \qquad (19.17)$$

As is well known, the Fourier transform of a truncated sinusoidal pulse is given by the sinc function, and is proportional to

$$\frac{\sin(\pi N_u \Delta\omega/\omega_1(\theta))}{\Delta\omega} , \qquad (19.18)$$

where $\Delta\omega = \omega - \omega_1(\theta)$, and we have assumed $|\Delta\omega| \ll \omega_1(\theta)$. It is not surprising that the delta function in (19.15) is now replaced by the square (because the power is proportional to the electric field squared) of the sinc function:

$$\frac{d\mathcal{P}}{d\Omega d\omega} = \frac{d\mathcal{P}}{d\Omega} \frac{\sin(\pi N_u \Delta\omega/\omega_1(\theta))^2}{\pi^2 N_u \Delta\omega^2/\omega_1(\theta)} , \qquad (19.19)$$

where $d\mathcal{P}/d\Omega$ is given by Eq. (19.14) and the additional factors on the right-hand side are chosen from the requirement that the integration of $d\mathcal{P}/d\Omega d\omega$ over the frequency ω should give $d\mathcal{P}/d\Omega$. It follows from this equation that the spectral width of the undulator radiation in a given direction θ is inversely proportional to the number of periods, $\Delta\omega \sim \omega_1(\theta)/N_u$.

For an undulator of finite length it makes more sense to talk about the total radiated energy $\mathcal{W}$ and its angular spectrum $d\mathcal{W}/d\Omega d\omega$ rather than the radiation power $\mathcal{P}$ and the angular spectrum $d\mathcal{P}/d\Omega d\omega$. The total energy radiated from the length of the undulator is obtained by multiplying the power by the time of passage through the undulator, approximately equal to L_u/c where $L_u = 2\pi N_u/k_u$ is the length of the undulator. In particular, $d\mathcal{W}/d\Omega d\omega$ is given by the right-hand side of Eq. (19.19) multiplied by L_u/c.

19.4 Wiggler Radiation for $K \gtrsim 1$

The undulator radiation in a wiggler with large K can be derived using the same approach as for $K \ll 1$ case — using the Lorentz transformation to the particle frame of reference. However, calculations become much more involved when $K \gtrsim 1$, so we will limit our analysis here to the derivation of the radiation frequency and a discussion of some qualitative properties of the spectrum. We assume that $K \ll \gamma$.

First, we need to calculate the averaged velocity $\bar{v}_z$ of the particle moving along the z-axis of the wiggler. The x velocity can be found from Eq. (19.3),

Fig. 19.2 Electric field versus time (in arbitrary units) for undulator radiation in the forward direction with $N_u = 10$, $K = 0.1$ and $\gamma = 10$. The time is normalized by the period $T_0 = 2\pi/\omega_1(0)$

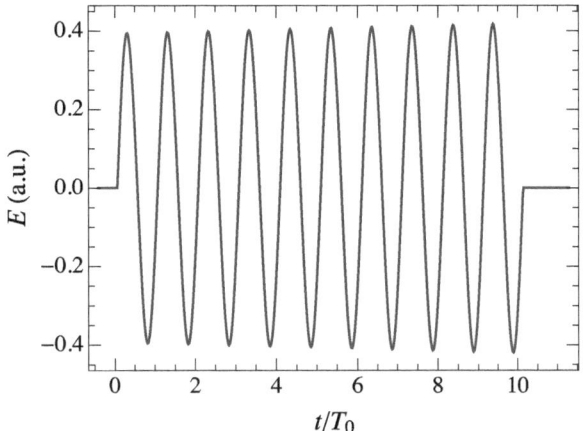

$$v_x = \frac{dx}{dt} = -\frac{Kc}{\gamma} \sin(k_u ct), \qquad (19.20)$$

which gives for v_z

$$v_z = \sqrt{v^2 - v_x^2} \approx v\left(1 - \frac{v_x^2}{2c^2}\right) = v\left[1 - \frac{K^2}{2\gamma^2} \sin^2(k_u ct)\right], \qquad (19.21)$$

where we have used the approximations $|v_x| \ll v$ and $v \approx c$. Averaging Eq. (19.21) over time, we obtain

$$\bar{v}_z = v\left(1 - \frac{K^2}{4\gamma^2}\right). \qquad (19.22)$$

This is the velocity of the reference frame in which the particle, on average, is at rest. Given that $K \ll \gamma$, we see that this velocity is still close to the speed of light, but the effective gamma factor corresponding to $\bar{v}_z$, which we denote by γ_z, is smaller than the original γ:

$$\gamma_z = \frac{1}{\sqrt{1 - \bar{v}_z^2/c^2}} \approx \left[1 - \frac{v^2}{c^2}\left(1 - \frac{K^2}{2\gamma^2}\right)\right]^{-1/2} \approx \left(1 - \frac{v^2}{c^2} + \frac{K^2}{2\gamma^2}\right)^{-1/2}$$

$$= \left(\frac{1}{\gamma^2} + \frac{K^2}{2\gamma^2}\right)^{-1/2} = \frac{\gamma}{\sqrt{1 + K^2/2}}. \qquad (19.23)$$

When we make the Lorentz transformation to the particle frame of reference, the frequency of the electromagnetic plane wave Lorentz transformed from the wiggler field will be $\omega' = \gamma_z k_u c$; correspondingly γ_z should replace γ in Eq. (19.10) for the radiation frequency in the lab frame,

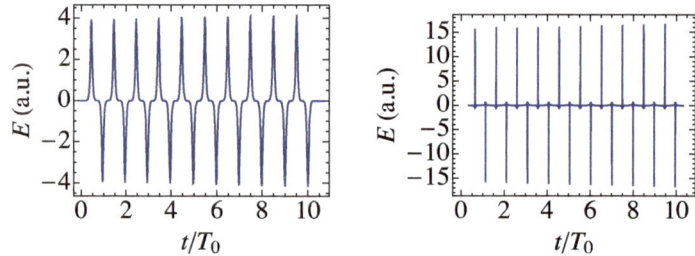

Fig. 19.3 Electric field versus time (in arbitrary units) for an undulator radiation with $K = 1$ (left panel) and $K = 4$ (right panel). The time is normalized by $T_0 = 2\pi/\omega_0$. The undulator has $N_u = 10$ periods and $\gamma = 10$

Fig. 19.4 Undulator spectrum for $K = 4$ (in the limit $N_u \gg 1$). The frequency ω_0 is given by Eq. (19.25)

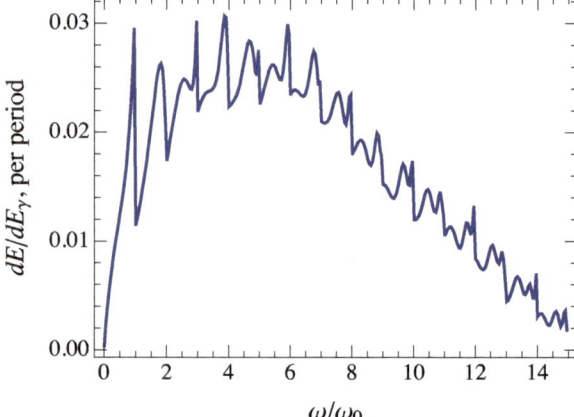

$$\omega = \frac{2\gamma_z^2 k_u c}{1 + \gamma_z^2 \theta^2} = \frac{2\gamma^2 k_u c}{1 + K^2/2 + \gamma^2 \theta^2} \, . \tag{19.24}$$

In particular, for $\theta = 0$, that is the frequency in the forward direction, we obtain

$$\omega_0 = \frac{2k_u c \gamma^2}{1 + K^2/2} \, . \tag{19.25}$$

This, however, is not the maximum frequency of the radiation as it was in the case of $K \ll 1$. The reason for this is illustrated by Fig. 19.3 where we show the electric field of the wiggler radiation detected in the forward direction for the two cases $K = 1$ and $K = 4$. The repetition rate of the pulses is equal to the frequency ω_0, but for larger values of K the spikes of the electric field become narrower which leads to a rich content of *higher harmonics* in the radiation spectrum. Indeed, as shown in Fig. 19.4 for $K = 4$, the undulator spectrum (integrated over all angles Ω) extends to frequencies much higher than ω_0.

Fig. 19.5 The spectrum of the undulator radiation given by Eq. (19.26)

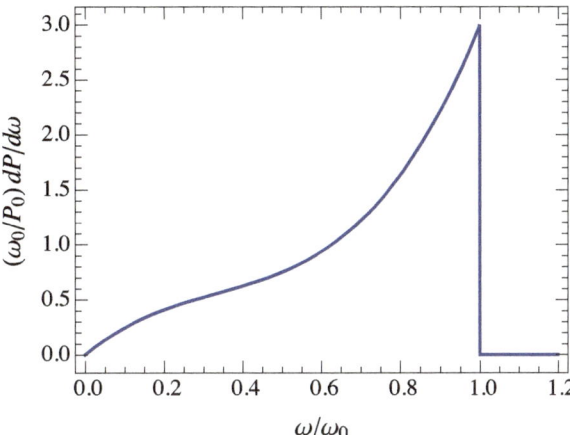

Worked Examples

Problem 19.1 *Integrate equation* (19.15) *over the solid angle and show that the intensity of the radiation per unit frequency is*

$$\frac{d\mathcal{P}}{d\omega} = \frac{3\mathcal{P}_0}{\omega_0} \frac{\omega}{\omega_0} \left[1 - 2\frac{\omega}{\omega_0} + 2\left(\frac{\omega}{\omega_0}\right)^2 \right] \tag{19.26}$$

for $\omega < \omega_0$, and zero for $\omega > \omega_0$. The plot of this function is shown in Fig. 19.5.

Solution: Expressing the elementary solid angle as $d\Omega = \theta d\theta d\phi$ (where we approximated $\sin\theta \approx \theta$) we can write $d\mathcal{P}/d\Omega d\omega = \theta^{-1} d\mathcal{P}/d\theta d\phi d\omega$. We then first integrate it over ϕ to obtain

$$\frac{d\mathcal{P}}{\theta d\theta d\omega} = \frac{Z_0}{4\pi^2} \frac{q^4 \gamma^4 B_0^2}{m^2} \delta\left(\omega - \frac{2\gamma^2 k_u c}{1 + \gamma^2\theta^2}\right) \int_0^{2\pi} d\phi \frac{(1 + \gamma^2\theta^2)^2 - 4\gamma^2\theta^2 \cos^2\phi}{(1 + \gamma^2\theta^2)^5}$$

$$= \frac{Z_0}{2\pi} \frac{q^4 \gamma^4 B_0^2}{m^2} \frac{(1 + \gamma^2\theta^2)^2 - 2\gamma^2\theta^2}{(1 + \gamma^2\theta^2)^5} \delta\left(\omega - \frac{2\gamma^2 k_u c}{1 + \gamma^2\theta^2}\right). \tag{19.27}$$

In the second step we integrate this equation over the angle θ,

$$\frac{d\mathcal{P}}{d\omega} = \int_0^\infty d\theta \frac{d\mathcal{P}}{d\theta d\omega},$$

where we have extended the upper limit of the integration range to infinity because the dominant contribution to the integral comes from small angles $\sim 1/\gamma$. Integration of the delta function in Eq. (19.27) gives an inverse derivative of its argument with respect to θ with the rule that the angle θ should be expressed through the frequency ω from the Eq. (19.10). We then obtain

$$\frac{dP}{d\omega} = \frac{Z_0}{2\pi} \frac{q^4 \gamma^4 B_0^2}{m^2} \frac{(1 + \gamma^2 \theta^2)^2}{2\omega_0 \gamma^2} \frac{(1 + \gamma^2 \theta^2)^2 - 2\gamma^2 \theta^2}{(1 + \gamma^2 \theta^2)^5}$$

$$= \frac{Z_0}{4\pi} \frac{q^4 \gamma^2 B_0^2}{m^2 \omega_0} \left(\frac{\omega}{\omega_0}\right)^3 \left[\left(\frac{\omega_0}{\omega}\right)^2 - 2\frac{\omega_0}{\omega} + 2\right]$$

$$= \frac{3P_0}{\omega_0} \frac{\omega}{\omega_0} \left[1 - 2\frac{\omega}{\omega_0} + 2\left(\frac{\omega}{\omega_0}\right)^2\right],$$

with $P_0 = Z_0 q^4 \gamma^2 B_0^2 / 12\pi m^2$. From the definition, ω_0 is at $\theta = 0$, when the frequency is maximized, so $\omega \leq \omega_0$.

Chapter 20
Formation Length of Radiation and Coherent Effects

The radiation process is not instantaneous — it requires some time and free space around the orbit for the radiation to be formed. In this chapter we estimate the longitudinal extent and transverse size of the free space volume needed for the synchrotron radiation. We then analyze the radiation of a bunch of many particles.

20.1 Longitudinal Formation Length

It takes some time and space for a moving charge to generate radiation. Let us take a closer look at the derivation in Sect. 18.1 and try to figure out what fraction of the length of the orbit is involved in the formation of the synchrotron pulse.

In Eq. (18.10) the variable $\xi = c\tau/\rho$ is related to the retarded time τ. It follows from this equation that the characteristic width of the electromagnetic pulse in variable ξ is $\Delta\xi \sim \gamma^{-1}$, which corresponds to the time duration $\tau \sim \rho/c\gamma$. Hence the length of the orbit necessary for the formation of the radiation pulse, which we will call the *formation length* l_f, is

$$l_f \sim c\tau \sim \frac{\rho}{\gamma}\,. \tag{20.1}$$

In quantum language, the formation length is needed to convert a virtual photon carried by the electromagnetic field of a particle into a real photon.

How does this formation length agree with the duration of the radiation pulse $\sim \rho/\gamma^3 c$? Since the charge is moving with the velocity $v \approx c(1 - 1/2\gamma^2)$, the relative velocity with which the electromagnetic field propagating with the speed of light overtakes the charge is $\Delta v \sim c/\gamma^2$. During the formation time τ the radiated field propagates ahead of the charge at the distance $\tau \Delta v \sim \rho/\gamma^3$, which explains the duration of the pulse (18.12).

© Springer International Publishing AG, part of Springer Nature 2018
G. Stupakov and G. Penn, *Classical Mechanics and Electromagnetism in Accelerator Physics*, Graduate Texts in Physics,
https://doi.org/10.1007/978-3-319-90188-6_20

Fig. 20.1 The orbit consists
of a circular arc limited by
angles $-\varphi_2 < \varphi < \varphi_1$. The
horizontal arrow shows the
direction in which the
radiation is observed

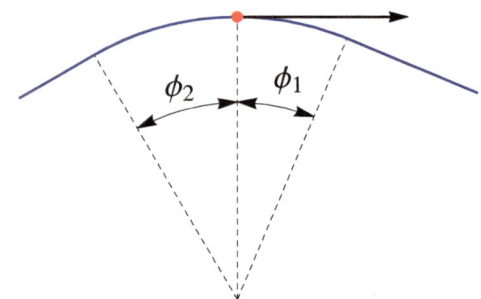

The concept of the formation length has important practical applications. In accelerators particles emit synchrotron radiation when passing through dipole magnets of finite length, where their trajectory is an arc of a circle. Outside of the magnet the trajectories are represented by segments of straight lines for which the formal value of the bending radius is infinity. The question then arises: under what conditions can one apply the results of Chap. 18 derived for a circular orbit to the case of a magnet of finite length? The answer to this question is that the circular orbit approximation is valid if the magnet length L is several times longer than l_f. Because the change of angle of a charged particle passing through a magnet is roughly equal to L/ρ (we assume that the angle is small), this means that the angle must be several times larger than $1/\gamma$.

Radiation from a magnet that is shorter than l_f has very different properties than what we have calculated in Chap. 18. Even for longer magnets, the radiation coming from the region within a distance $\sim l_f$ from the edges of the magnet will be different. Some of the properties of radiation from short magnets can be explained using the time profile of the radiation pulse shown in Fig. 18.2. Recall that each point on this plot has a corresponding position on the orbit from which the field at this point was emitted. This position is determined by the value of the variable ξ approximately equal to the angle with the vertical axis.

Let us now consider an orbit that consists of a circular arc with the angular extension $-\varphi_2 < \varphi < \varphi_1$ as shown in Fig. 20.1; outside of the arc the particle moves along straight lines (tangential to the end points of the arc) with a constant velocity. Since there is no acceleration in the straight parts of the orbit, the radiation pulse shown in Fig. 18.2 will be truncated: the value of the radiation field $\hat{B}$ becomes zero for $\xi < -\varphi_2$ or $\xi > \varphi_1$, while it remains the same for the points on the arc where $-\varphi_2 < \xi < \varphi_1$. Recalling the relation $\zeta = \gamma \xi$ in Eq. (18.11), we conclude that the radiation pulse for a short magnet is given by the same Eq. (18.11), in which ζ is now constrained by $-\gamma \varphi_2 < \zeta < \gamma \varphi_1$. An example of the pulse shape for $\gamma \varphi_2 = 0.7$ and $\gamma \varphi_1 = 0.5$ is shown in Fig. 20.2. In reality, the discontinuities of the field at the front and the tail will be somewhat smeared out due to the fact that the magnetic fields do not abruptly drop to zero at the entrance to and the exit from the magnet. Instead the

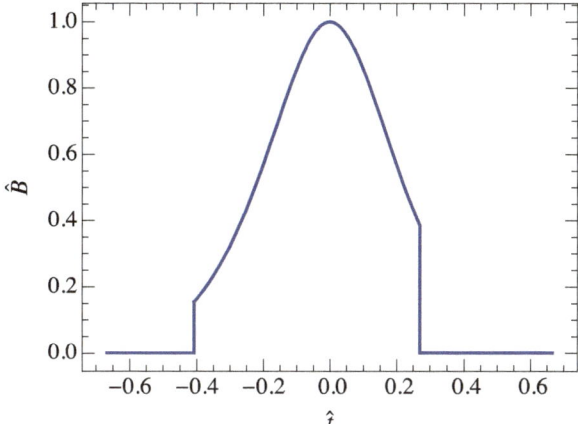

Fig. 20.2 The radiation pulse of the electromagnetic field for a short magnet with $\gamma\varphi_2 = 0.7$ and $\gamma\varphi_1 = 0.5$. The variables $\hat{B}$ and $\hat{t}$ are the same as in Fig. 18.2. The shape of the central part of the pulse is the same as in Fig. 18.2; the field drops to zero outside of this pulse

fields decrease over some finite extent, usually related to the magnetic aperture. This region of non-uniform field is called the fringe field[1].

The estimate for the formation length given in Eq. (20.1) is valid for the formation of the main part of the radiation pulse which carries the bulk of the electromagnetic energy. The formation length is different for the low frequency part of the radiation, with frequencies $\omega \ll \omega_c$. Low frequencies are produced by the long tails of the electromagnetic pulse in Fig. 18.2 and require a longer formation length $l_f(\omega)$. To estimate $l_f(\omega)$ we will analyze Eq. (18.14) and find the length $\Delta\xi$ of the region that makes a dominant contribution to the integral for a given frequency ω. We first note that for $\omega \ll \omega_c$ one can neglect the term with γ^{-2} in the exponent of Eq. (18.14), and the integral becomes

$$\int_{-\infty}^{\infty} \xi e^{i\omega\rho\xi^3/6c} d\xi \,, \tag{20.2}$$

from which it follows that $\Delta\xi \sim (c/\omega\rho)^{1/3}$. When ξ is much larger than this quantity, the function $e^{i\omega\rho\xi^3/6c}$ begins to rapidly oscillate thus suppressing the contribution from the region $|\Delta\xi| \gg (c/\omega\rho)^{1/3}$. Recalling that ξ represents the angle along the orbit, we estimate the formation length as

$$l_f(\omega) \sim \rho\Delta\xi \sim \rho^{2/3}\lambda^{1/3} \,, \tag{20.3}$$

where $\lambda = c/\omega$ is the reduced wavelength. For the critical frequency $\omega = \omega_c$ this formula reduces to the previous expression (20.1).

Using the result of Problem 18.3 that the angular spread of the synchrotron radiation at frequency $\omega \ll \omega_c$ is of order of $\Delta\psi \sim (\lambda/\rho)^{1/3}$, we can relate the formation

---

[1]The specific features of the radiation caused by the jumps in $\hat{B}$ at the entrance and the exit from the magnet are usually referred to as the *edge radiation*. It has a similarity with the transition radiation studied in Sect. 17.1.

length to the angular spread of the radiation $\Delta\psi$,

$$l_f \sim \frac{\lambda}{\Delta\psi^2} \, . \tag{20.4}$$

20.2   Transverse Formation Length

In addition to the requirement of having a long enough path length for the radiation formation, the charge also needs some space in the direction perpendicular to the orbit. We can estimate the needed transverse extension if we note that the radiation with frequency ω and angular spread $\Delta\psi$ has a transverse wavenumber $k_\perp \sim \omega\Delta\psi/c$. Any electromagnetic field with a characteristic value of $k_\perp$ extends transversely at least over the distance of one reduced transverse wavelength $k_\perp^{-1}$, hence the minimal transverse size needed for the formation of radiation is

$$l_\perp \sim \frac{\lambda}{\Delta\psi} \sim \rho^{1/3}\lambda^{2/3} \, , \tag{20.5}$$

where on the last step we have used the relation $\Delta\psi \sim (\lambda/\rho)^{1/3}$ from Problem 18.3. We will call $l_\perp$ the *transverse formation length*.

An interesting connection of $l_\perp$ to the properties of Gaussian beams can be traced if we recall the results of Sect. 13.2. For a Gaussian beam with an angular spread θ, the minimal transverse size (at the focal point) is the beam waist w_0 which, within a factor of two, is equal to λ/θ (see Eq. (13.20)). Hence $l_\perp$ is analogous to the laser beam waist w_0. Moreover, comparing Eq. (20.4) with Eq. (13.20) for Z_R we see that the longitudinal formation length is analogous to the Rayleigh length of a focused laser beam.

The practical importance of the transverse formation length is that synchrotron radiation can be suppressed by nearby metal walls of the vacuum chamber if they are located closer than $l_\perp$ — the so-called *shielding* effect. More specifically, the synchrotron radiation of a relativistic beam that moves on a circular orbit in a metal pipe having a transverse size a is suppressed in the frequency range where $a \lesssim l_\perp$. From Eq. (20.5) we find that the suppression occurs at long wavelength, $\lambda \gtrsim \sqrt{a^3/\rho}$. The suppression factor, defined as the ratio of the power radiated in the pipe to the free space radiation, depends on the cross section shape of the pipe. In Fig. 20.3 we plot the suppression factor for a model in which the circular orbit of the particle is located in the middle plane between two parallel perfectly conducting plates separated by a distance $2h$ (so that the distance from the orbit to each plate is h). The result shown in Fig. 20.3 is valid in the limit of small frequencies, $\omega \ll \omega_c$, when the free space radiation is given by Eq. (18.25a). Note that the horizontal axis in the plot is $\omega h^{3/2}\rho^{-1/2}/c \sim (h/l_\perp)^{3/2}$, and one can see that the suppression factor approaches zero when h becomes much smaller than $l_\perp$. Interestingly, the radiation is actually amplified by about 30% in the region of frequencies where $\omega h^{3/2}\rho^{-1/2}/c \approx 2$.

Fig. 20.3 Suppression factor for the intensity of the synchrotron radiation as a function of the frequency ω for the case of parallel conducting plates

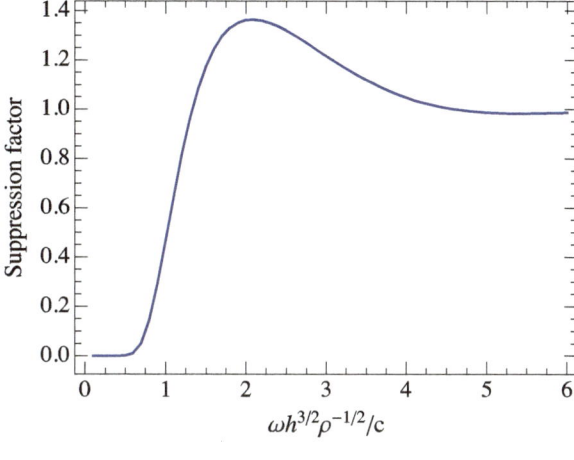

Fig. 20.4 Two particles, i and j, in a bunch emit separate pulses of electromagnetic radiation

This shielding effect plays a crucial role in electron accelerators with intense short electron bunches, acting to limit the beam energy loss going into radiation at long wavelengths.

20.3 Coherent Radiation

We now consider radiation of a bunch of particles. First, we neglect the transverse size of the bunch and consider the beam as infinitely thin with the longitudinal distribution function $\lambda(s)$. This function gives the probability for a particle to be located at s; it is normalized so that $\int \lambda(s)ds = 1$.

When two particles radiate electromagnetic pulses as shown in Fig. 20.4, the pulses propagate to the observer with delays that are determined by the arrival times of the particles to the point on the orbit from which the radiation is emitted. Let us denote by $B(t)$ the magnetic field of the pulse at the detector emitted by a reference particle in the beam (for synchrotron radiation, the function $B(t)$ is calculated in Sect. 18.1). In what follows, we will also need the Fourier transform of this field,

$$\tilde{B}(\omega) = \int_{-\infty}^{\infty} dt\, B(t)e^{i\omega t} .\tag{20.6}$$

The total field $\mathcal{B}(t)$ radiated by the bunch is the sum of the pulses,

$$\mathcal{B}(t) = \sum_{i=1}^{N} B(t - t_i) \,, \tag{20.7}$$

where $t_i = s_i/c$ with s_i the position of the particle i, and N is the total number of particles in the bunch. The Fourier image of this field is

$$\tilde{\mathcal{B}}(\omega) = \int dt \, \mathcal{B} e^{i\omega t} = \sum_{i=1}^{N} \int dt \, B(t - t_i) e^{i\omega t} = \sum_{i=1}^{N} \tilde{B}(\omega) e^{i\omega t_i} \,. \tag{20.8}$$

The spectral intensity of the radiation is proportional to $|\tilde{\mathcal{B}}(\omega)|^2$ for which we have

$$|\tilde{\mathcal{B}}(\omega)|^2 = \left| \sum_{i=1}^{N} \tilde{B}(\omega) e^{i\omega t_i} \right|^2 = |\tilde{B}(\omega)|^2 \left(N + \sum_{i \neq j} e^{i\omega(t_i - t_j)} \right)$$

$$= N|\tilde{B}(\omega)|^2 + 2|\tilde{B}(\omega)|^2 \sum_{i<j} \cos \left(\omega \frac{s_i - s_j}{c} \right) \,. \tag{20.9}$$

The first term in this equation is the *incoherent* radiation — it is proportional to the number of particles in the beam N. The second one is the *coherent* radiation term. The number of summands in the last sum is $N(N-1)/2 \approx N^2/2$. Instead of doing summation over i and j we can average $\cos(\omega(s_i - s_j)/c)$ and multiply the result by $N(N-1)/2 \approx N^2/2$ assuming that s_i and s_j are distributed with the probability given by $\lambda(s)^2$:

$$2 \sum_{i<j} \cos \left(\omega \frac{s_i - s_j}{c} \right) \approx N^2 \int ds' ds'' \lambda(s') \lambda(s'') \cos \left(\omega \frac{s' - s''}{c} \right)$$

$$= N^2 F(\omega) \,, \tag{20.10}$$

where the *form factor* $F(\omega)$ is

$$F(\omega) = \int ds' ds'' \lambda(s') \lambda(s'') \cos \left(\omega \frac{s' - s''}{c} \right) \,. \tag{20.11}$$

Because the spectral intensity of the radiation $dW/d\omega d\Omega$ is proportional to $|\tilde{\mathcal{B}}(\omega)|^2$ (see Eq. (17.10)) we obtain

$$\left. \frac{dW}{d\omega d\Omega} \right|_{\text{bunch}} = \frac{dW_1}{d\omega d\Omega} \left[N + N^2 F(\omega) \right] \,, \tag{20.12}$$

where $dW_1/d\omega d\Omega$ refers to the radiation of a single particle.

---

[2] Here we implicitly assume that the positions of particles i and j are not correlated.

Equation (20.11) can also be written as

$$F(\omega) = \left| \int_{-\infty}^{\infty} ds\, \lambda(s) e^{i\omega s/c} \right|^2 , \tag{20.13}$$

which is easily established by writing the square of the absolute value as a product of the integral $\int_{-\infty}^{\infty} ds\, \lambda(s) e^{i\omega s/c}$ with its complex conjugate. For a Gaussian distribution function,

$$\lambda(s) = \frac{1}{\sqrt{2\pi}\sigma_z} e^{-s^2/2\sigma_z^2} , \tag{20.14}$$

we have

$$F(\omega) = e^{-(\omega\sigma_z/c)^2} . \tag{20.15}$$

We see that for reduced wavelengths that are longer than the bunch length, $\lambda \gtrsim \sigma_z$, $F(\omega)$ approaches unity and the coherent radiation term in Eq. (20.12) dominates because the power scales as the number of particles squared. However, this radiation can only occur at long wavelengths, and in many cases it is suppressed by the shielding effect. In the opposite limit, $\lambda \ll \sigma_z$, the form factor F is exponentially small and the coherent radiation is negligible.

20.4 Effect of the Transverse Size of the Beam

In the previous analysis we neglected the transverse size of the beam and tacitly assumed that the radiation propagates in the forward direction. We now take into account that the radiation can be emitted at an angle to the direction of motion of the beam and consider a 3D distribution of the beam illustrated in Fig. 20.5. The 3D distribution function is $\lambda(\boldsymbol{r})$ normalized so that $\int d^3r\, \lambda(\boldsymbol{r}) = 1$. From Fig. 20.5 it is seen that the delay between pulses radiated by two particles located in the bunch at coordinates $\boldsymbol{r}_i$ and $\boldsymbol{r}_j$ is equal to $\Delta t = (\boldsymbol{r}_i - \boldsymbol{r}_j) \cdot \boldsymbol{n}/c$, where $\boldsymbol{n}$ is the unit vector in the direction of radiation. The radiation field of the bunch (20.9) can now be written as

$$|\tilde{B}(\omega, \boldsymbol{n})|^2 = N|\tilde{B}(\omega, \boldsymbol{n})|^2 + 2|\tilde{B}(\omega, \boldsymbol{n})|^2 \sum_{i<j} \cos\left(\omega \frac{\boldsymbol{n} \cdot (\boldsymbol{r}_i - \boldsymbol{r}_j)}{c}\right) , \tag{20.16}$$

where the second argument $\boldsymbol{n}$ in the magnetic field indicates the dependence versus the direction of radiation.

Repeating the derivation from the previous section we arrive at the following result for the form factor F:

Fig. 20.5 Radiation pulses
that propagate at an angle ψ
to the direction of motion of
the beam

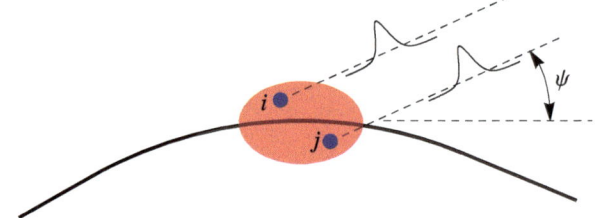

$$F(\omega, \boldsymbol{n}) = \int d^3 r' d^3 r'' \lambda(\boldsymbol{r}')\lambda(\boldsymbol{r}'') \cos\left(\omega \frac{\boldsymbol{n}\cdot(\boldsymbol{r}'-\boldsymbol{r}'')}{c}\right), \qquad (20.17)$$

which can also be written as the square of the absolute value of the three dimensional
Fourier transform of the distribution function,

$$F(\omega, \boldsymbol{n}) = \left|\int d^3 r \lambda(\boldsymbol{r}) e^{i\omega \boldsymbol{n}\cdot\boldsymbol{r}/c}\right|^2. \qquad (20.18)$$

The calculation of the form factor for a highly simplified beam distribution in
Problem 20.1 shows that the coherent radiation is suppressed if

$$\sigma_r \gtrsim \lambda/\psi, \qquad (20.19)$$

that is if the transverse size of the beam is larger than the transverse formation size of
the radiation. As an order of magnitude estimate, this conclusion is valid for a broad
class of other distribution functions.

Worked Examples

Problem 20.1 *Calculate the integral equation* (20.18) *for a "pancake" distribution*

$$\lambda(\boldsymbol{r}) = \delta(z)\frac{1}{2\pi\sigma_r^2}e^{-(x^2+y^2)/2\sigma_r^2}.$$

   Solution: We need to calculate the integral

$$\int d^3 r \lambda(\boldsymbol{r}) e^{i k\boldsymbol{n}\cdot\boldsymbol{r}},$$

with $k = \omega/c$ and the vector $\boldsymbol{n}$ directed at angle ψ to the z axis. By symmetry of the
problem, we can choose $\boldsymbol{n}$ to lie in the x-z plane, $\boldsymbol{n} = (\sin\psi, 0, \cos\psi)$. For the form
factor F we obtain:

$$F = \left| \int dx\, dy\, dz\, \frac{\delta(z)}{2\pi\sigma_r^2} e^{-(x^2+y^2)/2\sigma_r^2} e^{ik(x \sin \psi + z \cos \psi)} \right|^2$$

$$= \frac{1}{2\pi\sigma_r^2} \left| \int dx\, e^{-x^2/2\sigma_r^2} e^{ikx \sin \psi} \right|^2$$

$$= e^{-(k^2\sigma_r^2/2) \sin^2 \psi} \ .$$

For small angles, the coherent radiation is suppressed if $\lambda \lesssim \sigma_r \psi$.

Chapter 21
Topics in Laser-Driven Acceleration

A focused laser beam can easily produce an extremely high electric field at the focal point. For example, a laser beam with an energy of 1 J and pulse duration of 100 fs focused to a spot size of $10\,\mu$m, has a maximum electric field of about 40 GV/cm which is many orders of magnitude larger than what is used in traditional accelerators. In this chapter we will discuss several topics related to the possibility of using laser fields for accelerating charged particles.

21.1 The Lawson–Woodward Theorem

In the simplest setup one can try to accelerate charged particles by sending them through a focal point of a laser beam, as shown in Fig. 21.1. This is referred to as *direct laser acceleration*, or DLA. Because the electric field in a paraxial laser beam is directed predominantly transversely to the direction of propagation, one can expect that the particle trajectories should cross the focal region at an angle so that the component of the transverse laser field along the particle trajectory is not zero. Nevertheless, it is not immediately clear what would be the net acceleration effect after the interaction ceases — the electromagnetic field in the laser beam oscillates with time, and a particle crossing the focal point would experience both accelerating and decelerating phases.

Unfortunately, as we will show below, no matter how we organize the interaction of the laser beam with the particles, in the *linear* approximation, there is no net acceleration in free space. This statement is often called the *Lawson–Woodward theorem*.

We will now explain the meaning of the linear approximation. In this approximation we calculate the energy gain ΔW of a particle passing through an external field $E(r, t)$ assuming that it moves along a straight line with a constant velocity v,

© Springer International Publishing AG, part of Springer Nature 2018

G. Stupakov and G. Penn, *Classical Mechanics and Electromagnetism in Accelerator Physics*, Graduate Texts in Physics, https://doi.org/10.1007/978-3-319-90188-6_21

Fig. 21.1 A particle
trajectory (straight line) is
tilted relative to the direction
of the laser beam (outlined
by the two hyperbolas)

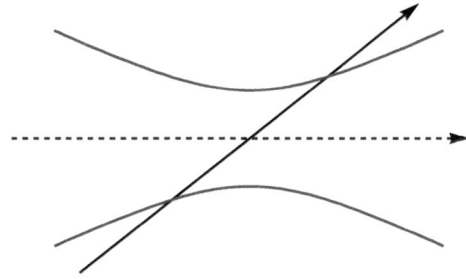

$$\Delta W = q \int_{-\infty}^{\infty} \boldsymbol{v} \cdot \boldsymbol{E}(\boldsymbol{r}_0 + \boldsymbol{v}t, t)dt \,. \tag{21.1}$$

Hence, in this approximation, we neglect the influence of the accelerating field on
the particle velocity and its orbit. This is a reasonable approximation for relativistic
particles that move with almost constant velocity close to the speed of light and due
to the relativistically increased inertia are difficult to deflect from a straight path.
This approximation becomes less accurate when the dimensionless electromagnetic
parameter a from Chap. 16 is much larger than unity.

Let us now prove that in free space far from material boundaries, and in the
absence of static fields, the above integral is equal to zero. The proof is based on
the fact that an arbitrary electromagnetic field in vacuum can be represented as a
superposition of plane electromagnetic waves propagating in all possible directions
(see Problem 13.1). Mathematically, this representation is written as

$$\boldsymbol{E}(\boldsymbol{r}, t) = \int d^3k \, \tilde{\boldsymbol{E}}(\boldsymbol{k}) e^{i\boldsymbol{k}\cdot\boldsymbol{r} - i\omega t} \,, \tag{21.2}$$

where $\boldsymbol{k}$ is the wavenumber of a plane electromagnetic wave, $\tilde{\boldsymbol{E}}(\boldsymbol{k})$ is the amplitude
of the electric field in the wave and $\omega = ck$.[1] The integration over $\boldsymbol{k}$ is carried over
the whole space. Substituting Eq. (21.2) into (21.1) we obtain

$$W = q\boldsymbol{v} \cdot \int_{-\infty}^{\infty} dt \int d^3k \, \tilde{\boldsymbol{E}}(\boldsymbol{k}) e^{i\boldsymbol{k}\cdot(\boldsymbol{r}_0 + \boldsymbol{v}t) - i\omega t}$$
$$= 2\pi \int d^3k \, q\boldsymbol{v} \cdot \tilde{\boldsymbol{E}}(\boldsymbol{k}) e^{i\boldsymbol{k}\cdot\boldsymbol{r}_0} \delta(\omega - \boldsymbol{k} \cdot \boldsymbol{v}) \,. \tag{21.3}$$

The argument in the delta function in the last integral is never equal to zero, because

$$\omega - \boldsymbol{k} \cdot \boldsymbol{v} = k(c - v \cos\alpha) > 0 \,, \tag{21.4}$$

---

[1] Here we ignore the fact that the frequency can actually be both positive and negative, $\omega = \pm ck$.
This does not change the proof.

where α is the angle between the directions of $\boldsymbol{v}$ and $\boldsymbol{k}$. Hence the integral (21.1) is equal to zero. This result can be explained as follows. As electromagnetic waves propagate with the speed of light they overtake particles moving with a velocity smaller than c. As a result, a particle is slipping through the accelerating and decelerating phases in the wave, and the total acceleration averages to zero.

There are several ways to overcome the limitations of the Lawson–Woodward theorem and achieve acceleration in an electromagnetic field. First, one can limit the interaction between the particle and the waves by introducing material boundaries and preventing the field from occupying the whole space. This would make the representation (21.2) invalid. We consider a model for such acceleration in the next section. Second, for particles of relatively small energy and intense laser fields one may need to take into consideration the wiggling of the orbit caused by the laser. This means that one has to drop the assumption of constant velocity and a straight orbit in Eq. (21.3). Such effects play an important role in the acceleration of not very relativistic electrons by a strong laser beam with parameter $a \gg 1$. Finally, one can introduce an external magnetic field and intentionally bend the orbit. One example of such acceleration — the so-called *inverse FEL* acceleration — will be considered in the last section of this chapter.

Other types of laser acceleration use some form of *indirect* acceleration, which incorporates materials that mediate the interaction between the laser fields and the particles to be accelerated; examples employ dielectrics, thin foils, and plasmas.

21.2 Laser Acceleration in Space with Material Boundaries

We will now calculate the energy gain of a charged particle passing through a focused laser field reflected back by a flat mirror as shown in Fig. 21.2, see Ref. [1]. The mirror is located at $z = 0$ and has a small hole for the passage of the particle. We assume that the hole does not perturb the laser field except for a small vicinity near the hole. In the calculations, we neglect interaction with the reflected part of the field, which turns out to be small (see Problem 21.2). A similar setup has been tested experimentally [2, 3].

We assume that the particle moves along a straight line parallel to the z axis with velocity v and an offset x_0. The z coordinate of the particle at time t is equal to $z_0 + vt$. The energy gain is then given by the following integral:

$$\Delta W = q \int_{-\infty}^{0} dt \, v E_z(x_0, 0, z_0 + vt, t) = q \int_{-\infty}^{0} dz \, E_z(x_0, 0, z, (z - z_0)/v).$$
(21.5)

Note that the integration extends from $z = -\infty$ to $z = 0$ because there is no field behind the metal mirror. The longitudinal component of the electric field in the laser focus was calculated in Problem 13.3, and is given by

Fig. 21.2 The laser beam is reflected by a metal surface; the particle passes through a hole in the metal

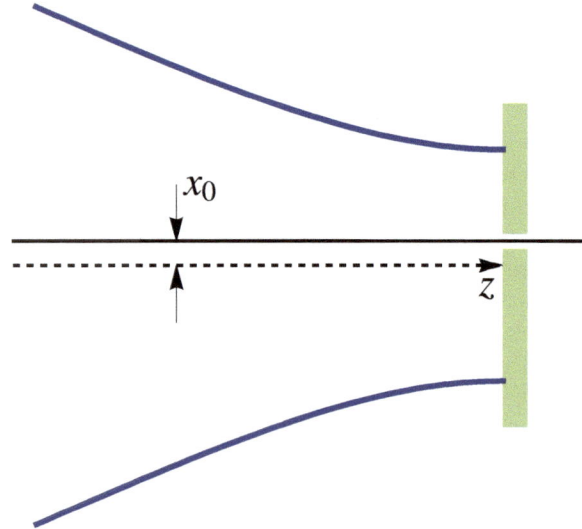

$$E_z(x, y, z, t) = -\frac{1}{ik}\frac{\partial E_x}{\partial x} = -\frac{2x}{ik}A(z)Q(z)e^{Q(z)\rho^2}e^{-i\omega t + ikz}, \qquad (21.6)$$

where A and Q are given by Eqs. (13.17) and (13.18). We need to take the real part of the field to calculate ΔW:

$$\Delta W = -\mathrm{Re}\left[\frac{2x_0 q}{ik}e^{ikz_0\beta^{-1}}\int_{-\infty}^{0}dz\, A(z)Q(z)e^{Q(z)x_0^2}e^{-ikz(\beta^{-1}-1)}\right]$$

$$= \mathrm{Re}\left[\frac{2E_0 x_0 q}{ikw_0^2}e^{ikz_0\beta^{-1}}\int_{-\infty}^{0}\frac{dz}{(1+2iz/kw_0^2)^2}e^{-x_0^2/w_0^2(1+2iz/kw_0^2)}e^{-ikz(\beta^{-1}-1)}\right].$$

$$(21.7)$$

Finally, switching to the integration variable $\xi = 2z/kw_0^2 = z/Z_R$ we obtain

$$\Delta W = -\mathrm{Re}\left[iE_0 x_0 q e^{ikz_0\beta^{-1}}\int_{-\infty}^{0}\frac{d\xi}{(1+i\xi)^2}e^{-x_0^2/w_0^2(1+i\xi)}e^{-i\xi k^2 w_0^2(\beta^{-1}-1)/2}\right]. \quad (21.8)$$

We first note that if there is no mirror and the particle interacts with the laser beam from $z = -\infty$ to $z = \infty$ the upper limit in this integral should be set to infinity. It can be proven that the resulting integral is equal to zero for $\beta \leq 1$. This is, of course, in agreement with the Lawson–Woodward theorem.

Returning to the integral (21.8) let us now consider an ultrarelativistic particle and set $\beta = 1$ (here the ultrarelativistic limit corresponds to $\gamma \gg kw_0$). Then the integration in (21.8) is easy to carry out, and the result is

Fig. 21.3 A particle trajectory in an undulator and the laser pulse co-propagating with the particle. The laser pulse is represented by a plane electromagnetic wave (blue vertical lines)

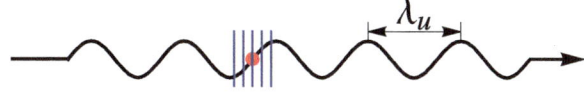

$$\Delta W = -\mathrm{Re}\left[i E_0 x_0 q e^{ikz_0} \int_{-\infty}^{0} \frac{d\xi}{(1+i\xi)^2} e^{-x_0^2/w_0^2(1+i\xi)} \right]$$
$$= \frac{E_0 q w_0^2}{x_0}(1 - e^{-x_0^2/w_0^2}) \cos(kz_0) . \tag{21.9}$$

The factor $\cos(kz_0)$ in this equation indicates that the sign of the energy change depends on the position of the charge relative to the phase of the laser field; for bunches of particles longer than the wavelength of the laser radiation such an interaction modulates the energy of the beam with the period equal to the laser wavelength.

Many more details of laser acceleration can be found in Ref. [4].

21.3 Inverse FEL Acceleration

We will now consider a laser-particle interaction when the particle is moving through an undulator as shown in Fig. 21.3. This kind of laser-beam interaction is called the *inverse FEL* acceleration, where FEL stands for Free-Electron Laser. To simplify calculations, we will assume that the undulator parameter K is small, $K \ll 1$.

We will represent the laser field by a plane electromagnetic wave propagating in the z direction, with electric field E_x given by

$$E_x(z, t) = E_0 \cos(\omega t - \omega z/c) , \tag{21.10}$$

where E_0 is the wave amplitude and ω is the laser frequency. The energy change is given by the following formula:

$$\Delta W = q \int dt\, \boldsymbol{E} \cdot \boldsymbol{v} = q \int dt\, E_x v_x . \tag{21.11}$$

The particle orbit in a small-K undulator has been calculated in Chap. 19, see Eq. (19.3), and the velocity v_x can be obtained by differentiating x with respect to time:

$$v_x = \frac{cK}{\gamma} \sin(k_u z) . \tag{21.12}$$

Using the approximation $z \approx z_0 + ct(1 - 1/2\gamma^2)$ gives

$$
\begin{aligned}
\Delta W &= q \int dt\, \boldsymbol{E} \cdot \boldsymbol{v} \\
&= q \int dt\, \frac{cK}{\gamma} \sin[k_u(z_0 + ct)] E_0 \cos\left[\omega t - kct\left(1 - \frac{1}{2\gamma^2}\right) - kz_0\right] \\
&\approx \frac{cqKE_0}{2\gamma} \int dt\, \sin\left[k_u ct - \omega t + \omega t\left(1 - \frac{1}{2\gamma^2}\right) + (k + k_u)z_0\right] \\
&\approx \frac{cqKE_0}{2\gamma} \int dt\, \sin\left[\left(k_u c - \omega\frac{1}{2\gamma^2}\right)t + (k + k_u)z_0\right] .
\end{aligned}
\tag{21.13}
$$

In this equation we discarded the term with the sum of the arguments in the sine function because it rapidly oscillates and adds only a small contribution to the result. The most effective acceleration occurs if

$$
\omega = 2\gamma^2 k_u c , \tag{21.14}
$$

when the time dependent part of the sine argument cancels. This means that the laser frequency should be equal to the frequency of the undulator radiation in the forward direction. In this case we find

$$
\Delta W = \frac{qKE_0 L_u}{2\gamma} \sin\left[(k + k_u)z_0\right] . \tag{21.15}
$$

Depending on the position of the particle z_0 relative to the phase of the laser, the energy gain can have both positive and negative signs. This will result in an energy modulation of the beam.

As a final note, we mention that the plane wave approximation is valid if the Raleigh length for the laser beam is much larger than the undulator length.

Worked Examples

Problem 21.1 *Prove that Eq. (21.8) with the upper limit of integration changed to infinity evaluates to zero for $\beta = 1$.*

Solution: For $\beta = 1$, the integral in the modified form of Eq. (21.8) becomes

$$
\int_{-\infty}^{\infty} \frac{d\xi}{(1 + i\xi)^2} e^{-x_0^2/\omega_0^2(1+i\xi)} = -\frac{i\omega_0^2}{x_0^2} e^{-x_0^2/\omega_0^2(1+i\xi)}\Big|_{\xi=-\infty}^{\infty} = 0 .
$$

Problem 21.2 *Estimate the contribution to ΔW in Eq. (21.8) from the reflected part of the laser field.*

Solution: To adapt Eq. (21.8) for the reflected wave, we have to change the sign of k to be $-k$ which implies that we also replace ξ with $-\xi$. There is one exception however, because the ξ/β term actually came from the frequency ω instead of k, so that sign does not change. Thus, instead of the quantity $\beta^{-1} - 1$ in the last exponent we have $\beta^{-1} + 1 \geq 2$. In the previous case, this term could become small as $\beta \to 1$. Because in the paraxial approximation a laser waist will have to satisfy $w_0 \gg 1/k$, there will be many oscillations for moderate ranges of the ξ parameter. While the final integral is difficult to simplify numerically, we can see that there will be a strong suppression of the contribution from the reflected wave to the total acceleration, even when stopping the integration at $\xi = 0$. At its largest, this contribution should be a factor of $1/(k^2 w_0^2) = 1/(2k Z_R)$ smaller than the acceleration from the forwards wave.

Problem 21.3 *Assume that you are given a laser with a given energy E_L, frequency ω and duration τ of the laser pulse. Optimize the parameters of the laser acceleration experiment shown in Fig. 21.2 to achieve the maximum energy gain for relativistic particles. Express the energy gain in terms of E_L, ω and τ.*

Solution: The energy gain is given by Eq. (21.9),

$$\Delta W = \frac{q E_0 w_0^2}{x_0} (1 - e^{-x_0^2/w_0^2}) \cos(k z_0) \,.$$

Depending on the sign of the charge, $k z_0$ should be either 0 or π. For either $x_0 \ll w_0$ or $x_0 \gg w_0$, $\Delta W \to 0$. The optimum x_0 corresponds to the maximum of the function $x_0^{-1}(1 - e^{-x_0^2/w_0^2})$, which is found by equating the derivative to zero,

$$\frac{2 x_0^2}{w_0^2} = e^{x_0^2/w_0^2} - 1 \,,$$

so that $x_0 \approx w_0$ (actually $x_0/w_0 \simeq 1.12$). We then see $|\Delta W| \propto E_0 w_0$. The laser power, P_L, is proportional to $E_0^2 w_0^2$, so $\Delta W \propto \sqrt{P_L}$. In terms of pulse energy, $P_L \approx E_L/\tau$ so shorter pulses are more effective.

Problem 21.4 *Take the following parameters of the inverse FEL acceleration experiment from Ref. [3]: beam energy 60 MeV, laser pulse length 0.55 ps, laser energy 0.65 mJ, laser focused spot size 200 μm, undulator period 1.8 cm, number of periods 3, $K = 0.65$; and estimate the amplitude of the energy modulation of the beam.*

Solution: The energy modulation for an inverse FEL is given by Eq. (21.15),

$$\Delta W = \frac{q E_0 K L_u}{2\gamma} \sin[(k + k_u) z_0] \,.$$

For the given parameters, $\gamma = 118$, the total undulator length $L_u = 0.054$ m, and the laser peak power is $P_L = W_L/\tau = 0.65 \text{ mJ}/0.55 \text{ ps} = 1.18 \text{ GW}$. The E field is given by

$$E_0 = \sqrt{P_L 4 Z_0/\pi}/w_0 = 6.9 \text{ GV/m} \,.$$

Plugging in above we find for the peak energy gain and loss

$$\Delta W = 1.0 \text{ MeV} \,.$$

References

1. A. Chao, Lecture Notes on Topics in Accelerator Physics. SLAC-PUB-9574 (2002)
2. T. Plettner, R.L. Byer, E. Colby, B. Cowan, C.M.S. Sears, J.E. Spencer, R.H. Siemann, Visible-laser acceleration of relativistic electrons in a semi-infinite vacuum. Phys. Rev. Lett. **95**, 134801 (2005)
3. C.M.S. Sears, E.R. Colby, R. Ischebeck, C. McGuinness, J. Nelson, R. Noble, R.H. Siemann, J. Spencer, D. Walz, T. Plettner, R.L. Byer, Production and characterization of attosecond electron bunch trains. Phys. Rev. ST Accel. Beams **11**, 061301 (2008)
4. Eric Esarey, Phillip Sprangle, Jonathan Krall, Laser acceleration of electrons in vacuum. Phys. Rev. E **52**, 5443 (1995)

Chapter 22
Radiation Damping Effects

In this chapter we will show how the radiation damping in electron and positron rings can be added to the Hamiltonian and Vlasov formalism, and calculate how radiation damping affects the energy, transverse actions and distribution function.

22.1 Radiation Damping in Equations of Motion

Relativistic electrons and positrons in accelerators radiate intensely when moving in a circular orbit of an accelerator. As we discussed in Chap. 15, for a transverse magnetic field the radiation is emitted in the forward direction within an angle $\sim 1/\gamma$. The radiation reaction force acts in the opposite direction, and, similar to a friction force, tends to slow down the motion of the particle. It maintains the energy balance in the process: the energy of the radiation is taken from the kinetic energy of the particle. It also causes damping of the betatron oscillations and affects the energy spread of the beam. In this section we will derive the equations that describe the effect of the radiation on the beam dynamics.

We consider relativistic particles moving with $\gamma \gg 1$ in a circular accelerator. Let $\mathcal{P}$ be the power of radiation (the energy emitted per unit time) for a given particle at a specified location in the ring. Since for a relativistic particle we have approximately $p = h/c$, where h is the energy, the quantity $\mathcal{P}/c$ is equal to the decrease of the momentum of the particle per unit time. As mentioned above, the force is acting in the direction opposite to the momentum, so we can write the change in the momentum components per infinitesimally small time dt as

G. Stupakov and G. Penn, *Classical Mechanics and Electromagnetism in Accelerator Physics*, Graduate Texts in Physics, https://doi.org/10.1007/978-3-319-90188-6_22

$$dp_x = -\frac{p_x}{p}\frac{\mathcal{P}}{c}dt = -\frac{v_x}{v}\frac{\mathcal{P}}{c}dt \approx -\frac{\mathcal{P}}{c^2}v_x dt = -\frac{\mathcal{P}}{c^2}dx = -\frac{\mathcal{P}}{c^2}\frac{dx}{ds}ds\,,$$

$$dp_y = -\frac{p_y}{p}\frac{\mathcal{P}}{c}dt \approx -\frac{\mathcal{P}}{c^2}\frac{dy}{ds}ds\,,$$

$$dh = -\mathcal{P}dt = -\mathcal{P}\frac{dt}{ds}ds\,. \tag{22.1}$$

As we learned in Chap. 5, without the radiation effects the particle dynamics in a circular accelerator is governed by the Hamiltonian (5.19). We now need to add the additional changes of the momentum and energy defined by Eq. (22.1) to the Hamiltonian equations. Because Eq. (22.1) only modifies the canonical momenta, the differential equations for the evolution of the coordinates remain Hamiltonian:

$$\frac{dx}{ds} = \frac{\partial K}{\partial p_x}\,, \quad \frac{dy}{ds} = \frac{\partial K}{\partial p_y}\,, \quad \frac{dt}{ds} = -\frac{\partial K}{\partial h}\,, \tag{22.2}$$

while the original Hamiltonian equations for the momentum and energy are amended by the terms from Eq. (22.1):

$$\frac{dp_x}{ds} = -\frac{\partial K}{\partial x} - \frac{\mathcal{P}}{c^2}\frac{\partial K}{\partial p_x}\,,$$

$$\frac{dp_y}{ds} = -\frac{\partial K}{\partial y} - \frac{\mathcal{P}}{c^2}\frac{\partial K}{\partial p_y}\,,$$

$$\frac{dh}{ds} = \frac{\partial K}{\partial t} + \mathcal{P}\frac{\partial K}{\partial h}\,. \tag{22.3}$$

In these equations, we replaced the derivatives dt/ds, dy/ds and dt/ds on the right-hand sides of Eq. (22.1) by their Hamiltonian expressions Eq. (22.2). We emphasize here that Eqs. (22.2) and (22.3) are not Hamiltonian any more, but it is convenient to keep writing them using the previously introduced canonical variables and the Hamiltonian function K.

An expression for the radiation power was derived in Chap. 18, Eq. (18.27),

$$\mathcal{P}(h,s) = \frac{2}{3}\frac{r_c\gamma^4 mc^3}{\rho^2(s)} = \frac{2}{3}\frac{q^2 r_c h^2}{m^3 c^3}B^2(s)\,, \tag{22.4}$$

where the bending radius for a relativistic particle is expressed through the magnetic field, $|\rho| = h/c|q B(s)|$, $B(s)$ is the magnetic field taken at the particle's position, and r_c is the classical radius. In Eq. (22.4) we indicated the explicit dependence of $\mathcal{P}$ versus the coordinate s and the energy h. For the nominal energy h_0, we will use the notation

$$\mathcal{P}_0(s) = \frac{2}{3}\frac{q^2 r_c h_0^2}{m^3 c^3}B^2(s)\,. \tag{22.5}$$

Averaging $\mathcal{P}_0$ over the ring and dividing it by the nominal energy $p_0 c$ defines a characteristic damping time τ_s for the ring,

$$\frac{1}{\tau_s} = \frac{\langle \mathcal{P}_0(s) \rangle}{p_0 c} = \frac{1}{\gamma m c^2} \frac{1}{cT} \oint ds \, \mathcal{P}_0(s) = \frac{2}{3} \frac{r_c \gamma^3}{T} \oint \frac{ds}{\rho^2(s)}, \tag{22.6}$$

where T is the revolution period and the angular brackets denote the averaging over the ring circumference. The time τ_s has the meaning of the time scale needed for a particle to lose all its initial energy to radiation, if this energy is not replenished. In a typical accelerator ring the damping time is much longer than the revolution period T and the period of betatron oscillations. This observation will be used in the next section.

22.2 Synchrotron Damping of Betatron Oscillations

In this section we will consider the effect of the synchrotron damping on betatron oscillations using the machinery developed in Sect. 22.1.

First we need to transform Eqs. (22.2) and (22.3) to the variables of the linearized Hamiltonian $\mathcal{H}$ from Eq. (6.10) used in our analysis of the betatron oscillations[1],

$$\mathcal{H} \approx -1 - \eta - \eta \frac{x}{\rho(s)} + \frac{1}{2} P_x^2 + \frac{1}{2} P_y^2 + \frac{x^2}{2\rho^2(s)} - \frac{e}{p_0} \frac{1}{2} G(s) \left(y^2 - x^2 \right). \tag{22.7}$$

Recall that $\mathcal{H}$ was obtained from K by division by p_0 with a simultaneous transition from p_x, p_y to $P_x = p_x / p_0$, $P_y = p_y / p_0$, and from h to h/p_0. As a result the two first pairs of Eqs. (22.2) and (22.3) become

$$\frac{dx}{ds} = \frac{\partial \mathcal{H}}{\partial P_x}, \qquad \frac{dP_x}{ds} = -\frac{\partial \mathcal{H}}{\partial x} - \frac{\mathcal{P}}{p_0 c^2} \frac{\partial \mathcal{H}}{\partial P_x},$$

$$\frac{dy}{ds} = \frac{\partial \mathcal{H}}{\partial P_y}, \qquad \frac{dP_y}{ds} = -\frac{\partial \mathcal{H}}{\partial y} - \frac{\mathcal{P}}{p_0 c^2} \frac{\partial \mathcal{H}}{\partial P_y}. \tag{22.8}$$

For relativistic motion, we can identify h/p_0 with $pc/p_0 = c(1 + \eta)$ (see Eq. (5.22)), which means that in the third equation in (22.3) we can use $c\eta$:

$$c \frac{d\eta}{ds} = \frac{\partial \mathcal{H}}{\partial t} + \frac{\mathcal{P}}{p_0 c} \frac{\partial \mathcal{H}}{\partial \eta} = -\frac{\mathcal{P}}{p_0 c} \left(1 + \frac{x}{\rho} \right), \tag{22.9}$$

where in the last equation we assumed $\partial \mathcal{H} / \partial t = 0$ and used Eq. (6.10) for the derivative $\partial \mathcal{H} / \partial \eta$. One immediately sees from this equation that η monotonically

---

[1] In our analysis we will assume that ρ does not depend on the transverse coordinates x and y ignoring the case of the so-called combined function magnets where such dependence exists.

decreases with time due to the continuous energy loss to the radiation. If this loss is not compensated the initial energy of the particle will be radiated away within a time of the order of τ_s. In reality the particle energy is replenished by RF cavities in the ring, with the corresponding term in the Hamiltonian given by Eq. (5.27). However, to avoid complications associated with a treatment of a time-dependent Hamiltonian we will adopt a simpler model for the energy source that compensates for the radiated energy. We will assume that this source is uniformly distributed in the ring and is equal to the average over the ring circumference energy loss as defined by Eq. (22.5):

$$ c\frac{d\eta}{ds} = -\frac{\mathcal{P}}{p_0 c}\left(1 + \frac{x}{\rho}\right) + \frac{\langle \mathcal{P}_0(s) \rangle}{p_0 c} . \tag{22.10} $$

As it turns out, our results obtained with the model of a continuous energy source are valid, to a good approximation, for real machines with localized RF cavities in the ring. Since we assume that the relative energy deviation η is small we can expand the difference $\mathcal{P} - \mathcal{P}_0$ keeping only the linear term in η,

$$ \mathcal{P} - \mathcal{P}_0(s) \approx p_0 c\eta \frac{\partial \mathcal{P}}{\partial h}\bigg|_{h=h_0} = 2\eta \mathcal{P}_0(s), \tag{22.11} $$

where we took into account the quadratic dependence of $\mathcal{P}$ versus h in Eq. (22.4). As a result, we obtain

$$ \frac{d\eta}{ds} = -\frac{\mathcal{P}_0(s)}{p_0 c^2}\left(2\eta + \frac{x}{\rho}\right), \tag{22.12} $$

where we have neglected second order terms such as $x\eta$ and η^2 as well as the quantity $\mathcal{P}_0(s) - \langle \mathcal{P}_0(s) \rangle$, which by definition is an oscillation that will average to zero during each turn around the ring. There are two terms on the right-hand side of this equation. The first one describes damping of the energy deviation. The mechanism of this damping is due to the monotonic increase of the energy loss with energy, $\mathcal{P} \propto h^2$; the energy loss is higher for particles with energy larger than the nominal energy, $h > h_0$, and smaller for energy $h < h_0$. The second term is a driving force due to the changes in the path length of the orbit with a nonzero offset x in comparison with the nominal one. In contrast to Eq. (22.9) this equation exhibits an equilibrium solution with $x = \eta = 0$.

Let us now assume that a particle has an initial energy deviation η. How does this energy deviation evolve with time? We know that when the tune is not an integer, an off-momentum particle oscillates around the modified closed orbit defined by Eq. (8.14): $x_0(s) = \eta D(s)$. Substituting this relation into Eq. (22.12) we obtain

$$ \frac{d\eta}{ds} = -\frac{\mathcal{P}_0(s)}{p_0 c^2}\left[\eta\left(2 + \frac{D(s)}{\rho}\right) + \frac{x - x_0(s)}{\rho}\right]. \tag{22.13} $$

As was mentioned at the end of the previous section, the synchrotron damping is a slow process that lasts for many revolution periods. Therefore, it makes sense to consider the cumulative effect of the damping at a given s over multiple turns. The term $x - x_0(s)$ will oscillate and average nearly to zero, so this term will be removed. It also makes sense to average Eq. (22.13) over the circumference of the ring. Recalling that $\mathcal{P}_0 \propto \rho^{-2}$ and using Eq. (22.6) it is easy to obtain

$$\langle \frac{d\eta}{ds} \rangle = -\frac{\eta}{c\tau_s}(2 + \mathcal{D}), \tag{22.14}$$

where $\mathcal{D}$ is the *damping partition number*,

$$\mathcal{D} = \left(\int \frac{ds}{\rho^2} \right)^{-1} \int \frac{ds}{\rho^3} D(s). \tag{22.15}$$

In many accelerators the parameter $\mathcal{D}$ is small and to the lowest approximation can be neglected. We then have energy perturbations, on average, exponentially decaying with a time constant equal to half of τ_s.

We will now consider the damping of vertical betatron oscillations in a ring due to the synchrotron radiation. We know that without the damping, when the system is Hamiltonian, the action J_y given by Eq. (7.8) is conserved. Due to the synchrotron radiation it will be slowly (over many revolution periods) decreasing with time. To find its damping time, we need to calculate the derivative dJ_y/ds using Eq. (22.8) and average it over the ring circumference (as we did above for $d\eta/ds$). The calculation is simplified if we note that when $\mathcal{P} = 0$, $dJ_y/ds = 0$, and that the damping term in Eq. (22.8) involves only P_y. Hence

$$\frac{dJ_y}{ds} = \frac{\partial J_y}{\partial P_y} \times \left[-\frac{\mathcal{P}_0(s)}{p_0 c^2} \frac{\partial \mathcal{H}}{\partial P_y} \right] = -\frac{\mathcal{P}_0(s)}{p_0 c^2}(\beta P_y + \alpha y)P_y \tag{22.16}$$

(to simplify the notation we dropped the index y in β and α). We then use Eq. (7.10) to obtain

$$\frac{dJ_y}{ds} = -\frac{\mathcal{P}_0(s)}{p_0 c^2} \left[2J_y(\sin\phi + \alpha\cos\phi)^2 - 2\alpha J_y\cos\phi(\sin\phi + \alpha\cos\phi) \right]$$

$$= -\frac{2\mathcal{P}_0(s)}{p_0 c^2} J_y \left[\sin^2\phi + \alpha\sin\phi\cos\phi \right]. \tag{22.17}$$

We now average this expression over the ring circumference. The averaging consists of two parts. First we note that at subsequent revolutions in the ring the phase ϕ at a given location s increases by the betatron phase advance $2\pi\nu$ (see Sect. 7.3). Assuming that the tune is not close to an integer or half-integer, the angular part of the expression (22.17) effectively averages over the angle ϕ in a small number of turns, such that the change in J_y can be ignored over this time scale. Therefore we can take $\sin^2\phi \to \frac{1}{2}$ and $\sin\phi\cos\phi \to 0$. After that we need to do one more averaging of $\mathcal{P}_0(s)$ over s, again neglecting the change in J_y during one revolution around the ring:

$$\left\langle \frac{dJ_y}{ds} \right\rangle = -\frac{\langle \mathcal{P}_0 \rangle}{p_0 c^2} J_y = -\frac{1}{c\tau_s} J_y \,. \tag{22.18}$$

We see that the vertical betatron oscillations decay exponentially with a time constant equal to τ_s, which we assume to be much greater than the revolution period T. We also ignore the role of the energy offset, which would add an overall factor of $(1 + 2\eta)$ that is a higher order correction that will be ignored.

Calculation of the damping of the betatron oscillations in the horizontal plane is more complicated. This complication comes from the fact that the betatron oscillations are coupled to the energy through the term $-\eta x/\rho$ in the Hamiltonian (6.10) and, in addition, the evolution of η is coupled to x through the term x/ρ in Eq. (22.12). The closed orbit for off-energy particles is also distorted. As before, we take into account that $dJ_x/ds = 0$ when $\mathcal{P} = 0$, and the damping comes through the variables P_x and η. However, to take into account the fact that there is an orbit distortion associated with η, we have to use the action J_x defined by Eq. (8.16) instead of Eq. (7.8) to find that

$$\begin{aligned}
\frac{dJ_x}{ds} &= \frac{\partial J_x}{\partial P_x} \times \left(-\frac{\mathcal{P}_0(s)}{p_0 c^2} \frac{\partial \mathcal{H}}{\partial P_x} \right) + \frac{\partial J_x}{\partial \eta} \frac{d\eta}{ds} \\
&= -\frac{\mathcal{P}_0(s)}{p_0 c^2} P_x \left[\beta(P_x - \eta D') + \alpha(x - \eta D) \right] \\
&\quad - \frac{\mathcal{P}_0(s)}{p_0 c^2} \left(2\eta + \frac{x}{\rho} \right) \frac{1}{\beta} \Big\{ -D(x - \eta D) \\
&\qquad - (\beta D' + \alpha D)\left[\beta(P_x - \eta D') + \alpha(x - \eta D) \right] \Big\} \,.
\end{aligned} \tag{22.19}$$

We also must generalize the transformation to action-angle coordinates, replacing Eqs. (7.9) and (7.10) with:

$$\begin{aligned}
x - \eta D &= \sqrt{2\beta J_x}\cos\phi \,, \\
P_x - \eta D' &= -\sqrt{\frac{2J_x}{\beta}}\,(\sin\phi + \alpha\cos\phi) \,.
\end{aligned} \tag{22.20}$$

We can use these expressions to recast the evolution of J_x in terms of action-angle variables:

$$\frac{dJ_x}{ds} = -\frac{\mathcal{P}_0(s)}{p_0 c^2} \left\{ 2J_x \left[\sin\phi(\sin\phi + \alpha\cos\phi) - \frac{D}{\rho}\cos^2\phi \right. \right. \tag{22.21}$$

$$\left. + \frac{1}{\rho}(\beta D' + \alpha D)\sin\phi\cos\phi \right]$$

$$\left. + \eta\sqrt{2\beta J_x}\left[-D'\sin\phi + \frac{1}{\beta}\left(2 + \frac{D}{\rho}\right)(-D\cos\phi + (\beta D' + \alpha D)\sin\phi) \right] \right\}.$$

We now perform the two-step averaging using the same approach as for the vertical action J_y above. At this point, we ignore any variations in η along the ring as being a higher order correction. Averaging over the phase ϕ leaves only two terms on the right-hand side,

$$\frac{dJ_x}{ds} = -\frac{\mathcal{P}_0(s)}{p_0 c^2} J_x \left(1 - \frac{D}{\rho}\right), \tag{22.22}$$

and then averaging over the circumference of the ring gives

$$\left\langle \frac{dJ_x}{ds} \right\rangle = -\frac{1}{c\tau_s} J_x (1 - \mathcal{D}). \tag{22.23}$$

We see that the horizontal betatron oscillations decay with the time constant $\tau_s/(1 - \mathcal{D})$.

From the results of this section it follows that the synchrotron radiation causes energy deviations and betatron oscillations to damp down. This however does not mean that in a real machine they completely disappear after several damping times. There are various driving sources for these oscillations, and the most prominent one is due to the so-called *quantum diffusion* effect. The effect is explained by the quantum nature of the synchrotron radiation and the recoil momentum that a particle receives when it emits a single synchrotron photon. The competition between the synchrotron damping and excitation of betatron and synchrotron oscillations establishes a dynamic equilibrium and in many cases determines the energy spread and horizontal emittance of the beam in circular electron accelerators. In the absence of vertical bends, the equilibrium vertical emittance can be much smaller and is often determined by magnetic field errors.

22.3 Vlasov Equation and Robinson's Theorem

In Chap. 10 we introduced the kinetic equation to describe the evolution of an ensemble of beam particles. While our initial formulation of the continuity equation (10.7) was general and valid for arbitrary equations of motion, the subsequent assumption of Hamiltonian motion leads to the Vlasov equation (10.11) and the even more elegant Eq. (10.14). Our goal now is to include the effect of the synchrotron damping, as described by Eqs. (22.8) and (22.9), into the formalism of the Vlasov equation.

The distribution function f now depends on 7 variables: $f(x, P_x, y, P_y, t, \eta, s)$. With an evident generalization of equation (10.11) for 3 degrees of freedom, the kinetic equation is

$$\frac{\partial f}{\partial s} + \frac{\partial}{\partial x}\left(\frac{dx}{ds}f\right) + \frac{\partial}{\partial P_x}\left(\frac{dP_x}{ds}f\right) + \frac{\partial}{\partial y}\left(\frac{dy}{ds}f\right) + \frac{\partial}{\partial P_y}\left(\frac{dP_y}{ds}f\right)$$
$$+ \frac{\partial}{\partial t}\left(\frac{dt}{ds}f\right) + \frac{\partial}{\partial \eta}\left(\frac{d\eta}{ds}f\right) = 0 . \tag{22.24}$$

Let us calculate the full (or convective) derivative df/ds:

$$\frac{df}{ds} = \frac{\partial f}{\partial s} + \frac{dx}{ds}\frac{\partial f}{\partial x} + \frac{dP_x}{ds}\frac{\partial f}{\partial P_x} + \frac{dy}{ds}\frac{\partial f}{\partial y} + \frac{dP_y}{ds}\frac{\partial f}{\partial P_y} + \frac{dt}{ds}\frac{\partial f}{\partial t} + \frac{d\eta}{ds}\frac{\partial f}{\partial \eta} . \tag{22.25}$$

Substituting the partial derivative $\partial f/\partial s$ from Eqs. (22.24) into (22.25) we obtain

$$\frac{df}{ds} = -f\left(\frac{\partial}{\partial x}\frac{dx}{ds} + \frac{\partial}{\partial P_x}\frac{dP_x}{ds} + \frac{\partial}{\partial y}\frac{dy}{ds} + \frac{\partial}{\partial P_y}\frac{dP_y}{ds} + \frac{\partial}{\partial t}\frac{dt}{ds} + \frac{\partial}{\partial \eta}\frac{d\eta}{ds}\right) . \tag{22.26}$$

The Hamiltonian part of the equations of motion (22.8) does not contribute to the right-hand side of Eq. (22.26). Following the notation from Chap. 10 of writing the non-Hamiltonian part of the kinematic equations in terms of damping functions F_i, as in (10.32), this can be written as

$$\frac{df}{ds} = -f\left(\frac{\partial F_x}{\partial P_x} + \frac{\partial F_y}{\partial P_y} + \frac{\partial F_\eta}{\partial \eta}\right) = 4f\frac{P_0(s)}{p_0 c^2} , \tag{22.27}$$

where

$$F_x = -\frac{\mathcal{P}}{p_0 c^2}\frac{\partial \mathcal{H}}{\partial P_x} \simeq -\frac{P_0(s)}{p_0 c^2}P_x ,$$
$$F_y = -\frac{\mathcal{P}}{p_0 c^2}\frac{\partial \mathcal{H}}{\partial P_y} \simeq -\frac{P_0(s)}{p_0 c^2}P_y ,$$
$$F_\eta = \frac{\mathcal{P}}{p_0 c^2}\frac{\partial \mathcal{H}}{\partial \eta} \simeq -\frac{P_0(s)}{p_0 c^2}\left(2\eta + \frac{x}{\rho}\right) , \tag{22.28}$$

where we have used Eqs. (22.8), (22.12), and the linearized Hamiltonian.

According to Eq. (22.27) the distribution function f at a phase space point moving with a particle grows exponentially with time. This happens because, due to the synchrotron radiation, the phase space volume occupied by a given ensemble of particles decreases. Since f is the particle density in the phase space, it grows inversely proportionally to the phase space volume. This effect is associated with the name of K. Robinson who pointed it out in [1]. As was already mentioned in the previous section, in reality, due to the various diffusion sources that we neglected in our analysis, this exponential growth eventually stops and a steady state is established.

Reference

1. Kenneth W. Robinson, Radiation effects in circular electron accelerators. Phys. Rev. **111**(2), 373–380 (1958)

Appendix A
Maxwell's Equations, Equations of Motion, and Energy Balance in an Electromagnetic Field

A.1 Maxwell's Equations

Classical electrodynamics in vacuum is governed by the Maxwell equations. In the SI system of units, the Maxwell equations are

$$\nabla \cdot \boldsymbol{D} = \rho \,, \tag{A.1a}$$

$$\nabla \cdot \boldsymbol{B} = 0 \,, \tag{A.1b}$$

$$\nabla \times \boldsymbol{E} = -\frac{\partial \boldsymbol{B}}{\partial t} \,, \tag{A.1c}$$

$$\nabla \times \boldsymbol{H} = \boldsymbol{j} + \frac{\partial \boldsymbol{D}}{\partial t} \,, \tag{A.1d}$$

where ρ is the charge density, $\boldsymbol{j}$ is the current density, $\boldsymbol{D} = \epsilon_0 \boldsymbol{E}$ and $\boldsymbol{H} = \boldsymbol{B}/\mu_0$. Traditionally $\boldsymbol{B}$ is called the magnetic induction, and $\boldsymbol{H}$ is called the magnetic field, but in this book we refer to $\boldsymbol{B}$ as the magnetic field. The Maxwell equations are linear: a sum of two solutions, $\boldsymbol{E}_1, \boldsymbol{B}_1$ and $\boldsymbol{E}_2, \boldsymbol{B}_2$, is also a solution corresponding to the sum of densities $\rho_1 + \rho_2$, $\boldsymbol{j}_1 + \boldsymbol{j}_2$.

For a point charge q moving along a trajectory $\boldsymbol{r} = \boldsymbol{r}_0(t)$ the charge density and the current density are

$$\rho(\boldsymbol{r}, t) = q\delta(\boldsymbol{r} - \boldsymbol{r}_0(t)) \,, \qquad \boldsymbol{j}(\boldsymbol{r}, t) = q\boldsymbol{v}(t)\delta(\boldsymbol{r} - \boldsymbol{r}_0(t)) \,, \tag{A.2}$$

with $\boldsymbol{v}(t) = d\boldsymbol{r}_0(t)/dt$.

To find a particular solution of the Maxwell equations in a volume, proper boundary conditions should be specified at the volume boundary. On a surface of a good conducting metal the boundary condition requires the tangential component of the electric field to be equal to zero, $\boldsymbol{E}_t|_S = 0$.

© Springer International Publishing AG, part of Springer Nature 2018
G. Stupakov and G. Penn, *Classical Mechanics and Electromagnetism in Accelerator Physics*, Graduate Texts in Physics,
https://doi.org/10.1007/978-3-319-90188-6

A.2 Wave Equations

In free space with no local charges and currents the electric field satisfies the wave equation,

$$\frac{1}{c^2}\frac{\partial^2 E}{\partial t^2} - \frac{\partial^2 E}{\partial x^2} - \frac{\partial^2 E}{\partial y^2} - \frac{\partial^2 E}{\partial z^2} = 0. \tag{A.3}$$

The same equation is valid for the magnetic field B. A particular solution of Eq. (A.3) is a sinusoidal wave characterized by the frequency ω and the wave vector k,

$$E = E_0 \sin(\omega t - k \cdot r), \tag{A.4}$$

where E_0 is a constant vector perpendicular to k, $E_0 \cdot k = 0$, and $\omega = ck$.

A.3 Vector and Scalar Potentials

It is often convenient to express the fields in terms of the *vector potential* $A(r, t)$ and the *scalar potential* $\phi(r, t)$:

$$E = -\nabla\phi - \frac{\partial A}{\partial t},$$
$$B = \nabla \times A. \tag{A.5}$$

Substituting these equations into Maxwell's equations, we find that the second and third equations are satisfied identically. We only need to take care of the first and fourth equations.

A.4 Energy Balance and the Poynting Theorem

The electromagnetic field has an energy and momentum associated with it. The energy density of the field (energy per unit volume) is

$$u = \frac{1}{2}(E \cdot D + H \cdot B) = \frac{\epsilon_0}{2}(E^2 + c^2 B^2). \tag{A.6}$$

The Poynting vector,

$$S = E \times H, \tag{A.7}$$

Fig. A.1 Point charges
shown by red dots move
inside volume V and interact
through the electromagnetic
field

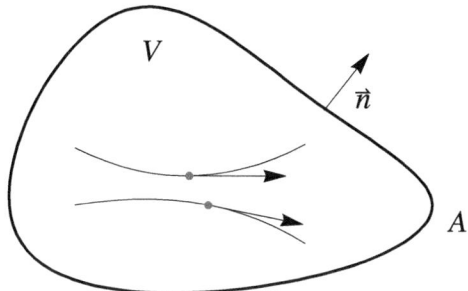

gives the energy flow (energy per unit area per unit time) in the electromagnetic field. Consider charges that move inside a volume V enclosed by a surface A, see Fig. A.1. The Poynting theorem states

$$\frac{\partial}{\partial t} \int_V u \, dV = - \int_V \boldsymbol{j} \cdot \boldsymbol{E} \, dV - \int_A \boldsymbol{n} \cdot \boldsymbol{S} \, dA \,, \tag{A.8}$$

where $\boldsymbol{n}$ is the unit vector normal to the surface and directed outward. The left-hand side of this equation is the rate of change of the electromagnetic energy due to the interaction with moving charges. The first term on the right-hand side is the work done by the electric field on the moving charges. The second term describes the electromagnetic energy flow from the volume through the enclosing surface.

A.5 Photons

The quantum view on electromagnetic radiation is that the electromagnetic field is represented by photons. Each photon carries the energy $\hbar\omega$ and the momentum $\hbar\boldsymbol{k}$, where the vector $\boldsymbol{k}$ is the wavenumber which points to the direction of propagation of the radiation, $\hbar = 1.05 \times 10^{-34}$ J· sec is the Planck constant divided by 2π, and $k = \omega/c$.

Appendix B
Lorentz Transformations and the Relativistic Doppler Effect

B.1 Lorentz Transformation and Matrices

Consider two coordinate systems, K and K'. The system K' is moving with velocity v in the z direction relative to the system K (see Fig. B.1). The coordinates of an event in both systems are related by the Lorentz transformation

$$
\begin{aligned}
x &= x', \\
y &= y', \\
z &= \gamma(z' + \beta ct'), \\
t &= \gamma(t' + \beta z'/c),
\end{aligned}
\tag{B.1}
$$

where $\beta = v/c$, and $\gamma = 1/\sqrt{1 - \beta^2}$.

The vector $(ct, \boldsymbol{r}) = (ct, x, y, z)$ is called a *4-vector*, and the above transformation is valid for any 4-vector quantity. The transformation from K to K' is also a Lorentz transformation, but the original frame K has a velocity $-v$ relative to the system K', so the inverse transformation is obtained from Eq. (B.1) by changing the sign of β:

Fig. B.1 Laboratory frame K and a moving frame K'

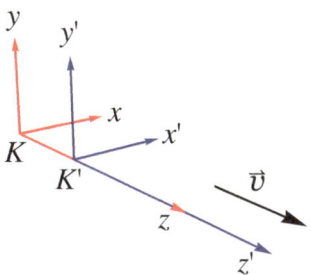

$$
\begin{aligned}
x' &= x\,, \\
y' &= y\,, \\
z' &= \gamma(z - \beta ct)\,, \\
t' &= \gamma(t - \beta z/c)\,.
\end{aligned}
\tag{B.2}
$$

The Lorentz transformation (B.1) can also be written in the matrix notation

$$
\begin{pmatrix} x \\ y \\ z \\ t \end{pmatrix}
=
\begin{pmatrix}
1 & 0 & 0 & 0 \\
0 & 1 & 0 & 0 \\
0 & 0 & \gamma & c\beta\gamma \\
0 & 0 & \frac{\beta\gamma}{c} & \gamma
\end{pmatrix}
\begin{pmatrix} x' \\ y' \\ z' \\ t' \end{pmatrix}
= L
\begin{pmatrix} x' \\ y' \\ z' \\ t' \end{pmatrix}\,,
\tag{B.3}
$$

where L denotes the 4×4 matrix in the middle of the equation. The advantage of using matrices is that consecutive transformations reduce to matrix multiplication.

B.2 Lorentz Contraction and Time Dilation

Two events occurring in the moving frame at the same point and separated by the time interval $\Delta t'$ will be measured by the lab observers as separated by Δt,

$$
\Delta t = \gamma \Delta t'\,.
\tag{B.4}
$$

This is the effect of relativistic *time dilation*.

An object of length l' aligned in the moving frame with the z' axis will have the length l in the lab frame:

$$
l = \frac{l'}{\gamma}\,.
\tag{B.5}
$$

This is the effect of relativistic *contraction*. The length in the direction transverse to the velocity is not changed.

B.3 Relativistic Doppler Effect

Consider a wave propagating in a moving frame K'. It has the time-space dependence:

$$
\propto \cos(\omega' t' - \boldsymbol{k}' \cdot \boldsymbol{r}')\,,
\tag{B.6}
$$

where ω' is the frequency and $\boldsymbol{k}'$ is the wavenumber of the wave in K'. An observer that measures this wave in the reference frame K will see the time-space dependence that is obtained from the Lorentz transformation of coordinates and time in Eq. (B.6):

$$\cos(\omega't' - \boldsymbol{k}' \cdot \boldsymbol{r}') = \cos(\omega'\gamma(t - \beta z/c) - k_x'x - k_y'y - k_z'\gamma(z - \beta ct))$$
$$= \cos(\gamma(\omega' + k_z'\beta c)t - k_x'x - k_y'y - \gamma(k_z' + \omega'\beta/c)z) . \quad \text{(B.7)}$$

We see that in the K frame this process is also a wave

$$\propto \cos(\omega t - \boldsymbol{k} \cdot \boldsymbol{r}) , \quad \text{(B.8)}$$

with the frequency and wavenumber

$$\begin{aligned}
k_x &= k_x' , \\
k_y &= k_y' , \\
k_z &= \gamma(k_z' + \beta\omega'/c) , \\
\omega &= \gamma(\omega' + \beta c k_z') .
\end{aligned} \quad \text{(B.9)}$$

Hence, the combination $(\omega, c\boldsymbol{k})$ is a 4-vector.

The above transformation is valid for any type of waves (electromagnetic, acoustic, plasma waves, etc.) Let us now apply it to electromagnetic waves in vacuum. For these waves we know that

$$\omega = ck . \quad \text{(B.10)}$$

Assume that an electromagnetic wave propagates at angle θ' in the frame K',

$$\cos\theta' = \frac{k_z'}{k'} , \quad \text{(B.11)}$$

and has a frequency ω' in that frame. What is the angle θ and the frequency ω of this wave in the lab frame? We can always choose the coordinate system such that $\boldsymbol{k} = (0, k_y, k_z)$, then

$$\tan\theta = \frac{k_y}{k_z} = \frac{k_y'}{\gamma(k_z' + \beta\omega'/c)} = \frac{\sin\theta'}{\gamma(\cos\theta' + \beta)} . \quad \text{(B.12)}$$

In the limit $\gamma \gg 1$ almost all angles θ' (except for the angles very close to π) are transformed to angles $\theta \sim 1/\gamma$. This explains why radiation of an ultrarelativistic beam goes mostly in the forward direction, within an angle of the order of $1/\gamma$.

A general expression for how the Lorentz transformation changes the angle θ of an electromagnetic wave with respect to the direction of the velocity is:

$$\cos\theta' = \frac{\cos\theta - \beta}{1 - \beta\cos\theta} , \qquad \sin\theta' = \frac{\sin\theta}{\gamma(1 - \beta\cos\theta)} . \quad \text{(B.13)}$$

For the frequency, a convenient formula relates ω with ω' and θ (not θ'). To derive it, we use the inverse Lorentz transformation

$$\omega' = \gamma(\omega - \beta c k_z) = \gamma(\omega - \beta c k \cos \theta) , \tag{B.14}$$

which gives

$$\omega = \frac{\omega'}{\gamma(1 - \beta \cos \theta)} . \tag{B.15}$$

Assuming a large γ and a small angle θ and using $\beta \approx 1 - 1/2\gamma^2$ and $\cos \theta = 1 - \theta^2/2$, we obtain

$$\omega = \frac{2\gamma\omega'}{1 + \gamma^2\theta^2} . \tag{B.16}$$

The radiation in the forward direction ($\theta = 0$) gets a large factor 2γ in the frequency transformation.

B.4 Lorentz Transformation of Fields

The electromagnetic field $(\boldsymbol{E}, \boldsymbol{B})$ is transformed from K' to K according to the following equations:

$$
\begin{aligned}
E_z &= E'_z , & \boldsymbol{E}_\perp &= \gamma \left(\boldsymbol{E}'_\perp - \boldsymbol{v} \times \boldsymbol{B}'\right) , \\
B_z &= B'_z , & \boldsymbol{B}_\perp &= \gamma \left(\boldsymbol{B}'_\perp + \frac{1}{c^2}\boldsymbol{v} \times \boldsymbol{E}'\right) ,
\end{aligned}
\tag{B.17}
$$

where $\boldsymbol{E}'_\perp$ and $\boldsymbol{B}'_\perp$ are the components of the electric and magnetic fields perpendicular to the velocity $\boldsymbol{v}$: $\boldsymbol{E}'_\perp = (E_x, E_y)$, $\boldsymbol{B}'_\perp = (B_x, B_y)$.

The electromagnetic potentials $(\phi/c, \boldsymbol{A})$ are transformed exactly as the 4-vector $(ct, \boldsymbol{r})$:

$$
\begin{aligned}
A_x &= A'_x , \\
A_y &= A'_y , \\
A_z &= \gamma \left(A'_z + \frac{v}{c^2}\phi'\right) , \\
\phi &= \gamma(\phi' + v A'_z) .
\end{aligned}
\tag{B.18}
$$

B.5 Lorentz Transformation and Photons

It is often convenient, even in classical electrodynamics, to consider electromagnetic radiation as a collection of photons. How do we transform the parameters of a photon from K' to K? The answer is rather evident: the combination $(\boldsymbol{k}, \omega)$ constitutes a 4-vector and is transformed according to Eq. (B.9). This is of course in agreement with the fact that the pair $(\hbar\boldsymbol{k}, \hbar\omega)$ is the momentum-energy 4-vector for the photon. The number of photons in K' to K is the same — it is a relativistic invariant.

Index

© Springer International Publishing AG, part of Springer Nature 2018
G. Stupakov and G. Penn, *Classical Mechanics and Electromagnetism
in Accelerator Physics*, Graduate Texts in Physics,
https://doi.org/10.1007/978-3-319-90188-6